THERAPEUTIC
DELIVERY SOLUTIONS

THERAPEUTIC DELIVERY SOLUTIONS

Edited by

CHUNG CHOW CHAN
CCC Consulting
Mississauga, Ontario, Canada

KWOK CHOW
Covar, Inc.
Mississauga, Ontario, Canada

BILL MCKAY
Medtronic, Inc.
Memphis, TN, USA

MICHELLE FUNG
Gordon and Leslie Diamond Health Care Centre
University of British Columbia
Vancouver, British Columbia, Canada

Published by John Wiley & Sons, Inc., Hoboken, New Jersey
Published simultaneously in Canada

For general information on our other products and services or for technical support, please contact
our Customer Care Department within the United States at (800) 762-2974, outside the United States
at (317) 572-3993 or fax (317) 572-4002.

Wiley also publishes its books in a variety of electronic formats. Some content that appears in print may
not be available in electronic formats. For more information about Wiley products, visit our web site at
www.wiley.com.

Library of Congress Cataloging-in-Publication Data:

Therapeutic delivery solutions / edited by Chung Chow Chan, Kwok Chow, Bill McKay, Michelle Fung.
    p. ; cm.
  Includes index.
  ISBN 978-1-118-11126-0 (cloth)
I. Chan, Chung Chow, editor.   II. Chow, Kwok, 1956– editor.   III. McKay, Bill, 1956– editor.
IV. Fung, Michelle, editor.
[DNLM: 1. Drug Delivery Systems–United States.   2. Cell- and Tissue-Based Therapy–methods–
United States.   QV 785]
  RS199.5
  615′.6–dc23
                                                                                        2014007293

Printed in the United States of America

ISBN: 9781118111260

10 9 8 7 6 5 4 3 2 1

CONTENTS

PREFACE

The technologies for the administration of therapeutic agents have been traditionally led by the pharmaceutical industry that develops drug molecules (both small and large molecules) in various dosage forms. The medical device industry has also evolved to apply its technologies to deliver drugs to various target sites.

Cellular therapy is now rapidly emerging as a new therapeutic solution platform, analogous to dosage form design and device development, in the last few decades. Under the Executive Order 13505 of March 9, 2009, in the United States, President Obama's Administration is committed to supporting and conducting ethically responsible, scientifically worthy human stem cell research, including human embryonic stem cell research. "National Institutes of Health Guidelines for Human Stem Cell Research" (Guidelines), effective July 7, 2009, applies to research using human embryonic stem cells and certain uses of human-induced pluripotent stem cells that have the potential to improve our understanding of human biology and aid in the discovery of new ways to prevent and treat illness. Researches in cellular therapy, for example, stem cells, have had very promising results as therapeutic solutions to diseased states and organ transplants.

This textbook provides a convergent link between traditional dosage form design, medical device development, and cellular therapeutics. It attempts to bring these three platforms of therapeutic delivery solution development together in one place to show the potential idiosyncrasies and common and dissimilar challenges that each platform faces to provide the best therapeutic delivery solution to the patient. Contemporary scientific and medical information as well as the newly emerging regulatory scientific information are discussed. This textbook will provide development scientists and medical professionals more options to develop a therapeutic agent to its fullest potential and create better and more creative therapeutic solutions.

The content of the book is grouped into five sections. Section 1 (consisting of Chapter 1) introduces the requirements and issues encountered in regulatory submissions in the pharmaceutical, cellular/gene products, and medical device industries. Section 2 (consisting of Chapters 2 and 3) explains in detail the traditional pharmaceutical drug therapy development. Section 3 (consisting of Chapters 4–6) provides an overview, current trends, and strategies of special medical device development. Section 4 (consisting of Chapters 7–9) introduces the reader to the latest advances and innovations in cellular and stem cell therapeutic delivery. Section 5 (consisting of Chapters 10–14) provides information on the analytical support needed for the research and development in Sections 2–4.

Chapter 1 provides an overview of the current regulatory requirements for the development of the three platforms of therapeutic solution and new FDA initiatives to ensure that innovative products reach the patients who need them and when they need them.

An overview of the approach and strategies for development of immediate release tablets after a drug candidate is selected is provided in Chapter 2. Chapter 3 discusses the strategies (with examples) for the development of low aqueous solubility drug products.

Chapter 4 starts with an overview, key trends, and drivers for drug delivery medical devices. Chapter 5 focuses on the local growth factor delivery to address metabolic bone disorders. "From glass syringes to feedback-controlled patch pumps", Chapter 6 discusses the amazing accomplishment for the pharmaceutical and medical device industries with the insulin pump to continuously deliver precise amounts of insulin 24 h a day.

Cell-based biologic therapies have a long history. Simple blood transfusions and tissue transplants are commonly utilized in medical practice. Chapter 7 reviews the history of islet transplantation, procedural issues, current outcomes, and future directions. Chapter 8 provides an overview of the latest developments of cell-based biologic therapies and discusses the future outlook for these novel treatment modalities, for example, cancer, infection, and autoimmune disorders. Chapter 9 reviews the history of stem cell research and development, sources of various stem cells (e.g., neonatal, adult, reprogrammed), technical and regulatory issues of stem cell therapy, and the prospect of industrialization of stem cell technology into future medical therapy.

Chapters 10 to 14 provide the analytical support needed in the development of the three platforms of therapeutic solution delivery. Chapter 10 summarizes the specifications setting and stability studies requirements for development work. Chapter 11 shows how LC–MS techniques have been used in all stages of the drug development process including discovery, preclinical, clinical, and manufacturing. Chapter 12 discusses the importance of biorelevant methods and how to achieve them. Chapter 13 provides information and importance of ICH guidelines for development and global harmonization. In the development of therapeutic solution, there will be situations when out of specification (OOS) or aberrant data are obtained. Chapter 14 looks at how the use of sound scientific judgment and good documentation can lead to a successful OOS/atypical result investigation in a case study according to current guidance.

CCC Consulting   CHUNG CHOW CHAN, PhD
Covar Inc.   KWOK CHOW, PhD
Medtronic Inc.   BILL McKAY, ME
University of British Columbia   MICHELLE FUNG, MD

CONTRIBUTORS

May Almukainzi, PhD, Faculty of Pharmacy and Pharmaceutical Sciences, University of Alberta, Edmonton, Alberta, Canada; Faculty of Pharmacy, Princess Nora Bint AbdulRahman University, Riyadh, Saudi Arabia

Ziliang Ao, MD, MSc, Ike Barber Human Islet Transplant Laboratory Surgery, Faculty of Medicine, University of British Columbia, Vancouver, British Columbia, Canada

James Blakemore, PhD, Cambridge Consultants Ltd., Science Park, Cambridge, UK

Nádia Araci Bou-Chacra, PhD, Faculty of Pharmaceutical Sciences, University of São Paulo, São Paulo, Brazil

Debra L. Bowen, MD, FACAAI, Pharma Science Consulting, Inc., Milford, NH, USA

Chung Chow Chan, PhD, CCC Consulting, Mississauga, Ontario, Canada

Kwok Chow, PhD, Covar, Inc., Mississauga, Ontario, Canada

Jared Diegmueller, MS, Medtronic Spinal and Biologics, Memphis, TN, USA

Man C. Fung, MD, MBA, MHCM, FACP, Janssen Pharmaceutical R&D (JNJ), Janssen Oncology, Raritan, NJ, USA

Michelle Fung, BASc, MD, MHSc, Department of Medicine, Faculty of Medicine, University of British Columbia, Vancouver, British Columbia, Canada

Sultan Ghani, BSc, BPharm, MS, Quality Management & Regulatory Affairs (QMRA), Getz Pharma Pvt. Limited, Karachi, Pakistan

Stephen G.F. Ho, MD, FRCPC, Department of Radiology, Faculty of Medicine, University of British Columbia, Vancouver, British Columbia, Canada

Klaudyne Hong, PhD, Temasek Bioscience Partners, NY, USA

James D. Johnson, PhD, Department of Surgery, Faculty of Medicine, University of British Columbia, Vancouver, British Columbia, Canada

Paul Keown, MD, Department of Medicine, Faculty of Medicine, University of British Columbia, Vancouver, British Columbia, Canada

Herman Lam, PhD, Wild Crane Horizon, Scarborough, Ontario, Canada

Raimar Löbenberg, PhD, Faculty of Pharmacy and Pharmaceutical Sciences, University of Alberta, Edmonton, Alberta, Canada

Bill McKay, ME, Medtronic Spinal and Biologics, Memphis, TN, USA

Mark Meloche, MD, Department of Surgery, Faculty of Medicine, University of British Columbia, Vancouver, British Columbia, Canada

Graydon Meneilly, MD, Department of Medicine, Faculty of Medicine, University of British Columbia, Vancouver, British Columbia, Canada

Breay W. Paty, MD, Department of Medicine, Faculty of Medicine, University of British Columbia, Vancouver, British Columbia, Canada

Steven Peckham, PhD, Medtronic Spinal and Biologics, Memphis, TN, USA

R. Jean Shapiro, MD, Department of Medicine, Faculty of Medicine, University of British Columbia, Vancouver, British Columbia, Canada

Iain Simpson, PhD, Cambridge Consultants Ltd., Science Park, Cambridge, UK

David Thompson, MD, Department of Medicine, Faculty of Medicine, University of British Columbia, Vancouver, British Columbia, Canada

Yu-Hong Tse, PhD, YH & MJ Consulting, Brampton, Ontario, Canada

Bill Van Antwerp, Zyomed Corp., Valencia, CA, USA

Bruce Vechere, PhD, Pathology & Laboratory Medicine and Surgery, Faculty of Medicine, University of British Columbia, Vancouver, British Columbia, Canada

Roderick B. Walker, Faculty of Pharmacy, Rhodes University, Grahamstown, South Africa

Garth Warnock, MD, MSc, Department of Surgery, Faculty of Medicine, University of British Columbia, Vancouver, British Columbia, Canada

ACKNOWLEDGMENT

We would like to thank all the authors and contributors who are leading scientists and physicians in the respective areas for their contributions to the chapters in this book.

SECTION 1

REQUIREMENTS AND ISSUES ENCOUNTERED IN REGULATORY SUBMISSIONS IN THE PHARMACEUTICAL, CELL THERAPY AND MEDICAL DEVICE INDUSTRIES

1

CHALLENGES TO QUALITY AND REGULATORY REQUIREMENT IN THE UNITED STATES—DRUGS, MEDICAL DEVICE, AND CELL THERAPY

CHUNG CHOW CHAN, SULTAN GHANI, IAIN SIMPSON, AND JAMES BLAKEMORE

1.1 OVERVIEW OF REGULATORY REQUIREMENTS FOR PHARMACEUTICAL, MEDICAL DEVICE, AND CELL THERAPIES

The technologies for the administration of therapeutic agents had been traditionally led by the pharmaceutical industry, which develops small drug molecules into various dosage forms. These developments have been followed by large-molecule pharmaceutical development (proteins, etc.), device development, and the new emerging cellular therapy. Recent breakthroughs in science and technology (ranging from sequencing of the human genome to advances in the application of nanotechnology to new medical products) are transforming the ability to treat diseases and bring with it new challenges in regulatory approval.

This chapter brings together the regulatory requirements for the development of the three platforms of therapeutic delivery solution (pharmaceutical, medical devices, and cellular therapeutic solutions) to illustrate the common/different strategies of regulating these three therapeutic deliveries and the current initiatives initiated in the United States and other countries. Note that the terms "drugs" and "pharmaceuticals" will be used interchangeably in this chapter. The common goal for all three platforms

Therapeutic Delivery Solutions, First Edition. Edited by Chung Chow Chan, Kwok Chow, Bill McKay, and Michelle Fung.
© 2014 John Wiley & Sons, Inc. Published 2014 by John Wiley & Sons, Inc.

of delivery is current Good Manufacturing Practices (CGMP). The detailed process of achieving the common goal of GMP is different in each therapeutic area. The summary of the common regulatory requirements and the different approaches to reach this goal are presented.

The evaluation and approval processes are being modernized by the Food and Drug Administration (FDA) in the United States and other global regulatory agencies to ensure that innovative products reach the patients who need them and when they need them. In the United States, this is being done through advancing Regulatory science, which is the science of developing new tools, standards, and approaches to assess the safety, efficacy, quality, and performance of FDA-regulated products [1].

In the United States, drug delivery is regulated by the Code of Federal Regulations (CFR). CFR is the codification of the general and permanent rules and regulations. This is published in the Federal Register by the executive departments and agencies of the Federal Government. It is divided into 50 titles that represent broad areas subject to Federal regulation.

Each title is divided into chapters, which usually bear the name of the issuing agency. Each chapter is further subdivided into parts that cover specific regulatory areas. Large parts may be subdivided into subparts. All parts are organized in sections, and most citations in the CFR are provided at the section level (http://www.gpo.gov/).

Title 21 of the CFR is reserved for Food and Drug under the rules of the FDA, Department of Health and Administrative Services. Title 21 contains the following three chapters:

- Chapter I—Food and Drug Administration, Department of Health and Human Services (Parts 1–1299)
- Chapter II—Drug Enforcement Administration, Department of Justice (Parts 1300–1321)
- Chapter III—Office of National Drug Control Policy (Parts 1400–1499)

1.2 REGULATORY REQUIREMENTS AND CHALLENGES FOR PHARMACEUTICAL, MEDICAL DEVICE, AND CELL THERAPIES

Title 21 Chapter 1 contains Parts 1–1299. The parts that are commonly encountered in the development of the three platforms of therapeutic delivery are listed below:

Part 3—Product Jurisdiction

Part 4—Current Good Manufacturing Practice Requirements for Combination Products (effective July 2013)

Part 11—Electronic Records; Electronic Signatures

Part 26—Mutual Recognition of Pharmaceutical Good Manufacturing Practice Reports, Medical Device Quality System Audit Reports, and Certain Medical

Device Product Evaluation Reports: United States and the European Community

Part 58—Good Laboratory Practice for Nonclinical Laboratory Studies

Part 210—Current Good Manufacturing Practice in Manufacturing, Processing, Packing, or Holding of Drugs; General

Part 211—Current Good Manufacturing Practice for Finished Pharmaceuticals

Part 312—Investigational New Drug Application

Part 600—Biological Products: General

Part 601—Biologic License Application

Part 610—General Biological Products Standards

Part 820—Quality System Regulation (Devices)

Part 814—Premarket Approval of Medical Devices

Part 1270—Human Tissue Intended for Transplantation

Part 1271—Human Cells, Tissues, and Cellular and Tissue-Based Products

In the United States, the regulatory requirements of the three platforms of drug delivery are implemented through three separate Centers in the FDA:

1. Center for Drug Evaluation and Research (CDER) for Pharmaceuticals. CDER's primary mission is to make certain that safe and effective drugs are available to the American people.
2. Center for Devices and Radiological Health (CDRH) for Medical Devices. CDRH is responsible for ensuring the safety and effectiveness of medical devices and eliminating unnecessary human exposure to man-made radiation from medical, occupational, and consumer products. It will advance public health and facilitate innovation to help bring novel technologies to market and make the medical devices that are already on the market safer and more effective.
3. Center for Biologics Evaluation and Research (CBER) for Cell Therapy. CBER regulates biological products for human use and protects and advances the public health by ensuring that biological products are safe and effective and available to those who need them.

Whether the item is a pharmaceutical agent, cell delivery agent, or medical device, it shares the common criteria in the regulatory approval of intended use of the product and CGMP. Pharmaceutical and cell therapy products share many common processes and techniques to provide relief to disease states of the patient. Device products are more varied and range from simple household products to highly sophisticated imaging products, which may provide other use in addition to providing relief to disease states. However, it still needs to fulfill the common criteria of intended use and be safe to the patients. As an example, a simple device product (Shoulder/Flex Massager) was used to "help relieve muscle pain" (intended use). However, because of incidents related to its safety (report of strangulation and death) at the time of its intended use, the product had been voluntarily recalled by the manufacturer [2].

1.2.1 Center for Drug Evaluation and Research

CDER enforces CGMP through Part 211 by implementing the regulatory sections tabulated in Table 1.1. Section 501(a)(2)(B) of the Food and Drug Act (FD&C Act) requires drugs, which include investigational new drug (IND) products, to comply with CGMP as follows:

> A drug…shall be deemed adulterated…if…the methods used in, or the facilities or controls used for, its manufacture, processing, packing, or holding do not conform to or are not operated or administered in conformity with current good manufacturing practice to assure that such drug meets the requirements of this Act as to safety and has the identity and strength, and meets the quality and purity characteristics, which it purports or is represented to possess.

Based on the statutory requirement for manufacturers to follow CGMP, FDA issued CGMP regulations for drug and biological products [3]. Although FDA stated at the time of issuance that the regulations applied to all types of pharmaceutical production, the preamble to the regulations indicated that FDA was considering proposing additional regulations governing drugs used in investigational clinical trials.

Because certain requirements in Part 211, which implement Section 501(a)(2) (B) of the FD&C Act, were directed at the commercial manufacture of products typically characterized by large, repetitive, commercial batch production (e.g., those regulations that address validation of manufacturing processes) and warehousing, they may not be appropriate to the manufacture of most investigational drugs used for Phase 1 clinical trials. Guidances on GMP requirements are now available for Phase 1–3 studies.

1.2.2 Center for Devices and Radiological Health

Medical devices employ a diversity of technologies to give a wide array of products in the healthcare sector. They range from simple devices such as bandages to life-maintaining active implantable devices such as insulin pump or heart pacemakers to sophisticated diagnostic imaging and surgical equipment. CDRH enforces CGMP through Part 820 by enforcing the regulatory requirements tabulated in Table 1.2. The quality system regulation of 820 govern the methods used in, and the facilities and controls used for, the design, manufacture, packaging, labeling, storage, installation, and servicing of all finished devices intended for human use. The requirements in this part are intended to ensure that finished devices will be safe and effective and otherwise in compliance with the FD&C Act.

Certain issues have arisen often relating to whether a product should be classified as a drug or a device. In Europe, the manufacturer is responsible for the classification of medical devices. In the United States, FDA is responsible for the classification of the medical devices. Accordingly, in the United States, a draft guidance document has been issued to focus particularly on when a product may be classified as a drug or a device [4].

TABLE 1.1 Regulatory sections of Part 211—current good manufacturing practice for finished pharmaceuticals

211.1	Scope
211.3	Definitions
211.22	Responsibilities of quality control unit
211.25	Personnel qualifications
211.28	Personnel responsibilities
211.34	Consultants
211.42	Design and construction features
211.44	Lighting
211.46	Ventilation, air filtration, air heating and cooling
211.48	Plumbing
211.50	Sewage and refuse
211.52	Washing and toilet facilities
211.56	Sanitation
211.58	Maintenance
211.63	Equipment design, size, and location
211.65	Equipment construction
211.67	Equipment cleaning and maintenance
211.68	Automatic, mechanical, and electronic equipment
211.72	Filters
211.80	General requirements
211.82	Receipt and storage of untested components, drug product containers, and closures
211.84	Testing and approval or rejection of components, drug product containers, and closures
211.86	Use of approved components, drug product containers, and closures
211.87	Retesting of approved components, drug product containers, and closures
211.89	Rejected components, drug product containers, and closures
211.94	Drug product containers and closures
211.100	Written procedures; deviations
211.101	Charge-in of components
211.103	Calculation of yield
211.105	Equipment identification
211.110	Sampling and testing of in-process materials and drug products
211.111	Time limitations on production
211.113	Control of microbiological contamination
211.115	Reprocessing
211.122	Materials examination and usage criteria
211.125	Labeling issuance
211.130	Packaging and labeling operations
211.132	Tamper-evident packaging requirements for over-the-counter (OTC) human drug products
211.134	Drug product inspection
211.137	Expiration dating
211.142	Warehousing procedures
211.150	Distribution procedures
211.160	General requirements
211.165	Testing and release for distribution

(Continued)

TABLE 1.1 (Cont'd)

211.166	Stability testing
211.167	Special testing requirements
211.170	Reserve samples
211.173	Laboratory animals
211.176	Penicillin contamination
211.180	General requirements
211.182	Equipment cleaning and use log
211.184	Component, drug product container, closure, and labeling records
211.186	Master production and control records
211.188	Batch production and control records
211.192	Production record review
211.194	Laboratory records
211.196	Distribution records
211.198	Complaint files
211.204	Returned drug products
211.208	Drug product salvaging

TABLE 1.2   Regulatory sections of Part 820—quality system regulation

820.1	Scope
820.3	Definitions
820.5	Quality system
820.20	Management responsibility
820.22	Quality audit
820.25	Personnel
820.30	Design controls
820.40	Document controls
820.50	Purchasing controls
820.60	Identification
820.65	Traceability
820.70	Production and process controls
820.72	Inspection, measuring, and test equipment
820.75	Process validation
820.80	Receiving, in-process, and finished device acceptance
820.86	Acceptance status
820.90	Nonconforming product
820.100	Corrective and preventive action
820.120	Device labeling
820.130	Device packaging
820.140	Handling
820.150	Storage
820.160	Distribution
820.170	Installation
820.180	General requirements
820.181	Device master record
820.184	Device history record
820.186	Quality system record
820.198	Complaint files
820.200	Servicing
820.250	Statistical techniques

If the classification of a product as a drug, device, biological product, or combination product is unclear or in dispute, the sponsor can file a request for designation (RFD) with FDA Office of Combination Products (OCP) in accordance with Part 3 of Title 21 of the Code of Federal Regulations (21 CFR Part 3) to obtain a formal classification determination for the product, as provided for under section 563 of the FD&C Act (21 USC 360bbb-2). In reviewing an RFD, the Agency considers the information provided in the RFD as well as other information available to the Agency at that time. Generally, the Agency will respond in writing within 60 days of the sponsor's RFD filing, identifying the classification of the product as a drug, device, biological product, or combination product. If the Agency does not provide a written response within 60 days, the sponsor's recommendation respecting the classification of the product is considered to be the final determination.

In the United States, FDA's determination of whether to classify a product as a drug or a device is based on the statutory definitions of these terms set forth in sections 201(g) and 201(h) of the FD&C Act, as applied to the scientific data concerning the product that are available to FDA at the time the classification determination is made.

1.2.2.1 Definition of Drug Section 201(g) of the FD&C Act defines the term "drug" as (A) articles recognized in the official United States Pharmacopoeia, official Homoeopathic Pharmacopoeia of the United States, or official National Formulary, or any supplement to any of them; (B) articles intended for use in the diagnosis, cure, mitigation, treatment, or prevention of disease in man or other animals; (C) articles (other than food) intended to affect the structure or any function of the body of man or other animals; and (D) articles intended for use as a component of any articles specified in clause (A), (B), or (C).

1.2.2.2 Definition of Device Section 201(h) of the FD&C Act defines the term "device" as …an instrument, apparatus, implement, machine, contrivance, implant, *in vitro* reagent, or other similar or related article, including any component, part, or accessory, which is:

1. recognized in the official National Formulary or the United States Pharmacopeia or any supplement to them,
2. intended for use in the diagnosis of disease or other conditions or in the cure, mitigation, treatment, or prevention of disease in man or other animals, or
3. intended to affect the structure or any function of the body of man or other animals and which does not achieve its primary intended purposes through chemical action within or on the body of man or other animals and which is not dependent upon being metabolized for the achievement of its primary intended purposes.

1.2.3 Center for Biologics Evaluation and Research

Human cells or tissue intended for implantation, transplantation, infusion, or transfer into a human recipient is regulated as a human cell, tissue, and cellular and tissue-based product or HCT/P. CBER regulates HCT/Ps under 21 CFR Parts 1270 and 1271. CBER enforces CGMP through Part 600, 601, and 610 (in addition to GMP Part 211) in Table 1.3, Table 1.4, and Table 1.5. CBER's role includes implementation of

TABLE 1.3   Regulatory sections of Part 600—biological products: general

600.2	Mailing addresses
600.3	Definitions
600.10	Personnel
600.11	Physical establishment, equipment, animals, and care
600.12	Records
600.13	Retention samples
600.14	Reporting of biological product deviations by licensed manufacturers
600.15	Temperatures during shipment
600.20	Inspectors
600.21	Time of inspection
600.22	Duties of inspector
600.80	Postmarketing reporting of adverse experiences
600.81	Distribution reports
600.90	Waivers

TABLE 1.4   Regulatory sections of Part 601—biologic license application

601.2	Applications for biologics licenses; procedures for filing
601.3	Complete response letter to the applicant
601.4	Issuance and denial of license
601.5	Revocation of license
601.6	Suspension of license
601.7	Procedure for hearings
601.8	Publication of revocation
601.9	Licenses; reissuance
601.12	Changes to an approved application
601.14	Regulatory submissions in electronic format
601.15	Foreign establishments and products: samples for each importation
601.20	Biologics licenses; issuance and conditions
601.21	Products under development
601.22	Products in short supply; initial manufacturing at other than licensed location
601.25	Review procedures to determine that licensed biological products are safe, effective, and not misbranded under prescribed, recommended, or suggested conditions of use
601.26	Reclassification procedures to determine that licensed biological products are safe, effective, and not misbranded under prescribed, recommended, or suggested conditions of use
601.27	Pediatric studies
601.28	Annual reports of postmarketing pediatric studies
601.29	Guidance documents
601.30–601.36	Diagnostic radiopharmaceuticals
601.30	Scope
601.31	Definition
601.32	General factors relevant to safety and effectiveness
601.33	Indications
601.34	Evaluation of effectiveness

(Continued)

TABLE 1.4　(Cont'd)

601.35	Evaluation of safety
601.40–601.46	Accelerated approval of biological products for serious or life-threatening illnesses
601.40	Scope
601.41	Approval based on a surrogate endpoint or on an effect on a clinical endpoint other than survival or irreversible morbidity
601.42	Approval with restrictions to assure safe use
601.43	Withdrawal procedures
601.44	Postmarketing safety reporting
601.45	Promotional materials
601.46	Termination of requirements
601.50	Confidentiality of data and information in an investigational new drug notice for a biological product
601.51	Confidentiality of data and information in applications for biologics licenses
601.70	Annual progress reports of postmarketing studies
601.90–601.95	Approval of biological products when human efficacy studies are not ethical or feasible
601.90	Scope
601.91	Approval based on evidence of effectiveness from studies in animals
601.92	Withdrawal procedures
601.93	Postmarketing safety reporting
601.94	Promotional materials
601.95	Termination of requirements

TABLE 1.5　Regulatory sections of Part 610—general biological product standards

610.1	Tests prior to release required for each lot
610.2	Requests for samples and protocols; official release
610.9	Equivalent methods and processes
610.10	Potency
610.11	General safety
610.11a	Inactivated influenza vaccine, general safety test
610.12	Sterility
610.13	Purity
610.14	Identity
610.15	Constituent materials
610.16	Total solids in serums
610.17	Permissible combinations
610.18	Cultures
610.20	Standard preparations
610.21	Limits of potency
610.30	Test for *mycoplasma*
610.40	Test requirements
610.41	Donor deferral
610.42	Restrictions on use for further manufacture of medical devices
610.44	Use of reference panels by manufacturers of test kits

(*Continued*)

TABLE 1.5   (Cont'd)

610.46	Human immunodeficiency virus (HIV) "lookback" requirements
610.47	Hepatitis C virus (HCV) "lookback" requirements
610.48	Hepatitis C virus (HCV) "lookback" requirements based on review of historical testing records
610.50	Date of manufacture
610.53	Dating periods for licensed biological products
610.60	Container label
610.61	Package label
610.62	Proper name; package label; legible type
610.63	Divided manufacturing responsibility to be shown
610.64	Name and address of distributor
610.65	Products for export
610.67	Barcode label requirements
610.68	Exceptions or alternatives to labeling requirements for biological products held by the strategic national stockpile

the regulation of preventive and therapeutic vaccines, blood and blood products, human cell and tissue-based products, gene therapies, and xenotransplantation (a procedure that uses a different species as a source of transplanted materials) [5].

No lot of any licensed product shall be released by the manufacturer prior to the completion of tests for conformity with standards applicable to such a product, which include tests for potency, sterility, purity, and identity (21 CFR Part 610, Subpart B). These requirements apply to all biological products, including autologous and single-patient allogeneic products, where a lot may be defined as a single dose.

Some Cellular and Gene Therapy (CGT) products may also contain, in addition to the active ingredient, one or more substances commonly referred to in the scientific literature as an "adjuvant." An adjuvant shall not be introduced into a product unless there is satisfactory evidence that it does not affect adversely the safety or potency of the product (21 CFR 610.15(a)).

Some of the challenges in the development of CGT products include the variability and complexity inherent in the components used to generate the final product, such as the source of cells (i.e., autologous or allogeneic), the potential for adventitious agent contamination, the need for aseptic processing, and the inability to "sterilize" the final product because it contains living cells. Distribution of these products can also be a challenge due to stability issues and the frequently short dating period of many cellular products, which may necessitate release of the final product for administration to a patient before certain test results are available.

1.2.4   Regulatory Submission Requirement

Each therapeutic delivery solution in the United States is regulated by different centers as mentioned earlier. Table 1.6 gives the summary of the regulating center and the documents that need to be filed for investigation and marketing.

TABLE 1.6 Summary of application type and designated regulating center

	Application type	Purpose	Regulating center
Clinical trials approval	IDE (investigation device exemption)	Approval to begin clinical evaluation of a device	CDRH
	IND (investigational new drug)	Approval to begin clinical evaluation of a drug	CDER or CBER (if biological drug)
Approval to market for a medical device or drug	PMA (premarket approval)	Permission to market a new medical device	CDRH
	510(k)	Premarket notification for a medical device substantially equivalent to an already marketed device	CDRH
	NDA (new drug application)	Permission to market a new drug	CDER
	ANDA (abbreviated (new drug application)	Permission to market a generic version of a drug comparable to an innovator drug product (already approved in the USA) in dosage form, strength, route of administration, quality, performance characteristics, and intended use.	CDER
	505 (b)(2)	Permission to market a drug product relying in part on data from existing reference drugs	CDER
	BLA (biologic license application)	Permission to market a new biologic drug	CBER

1.2.4.1 Small Molecule and Macromolecule Submission   Both small molecule and macromolecule drugs are under the jurisdiction of CDER and CBER respectively. Both classes of drugs will go through similar IND and new drug application (NDA) processes from its development to marketing. Generic drugs will go through the abbreviated new drug application (ANDA) process.

1.2.4.2 Medical Devices   Medical devices are classified into Class I, II, and III based upon the risk they are considered to present with the required level of regulatory control increasing from Class I to Class III.

Most Class I devices do not require premarket notification or approval and so are just subject to General Controls. Most Class II devices require Premarket Notification through a 510(k) process. Most Class III devices require Premarket Approval, for example, through the premarket approval (PMA) process. Device classification depends on the intended use of the device as well as its indications for use.

The FDA has classified around 1700 generic types of device which are grouped into 16 medical specialities or panels. Classification information is provided in a freely accessible database.

A device manufacturer can also request classification by the FDA. If the FDA concludes that the device is not substantially equivalent to a predicate device, then it will be designated as Class III unless the device manufacturer makes a de novo petition requesting the FDA to make a risk-based classification determination for the device. If the FDA grants the de novo petition, then the device will be reclassified from Class III to class II or I.

1.2.4.3 Medical Device 510(k) Premarket Notification   Some drug delivery devices aimed for general use are regulated as medical devices. For example, an autoinjector could be approved as a Class II device by the 510(k) route and then utilized with different drugs, each of which would be subject to its own submission as a combination product. But the fact that the autoinjector already has 510(k) approval should reduce the burden of review for the combination product.

This is the main route of approval for Class II devices and is based on showing that a new device is substantially equivalent to a predicate device, that is, that it is at least as safe and effective as an already marketed device.

1.2.4.4 Medical Device Premarket Approval (PMA)   This is an FDA route for approval for Class III devices and involves a detailed scientific and regulatory review to evaluate the safety and effectiveness of the device. Given the greater depth of review, the period is 180 days, although in practice, the review period can be much longer due to the need to provide additional information to the FDA. The process also requires Quality System Regulation (QSR) inspection prior to product approval and launch.

1.2.4.5 Medical Device Quality System Regulation   Class II and III device manufacturers need to comply with Quality System Regulation 21 CFR 820 (see Table 1.2 for summary). This is based on an early version of ISO 9001 (1994) with additional requirements for design and process validation and transfer.

1.2.5 FDA Compliance Program

FDA Compliance Programs are set up to provide instructions to FDA personnel for conducting activities to evaluate industry compliance with the FD&C Act and other laws administered by FDA [6]. These compliance programs neither create nor confer any rights for, or on, any person and do not operate to bind FDA or the public. Alternative approaches may be used as long as they satisfy the requirements of applicable statutes and regulations.

FDA Compliance Programs are organized by the following program areas:

- Biologics (CBER)
- Bioresearch Monitoring (BIMO)
- Devices/Radiological Health (CDRH)
- Drugs (CDER)
- Food and Cosmetics (CFSAN)
- Veterinary Medicine (CVM)

Compliance programs that affect the three therapeutic areas in CBER, BIMO, CDRH, and CDER are tabulated in Table 1.7, Table 1.8, Table 1.9, and Table 1.10.

TABLE 1.7   Compliance programs of CBER

Program no.	CBER compliance program title
7341.002	Inspection of Human Cells, Tissues, and Cellular and Tissue-Based Products (HCT/Ps)
7341.002A	Inspection of tissue establishments (covers human tissue recovered before 5/25/2005)
7342.001	Inspection of licensed and unlicensed blood banks, brokers, reference laboratories, and contractors
7342.002	Inspection of source plasma establishments, brokers, testing laboratories, and contractors
7342.007	Imported CBER-regulated products
7342.008	Inspection of licensed *in vitro* diagnostic (IVD) devices regulated by CBER
7345.848	Inspection of biological drug products (PDF—570 kb)
	Replaces 7342.006—inspection of plasma derivatives of human origin, 7345.001—inspection of licensed allergenic products, 7345.002—inspection of licensed vaccines

TABLE 1.8   Compliance program in BIMO

Program no.	BIMO compliance program title
7348.001	*In vivo* bioequivalence
7348.808	Good laboratory practice (nonclinical laboratories)
7348.808A	Good laboratory practice program (nonclinical laboratories) EPA data audit inspections
7348.809	Institutional review board
7348.809A	Radioactive drug research committee
7348.810	Sponsors, contract research organizations, and monitors
7348.811	Clinical investigators

TABLE 1.9   Compliance program in CDRH

Program no.	CDRH compliance program title
7382.845	Inspection of medical device manufacturers
7383.001	Medical device premarket approval and postmarket inspections
7385.014	Mammography facility inspections
7386.001	Inspection and field testing of radiation-emitting electronic products
7386.003	Field compliance testing of diagnostic medical X-ray equipment Attachments A-M
7386.003a	Inspection of domestic and foreign manufacturers of diagnostic X-ray equipment
7386.006	Compliance testing of electronic products at WEAC
7386.007	Imported electronic product
7386.008	Medical device and radiological health use control and policy implementation
7386.009	Emergency planning and response activities: Part VI

TABLE 1.10   CDER compliance program

Program no.	CDER compliance program title
7348.001	*In vivo* bioequivalence
7348.809A	Radioactive drug research committee
7346.832	Preapproval inspections/investigations
7346.843	Postapproval audit inspections
7352.002	Unapproved new drugs (marketed, human, prescription drugs only)
7352.004	*In vitro* method development and validation for generic drugs
7353.001	Postmarketing adverse drug experience (PADE) reporting inspections
7356.002	Drug manufacturing inspections
7356.002A	Sterile drug process inspections
7356.002B	Drug repackers and relabelers
7356.002C	Radioactive drugs
7356.002E	Compressed medical gases
7356.002F	Active pharmaceutical ingredients
7356.002M	Inspections of licensed biological therapeutic drug products
7356.002P	Positron emission tomography
7356.008	Drug quality sampling and testing—human drugs
7356.014	Drug listing
7356.014A	Drug listing—labeling review
7356.020	Compendial monograph evaluation and development (CMED)
7356.020A	Compendial method assessment
7356.021	Drug quality reporting system (DQRS) (MedWatch reports) NDA field alert reporting (FAR)
7356.022	Enforcement of the prescription drug marketing act (PDMA)
7361.003	OTC drug monograph implementation
7363.001	Fraudulent drugs

1.3   INITIATIVES IN THE PHARMACEUTICAL, MEDICAL DEVICE, AND CELL THERAPY REGULATORY REQUIREMENTS

In recent years, threats from adulteration (including economically motivated adulteration) of medical products is real. The consequences, throughout the world, have been tragic: Glycerin used in the manufacture of fever medicine and cough syrup and teething products was adulterated with diethylene glycol (DEG) resulting in the deaths of children in Haiti, Panama, and Nigeria. In 2007, pet food adulterated with the industrial chemical melamine sickened several thousand pets in the United States. That same contaminant was added to infant formula in China, fatally poisoning six babies in China and making 300,000 others gravely ill. In 2008, heparin contamination crisis in the United States was associated with several deaths and cases of serious illness.

FDA and other global regulatory agencies are playing an increasingly integral role, not just dedicated to ensuring safe and effective products, but also to promote public health and participate more actively in the scientific research enterprise directed toward new treatments and interventions. The global regulatory agencies are also modernizing its evaluation and approval processes by utilizing regulatory science to ensure that innovative products reach the patients who need them, when they need them.

Regulatory science is defined as the science of developing new tools, standards, and approaches to assess the safety, efficacy, quality, and performance of regulated products. Regulatory science is the foundation of FDA decision-making. Both the knowledge generated in developing new tools and the tools themselves have the potential to inform a broad range of health-related advances, involving numerous diseases and conditions. For example, a project to explore how to characterize and predict undesired immune responses that can alter or block the effects of recombinant proteins and monoclonal antibodies can demonstrate relevance to the treatment of cancer, rheumatoid arthritis, and other diseases. The knowledge generated from such studies may well be applicable across entire classes of medical products and could help better ensure that such medicines are both safe and effective.

Regulatory science does not take place only in laboratories. It involves scientific tools and information-gathering and analytical systems to study data, people, health systems, and communities. To be most effective, advances in regulatory science must be fully integrated into the entire product development process.

Outreach and collaborative efforts are integral to predicting the failure or success of new discoveries and technologies early in development and reducing product development costs. Advances in regulatory science will help make the evaluation and approval process more efficient, helping to deliver safe new products to patients faster and strengthening the ability to monitor product use and improve performance, thus enhancing patient outcomes.

To successfully achieve the mission to promote and protect the public health requires a right balance between innovation and safety. Regulatory science should not stifle innovation, but rather encourage innovation while maintaining a commitment to safety and effectiveness.

The Chemistry, Control, and Manufacturing issues faced by the development of pharmaceutical, medical device, and cell therapy delivery solutions are similar philosophically. The pharmaceutical and cell therapy deliveries follow more similar regulatory interpretation. However, unlike a drug whose active ingredient does not change and whose inherent flaws cannot generally be fixed, a device can be improved through changes to its design or composition at any time. As a result, regulatory initiatives and review processes of a medical device will follow a similar philosophy but will differ in detail and implementation.

1.3.1   FDA Initiative in Pharmaceutical and Cell Therapy Delivery

Biomedical research has dramatically expanded the understanding of biology and disease. However, the development of new therapies is in decline, and the cost of bringing them to market has increased significantly. Every opportunity to improve the effectiveness and outcomes of healthcare and address growing threats to the strength and innovation of the biotechnology industry ensures that the best medical treatments are made available to patients in a timely manner. The following are some of the challenges and initiatives taken by FDA to modernize product development to improve the speed, efficiency, predictability, capacity, and quality, from development to manufacturing.

1.3.1.1 Expedited Programs for Serious Conditions   Speeding the development and availability of drugs that treat serious diseases are in everyone's interest, especially when the drugs are the first available treatment or have advantages over existing treatments. FDA has developed four programs to making such drugs available as rapidly as possible: Fast Track, Breakthrough Therapy, Accelerated Approval, and Priority Review [7].

The following summary describes each program, how they differ, and how they complement each other:

Fast track   Fast track is a process designed to facilitate the development and expedite the review of drugs to treat serious conditions and fill an unmet medical need. The purpose is to get important new drugs to the patient earlier. Fast Track addresses a broad range of serious conditions.

Determining whether a condition is serious is a matter of judgment, but generally is based on whether the drug will have an impact on such factors as survival, day-to-day functioning, or the likelihood that the condition, if left untreated, will progress from a less severe condition to a more serious one. AIDS, Alzheimer's, heart failure, and cancer are obvious examples of serious conditions. However, diseases such as epilepsy, depression, and diabetes are also considered to be serious conditions.

Filling an unmet medical need is defined as providing a therapy where none exists or providing a therapy that may be potentially better than available therapy.

Any drug being developed to treat or prevent a condition with no current therapy obviously is directed at an unmet need. If there are available therapies, a fast track drug must show some advantage over available therapy, such as:

- Showing superior effectiveness, effect on serious outcomes, or improved effect on serious outcomes
- Avoiding serious side effects of an available therapy
- Improving the diagnosis of a serious condition where early diagnosis results in an improved outcome
- Decreasing the clinically significant toxicity of an available therapy that is common and causes discontinuation of treatment
- Ability to address emerging or anticipated public health needs

A drug that receives Fast Track designation is eligible for some or all of the following:

- More frequent meetings with FDA to discuss the drug's development plan and ensure collection of appropriate data needed to support drug approval
- More frequent written correspondence from FDA about such things as the design of the proposed clinical trials and use of biomarkers
- Eligibility for Accelerated Approval and Priority Review, if relevant criteria are met
- Rolling Review, which means that a drug company can submit completed sections of its Biological License Application (BLA) or New Drug Application (NDA) for review by FDA, rather than waiting until every section of the application is completed before the entire application can be reviewed. BLA or NDA review usually does not begin until the drug company has submitted the entire application to the FDA

Fast Track designation must be requested by the drug company. The request can be initiated at any time during the drug development process. FDA will review the request and make a decision within 60 days based on whether the drug fills an unmet medical need in a serious condition.

Once a drug receives Fast Track designation, early and frequent communication between the FDA and a drug company is encouraged throughout the entire drug development and review process. The frequency of communication assures that questions and issues are resolved quickly, often leading to earlier drug approval and access by patients.

Breakthrough therapy   Breakthrough Therapy designation is a process designed to expedite the development and review of drugs that are intended to treat a serious condition, and preliminary clinical evidence indicates that the drug may demonstrate substantial improvement over available therapy on a clinically significant endpoint(s).

To determine whether the improvement over available therapy is substantial is a matter of judgment and depends on both the magnitude of the treatment effect, which could include duration of the effect, and the importance of the observed clinical outcome. In general, the preliminary clinical evidence should show a clear advantage over available therapy.

For purposes of Breakthrough Therapy designation, clinically significant endpoint generally refers to an endpoint that measures an effect on irreversible morbidity or mortality (IMM) or on symptoms that represent serious consequences of the disease. A clinically significant endpoint can also refer to findings that suggest an effect on IMM or serious symptoms, including:

- An effect on an established surrogate endpoint
- An effect on a surrogate endpoint or intermediate clinical endpoint considered reasonably likely to predict a clinical benefit (i.e., the accelerated approval standard)
- An effect on a pharmacodynamic biomarker(s) that does not meet criteria for an acceptable surrogate endpoint, but strongly suggests the potential for a clinically meaningful effect on the underlying disease
- A significantly improved safety profile compared to available therapy (e.g., less dose-limiting toxicity for an oncology agent), with evidence of similar efficacy

A drug that receives Breakthrough Therapy designation is eligible for the following:

- All Fast Track designation features
- Intensive guidance on an efficient drug development program, beginning as early as Phase 1
- Organizational commitment involving senior managers

Breakthrough Therapy designation is requested by the drug company. If a sponsor has not requested breakthrough therapy designation, FDA may suggest that the sponsor consider submitting a request if (1) after reviewing submitted data and information (including preliminary clinical evidence), the Agency thinks the drug development program may meet the criteria for Breakthrough Therapy designation and (2) the remaining drug development program can benefit from the designation.

Ideally, a Breakthrough Therapy designation request should be received by FDA no later than the end-of-phase-2 meetings if any of the features of the designation are to be obtained. Because the primary intent of Breakthrough Therapy designation is to develop evidence needed to support approval as efficiently as possible, FDA does not anticipate that Breakthrough Therapy designation requests will be made after the submission of an original BLA or NDA or a supplement. FDA will respond to Breakthrough Therapy designation requests within 60 days of receipt of the request.

Accelerated approval   When studying a new drug, it can sometimes take many years to learn whether a drug actually provides a real effect on how a patient survives, feels, or functions. A positive therapeutic effect that is clinically meaningful in the context of a given disease is known as "clinical benefit." It may take an extended period of time to measure a drug's intended clinical benefit. Therefore, in 1992 FDA instituted the Accelerated Approval regulations to allow drugs for serious conditions that filled an unmet medical need to be approved based on a surrogate endpoint. Using a surrogate endpoint enabled the FDA to approve these drugs faster.

Section 901 of the Food and Drug Administration Safety Innovations Act (FDASIA) in 1992 amended the Federal Food, Drug, and Cosmetic Act (FD&C Act) to allow the FDA to base accelerated approval for drugs for serious conditions that fill an unmet medical need on whether the drug has an effect on a surrogate or an intermediate clinical endpoint.

A surrogate endpoint used for accelerated approval is a marker—a laboratory measurement, radiographic image, physical sign, or other measure that is thought to predict clinical benefit, but is not itself a measure of clinical benefit. Likewise, an intermediate clinical endpoint is a measure of a therapeutic effect that is considered reasonably likely to predict the clinical benefit of a drug, such as an effect on irreversible morbidity and mortality (IMM).

The FDA bases its decision on whether to accept the proposed surrogate or intermediate clinical endpoint on the scientific support for that endpoint. Studies that demonstrate a drug's effect on a surrogate or intermediate clinical endpoint must be "adequate and well controlled" as required by the FD&C Act.

Using surrogate or intermediate clinical endpoints can save valuable time in the drug approval process. For example, instead of having to wait to learn if a drug actually extends survival for cancer patients, the FDA may approve a drug based on evidence that the drug shrinks tumors, because tumor shrinkage is considered reasonably likely to predict a real clinical benefit. In this example, an approval based upon tumor shrinkage can occur far sooner than waiting to learn whether patients actually lived longer. The drug company will still need to conduct studies to confirm that tumor shrinkage actually predicts that patients will live longer. These studies are known as Phase 4 confirmatory trials.

Where confirmatory trials verify clinical benefit, FDA will generally terminate the requirement. Approval of a drug may be withdrawn or the labeled indication of the drug changed if trials fail to verify clinical benefit or do not demonstrate sufficient clinical benefit to justify the risks associated with the drug (e.g., show a significantly smaller magnitude or duration of benefit than was anticipated based on the observed effect on the surrogate).

Priority review   Prior to approval, each drug marketed in the United States must go through a detailed FDA review process. In 1992, under the Prescription Drug User Act (PDUFA), FDA agreed to specific goals for improving the drug review time and created a two-tiered system of review times—Standard Review and Priority Review. A Priority Review designation means FDA's goal is to take action on an application within 6 months (compared to 10 months under standard review).

A Priority Review designation will direct overall attention and resources to the evaluation of applications for drugs that, if approved, would be significant improvements in the safety or effectiveness of the treatment, diagnosis, or prevention of serious conditions when compared to standard applications.

Significant improvement may be demonstrated by the following examples:

- Evidence of increased effectiveness in treatment, prevention, or diagnosis of conditions
- Elimination or substantial reduction of a treatment-limiting drug reaction
- Documented enhancement of patient compliance that is expected to lead to an improvement in serious outcomes
- Evidence of safety and effectiveness in a new subpopulation

FDA decides on the review designation for every application. However, an applicant may expressly request priority review as described in the Guidance for Industry Expedited Programs for Serious Conditions—Drugs and Biologics. It does not affect the length of the clinical trial period. FDA informs the applicant of a Priority Review designation within 60 days of the receipt of the original BLA, NDA, or efficacy supplement. Designation of a drug as "Priority" does not alter the scientific/medical standard for approval or the quality of evidence necessary.

Fast Track, Breakthrough Therapy, Accelerated Approval, and Priority Review are approaches intended to make therapeutically important drugs available at an earlier time. They do not compromise the standards for the safety and effectiveness of the drugs that become available through this process.

1.3.1.2   Greater Availability of Generic Drugs   Generic drugs make up more than 70% of the prescriptions filled in the United States as well as other countries in the world and usually is the only solution to affordable treatment. However, many products do not have generic alternatives even though patents for the reference products have expired. More generic products could be made available if the difficulty in determining bioequivalence for some products could be overcome. Metered-dose inhalers, dry-powder inhalers, certain topical products, and products that are not systemically absorbed present challenges in determining bioequivalence. Developing validated methods for determining bioequivalence for these products so that quality, lower-cost generic products can become more widely available are being pursued.

Generic Drug User Fee Amendments of 2012 (GDUFA) provides user fees for FDA to ensure timely review of applications for generic drugs. GDUFA is designed to speed access to safe and effective generic drugs to the public and reduce costs to industry. The law requires industry to pay user fees to supplement the costs of reviewing generic drug applications and inspecting facilities. Additional resources enable FDA to reduce backlog of pending applications, cut the average time required to review generic drug applications for safety, and increase risk-based inspections.

GDUFA is built on the success of the Prescription Drug User Fee Act (PDUFA). Over the past 20 years, PDUFA has ensured a more predictable, consistent, and streamlined premarket program for industry and helped speed access to new, safe, and effective prescription drugs for patients. GDUFA will also enhance global supply chain safety by requiring that generic drug facilities and sites around the world self-identify.

The GDUFA Regulatory Science Plan had identified 13 research topics for further study and ranged from quality-by-design (QbD) and postmarketing surveillance to bioequivalence (BE) and pharmacokinetic (PK) evaluation of complex dosage forms [8].

FDA had also issued draft guidances on developing and approving biosimilars, using a risk-based "totality-of-the-evidence" approach. The guidance to industry is contained in three documents and represents FDA's interpretation of the Biologics Price Competition and Innovation Act of 2009 (BPCI Act), which creates an abbreviated licensure pathway for biological products shown to be biosimilar to or interchangeable with an FDA-licensed biological reference product and was part of the Patient Protection and Affordable Care Act.

The first document, "Scientific Considerations in Demonstrating Biosimilarity to a Reference Product," explains the evaluation approach, intended to help companies submitting new 351(k) applications for demonstrating biosimilarity. This document includes recommendation for a gradual or "stepwise" approach in the development of biosimilar products, which include a comparison of the proposed product and the reference product with respect to structure, function, animal toxicity, human pharmacokinetics (PK) and pharmacodynamics (PD), clinical immunogenicity, and clinical safety and effectiveness.

Once a product is determined to be biosimilar, it will be eligible for a separate interchangeability determination. To meet the higher standard of "interchangeability," an applicant must provide sufficient information to demonstrate biosimilarity and also to demonstrate that the biological product can be expected to produce the same clinical result as the reference product in any given patient. Interchangeable products may be substituted for the reference product without the intervention of the prescribing healthcare provider.

The second draft guidance document, "Quality Considerations in Demonstrating Biosimilarity to a Reference Protein Product," provides an overview of analytical factors for drug developers to consider when assessing biosimilarity between a proposed therapeutic protein product and a reference product. Those factors include the expression system, the manufacturing process, an assessment of physicochemical properties, functional activities, receptor binding and immunochemical properties, impurities, the reference product and reference standards, and stability. This guidance expects that the expression construct for a proposed biosimilar product will encode the same primary amino acid sequence as its reference product.

The third guidance document, "Biosimilars: Questions and Answers Regarding Implementation of the Biologics Price Competition and Innovation (BPCI) Act of 2009," answers common questions about biosimilar product development in a question-and-answer format. Questions are intended to address concerns arising in the early stages of product development, including requesting meetings with the FDA, addressing differences in formulation from the reference product, and requesting exclusivity.

Once applications are received for approval of a biosimilar drug, FDA has committed to reviewing them within 10 months under the fifth authorization of the Prescription Drug User Fee Act (PDUFA).

The BPCI Act also includes:

- A 12-year exclusivity period from the date of first licensure of the reference product, during which approval of a 351(k) application referencing that product may not be made effective
- A 4-year exclusivity period from the date of first licensure of the reference product, during which a 351(k) application referencing that product may not be submitted
- An exclusivity period for the first biological product determined to be interchangeable with the reference product for any condition of use, during which a second or subsequent biological product may not be determined interchangeable with that reference product

1.3.1.3 Other Product Development Initiatives

Modernized manufacturing and product quality   FDA and other global regulatory agencies are leading efforts on "Quality by Design (QbD)," which applies regulatory science to modernize the understanding and control of medical product manufacturing processes. Advances in regulatory science will not only ensure better quality, but could also lower development and manufacturing costs. In the United States, areas of investigation supported by FDA include (1) continuous processing, in which materials constantly flow in and out of the equipment and reduce overall manufacturing time and cost; (2) the use of process analytical technology (PAT) to monitor and control manufacturing processes as opposed to just testing products; and (3) new statistical approaches to detect changes in process or product quality. Applying these approaches will help control complex manufacturing processes, enhance their efficiency, and provide more reliable products to patients. In addition, new technologies such as flexible manufacturing facilities and the use of modular and disposable equipment can speed production of products in routine and emergency situations.

National Vaccine Plan   The National Vaccine Plan was initially created in 1994 to provide a strategic approach for maximizing the impact of vaccines on the health of U.S. populations [9]. In 2010, the National Vaccine Plan was updated to reflect the priorities, opportunities, and challenges of today's science and the national immunization program, and it provides a guiding vision for vaccines and immunization in the United States for the decade 2010–2020 with the following five goals.

Goal 1: Develop new and improved vaccines.

Goal 2: Enhance the vaccine safety system.

Goal 3: Support communications to enhance informed vaccine decision-making.

Goal 4: Ensure a stable supply of, access to, and better use of recommended vaccines in the United States.

Goal 5: Increase global prevention of death and disease through safe and effective vaccination.

Two areas to note in the development of new and improved vaccines are as follows.

Development of the next generation of influenza vaccines   Scientists at NIH's National Institute of Allergy and Infectious Diseases (NIAID) have recently devised a new strategy for the development of more broadly protective vaccines for influenza, an approach that represents a promising step forward toward a universal influenza vaccine. Since influenza viruses change rapidly, influenza vaccines are updated and produced annually to protect against the virus strains that will be most common that year. In animal studies, researchers at NIH/NIAID were able to elicit an immune response to sites within influenza viruses that are shared across different influenza strains and that typically do not change very much over time, despite ongoing mutations in the virus. This is one of the many strategies toward the development of a safe and effective universal influenza vaccine, which would potentially eliminate the need for a new seasonal influenza vaccine each year and could remove the threat of an influenza pandemic.

SMART Vaccines   As technological opportunities emerge and patterns of disease change over time, it is difficult to decide how best to invest in new vaccine development and introduce new vaccines into routine and campaign immunization programs. In 2012, the Institute of Medicine (IOM), with National Vaccine Program Office (NVPO), began developing a decision support tool for prioritizing vaccine targets for development and use. They developed a software called Strategic Multi-Attribute Ranking Tool for Vaccines (SMART Vaccines).

The SMART Vaccines software makes it possible for decision-makers to develop and test hypotheses and assumptions, weigh competing values, and explore alternative scenarios and vaccine attributes to assist in setting priorities for vaccine targets for development and introduction. Users can take into account multiple factors, including health, economic, demographic, scientific, and policy considerations and can assess their relative rank among a range of factors. The tool allows the flexibility of factoring in values such as aiming to eradicate or eliminate a disease. Users are also able to generate information on cost-effectiveness, premature deaths averted, and gains in worker productivity, among other topics of importance to vaccine development and introduction. Using this model, SMART Vaccines has the potential not only to guide discussions regarding vaccine goals but also to provide a common platform for determining priority areas of national and global interests. The SMART Vaccines software is now available to the public for download and use online through the National Academy of Sciences website at http://www.nap.edu/smartvaccines.

New approaches to evaluate product efficacy in vaccine   It is not always possible to test whether a vaccine or treatment will work against a new or emerging infectious disease or against a terrorist threat because the threat may be rare or even nonexistent at the time the therapy needs to be developed. Animal testing is often the only available option, but many diseases lack good animal models, and animal studies are technically difficult to conduct and typically limited in size. Therefore, regulatory science will help to develop and validate improved predictive models. Regulatory science can also support the identification and validation of surrogate measures of product efficacy. For example, FDA's definition and acceptance of a serum hemagglutination inhibition antibody titer, which helps predict the efficacy of influenza vaccines, took years off the time required to approve new flu vaccines and, as a result, helped to double the number and capacity of U.S. licensed flu vaccine makers. Such biomarkers (e.g., responses in blood tests and other measurements or medical images) that predict efficacy are not yet available for most terrorism threats, emerging pathogens, or major global infectious diseases. Efforts to develop, refine, and validate new biomarkers can lower development costs and improve and speed the development of safe and effective products for unmet public health needs.

More flexible and agile approaches to product development and manufacturing of vaccines and biotech products   Knowledge of genetic sequences enable production of DNA and recombinant vaccines or needed treatments and diagnostic tests more quickly and safely without using the pathogen in manufacturing.

The use of platform technologies of this sort may offer the potential to scale up production more rapidly. For example, several technologies could potentially allow production of large amounts of new influenza vaccines for a pandemic in weeks rather than months. Platform technologies may also be applicable across broader ranges of products. For example, the same virus-like particle, live vector, DNA vaccine, or recombinant protein expression system could be used as the basis to rapidly develop and produce different, distinct vaccines intended to protect against illnesses such as flu, plague, SARS, or TB. Even stronger commonalities apply across technologies that can be used for detection or diagnosis, such as high-throughput assays for antibody, antigen, and nucleic acid detection.

Regulatory science helps to evaluate multiuse technologies and products including new methodologies for measuring product quality, potency, safety, and effectiveness.

1.3.2   Initiative of the Medical Device Delivery System

There has been a lot of discussion about balancing innovation and safety—whether there is a need to have more regulation of medical devices to assure safety and effectiveness—or whether there is a need to have less regulation of medical devices to foster innovation. In the United States, the FDA's medical device initiative, Innovation Pathway, establishes how innovation and safety and effectiveness do not have to exist on opposite ends of a swinging pendulum. They can be complementary and mutually supporting.

1.3.2.1   Expedited Access Premarket Approval Application for Unmet Medical Needs for Life Threatening or Irreversibly Debilitating Diseases or Conditions ("Expedited Access PMA" or "EAP") Program   The program features earlier and more interactive engagement with FDA staff—including the involvement of senior management and a collaboratively developed plan for collecting the scientific and clinical data to support approval [10a].

To be eligible for participation in the program, the medical device must fulfill the following criteria:

- Be intended to treat or diagnose a life-threatening or irreversibly debilitating disease or condition
- Represent one of the following:
  1. no approved alternative treatment/diagnostic exists, or
  2. a breakthrough technology that provides a clinically meaningful advantage over existing technology, or
  3. offers a significant, clinically meaningful advantage over existing approved alternatives, or
  4. availability is in the patient's best interest
- Have an acceptable data development plan that has been approved by the FDA

The EAP builds on the Innovation Pathway pilot, which is described in the following, and the FDA's experience with expedited review programs for pharmaceuticals, including Accelerated Approval and Breakthrough Therapies.

In addition to the EAP, a separate draft guidance is published that outlines FDA's current policy on when data can be collected after product approval and what actions are available to the FDA if approval conditions, such as postmarket data collection, are not met. Also included in the guidance is advice on the use of surrogate or independent markers to support approval, similar to the data points used for accelerated approval of prescription drugs [10b].

1.3.2.2 Innovation Pathway

The goal of Innovation Pathway is to reduce the overall time and cost it takes for the development, assessment, and review of safe and effective medical devices that address unmet medical needs so these devices can get to the patients who need them sooner without jeopardizing patient safety. It will promote high-quality regulatory science and help FDA better prepare and respond to transformative technologies and scientific breakthroughs [10c].

Innovation Pathway will be developing and rapidly testing new approaches to premarket review including the use of a decision support tool that will help assure that the regulatory decisions are more transparent and consistent. Such a tool can help decide whether there is sufficient evidence to allow the device to be studied for the first time in humans. An example of the Initiative Pathway is its application to products for patients with end stage renal disease—ESRD.

Because these are novel technologies, it is likely to raise new scientific and regulatory challenges. Key features of this pathway will be identifying and resolving these issues early by leveraging scientific expertise outside of the agency from the Network of Experts.

Clinical trial protocols would be developed by the sponsor and CDRH through an interactive process and have flexibility built in to allow for repeat testing and redesign.

Front-loading resources will reduce unnecessary delays and review these devices for approval in roughly half the time it takes for the typical premarket approval, or

PMA, application. However, devices that utilize the Innovation Pathway must still adhere to the regulatory standards for new applications. Just because a device is accepted into the pathway does not mean it is destined for approval.

Another initiative by CDRH to strengthen device research is the creation of a voluntary third-party certification program for medical device test centers across the country. Eligible test centers would have expertise in both device design and the conduct of high-quality clinical studies.

Unlike a drug whose active ingredient does not change and whose inherent flaws cannot generally be fixed, a device can be improved through changes to its design or composition at any time. By providing incentives to universities and other institutions in a competitive way to combine expertise in developing and in assessing devices, they can help find and fix problems earlier. Additionally, since certified test centers have well-established safety records, they will be permitted to conduct first-in-human studies at an earlier stage in device development. As a result, the device development process would become more predictable, safer, and less costly.

1.3.2.3 *Training of New Regulatory Scientists in Medical Device* In the United States and other countries, unlike the pharmaceutical industry, the education system has few programs in the device development. To train future innovators, regulators, academia, industry, and the healthcare community will need to work together to develop a publicly available core curriculum in device design, testing, regulatory processes, and postmarketing surveillance.

1.3.2.4 *Acceptability of Data* Device manufacturers had been conducting much device research overseas. However, there are difficulties in the United States to accept these data. Clear guidance from the FDA on criteria and circumstances under which data developed overseas could be used to support device submissions will result in better data and less of a need to conduct additional clinical studies. This situation will also provide a smoother review, less cost to companies, and fewer risks to patients from investigational devices.

1.3.2.5 *Human Factors and Usability Engineering to Optimize Medical Device Design* Human factors engineering (HFE) and usability engineering (UE) is the study to understand and optimize how people interact with technology. HFE/UE are important to the development of medical devices and include three major components of the device-user system: (1) device users, (2) device use environments, and (3) device user interfaces.

The process of eliminating or reducing design-related use problems for medical devices that contribute to or cause unsafe or ineffective medical treatment is part of a process for controlling overall risk. Where harm could result from "use errors," the dynamics of user interaction are safety-related and should be components of risk analysis and risk management.

Medical devices should be designed so that the devices are safe and reliable for their intended uses. To achieve this goal, the possibilities of hazards arising from use of and failures of the device and its components should be evaluated.

Hazards traditionally considered in risk analysis include:

- Chemical hazards (e.g., toxic chemicals)
- Mechanical hazards (e.g., kinetic or potential energy from a moving object)
- Thermal hazards (e.g., high-temperature components)
- Electrical hazards (e.g., electrical shock, electromagnetic interference (EMI))
- Radiation hazards (e.g., ionizing and nonionizing)
- Biological hazards (e.g., allergic reactions, bioincompatibility, and infection)

These hazards most often result from instances of device or component failure that are not dependent on how the user interacts with the device.

In addition, hazards for medical devices that are associated with device use should also be considered and are referred to as use-related hazards. These hazards include use errors involving failure to perceive, read, interpret, or recognize and act on information from monitoring or diagnostic testing devices and improper treatment (e.g., ineffective or dangerous therapy) for devices that provide medical treatment.

Use-related hazards occur for one or more of the following reasons:

- Device use requires physical, perceptual, or cognitive abilities that exceed the abilities of the user.
- The use environment affects operation of the device and this effect is not recognized or understood by the user.
- The particular use environment impairs the user's physical, perceptual, or cognitive capabilities when using the device to an extent that negatively affects the user's interactions with the device.
- Device use is inconsistent with user's expectations or intuition about device operation
- Devices are used in ways that were not anticipated.
- Devices are used in ways that were anticipated but inappropriate and for which adequate controls were not applied.

HFE/UE should be incorporated into device design, development, and risk management processes. Three central steps are essential for performing a successful HFE/UE analysis:

- Identify anticipated use-related hazards and unanticipated use-related hazards and determine how hazardous use situations occur
- Develop and apply strategies to mitigate or control use-related hazards
- Demonstrate safe and effective device use through human factors' validation

From the regulatory perspective, the risk analysis that fulfills Quality System requirements should include use error [11]. To establish the design input for the user interface and carry out design verification, human factors activities conducted throughout the development process can include task/function analyses, user studies, prototype tests, and mock-up reviews. Formative and validation testing fulfill the requirements to test the device under realistic conditions. Validation testing should be used to demonstrate that the potential for use error has been minimized.

1.4 CURRENT GOOD MANUFACTURING PRACTICE REQUIREMENTS FOR COMBINATION PRODUCTS

The recent breakthrough in science and technology had transformed the ability to treat disease and physiological disorders. As a result, the therapeutic solutions available for a disease state and physiological disorder may have different options available. Using diabetes mellitus as an example, a diabetic patient can be treated using traditional therapy (e.g., antidiabetic oral formulation or subcutaneous insulin), a medical device (insulin pump device), or a combination of a traditional pharmaceutical/medical device and cellular therapy (pancreatic cell transplant). The development efforts of each of these individual aspects are discussed in detail in Chapters 2, 3, 6, 7, 8, and 9. Combination products are getting more commonly used for therapeutic solutions. The regulatory requirements for the combination products are covered from Parts 3 through 1271.

1.4.1 Definition of Combination Products

A combination product is a product comprised of any combination of a drug and a device; a device and a biological product; a biological product and a drug; or a drug, a device, and a biological product. A combination product includes the following:

1. A product comprised of two or more regulated components, that is, drug/device, biologic/device, drug/biologic, or drug/device/biologic, that are physically, chemically, or otherwise combined or mixed and produced as a single entity (single-entity combination products)
2. Two or more separate products packaged together in a single package or as a unit and comprised of drug and device products, device and biological products, or biological and drug products (copackaged combination products)
3. A drug, device, or biological product packaged separately that according to its investigational plan or proposed labeling is intended for use only with an approved individually specified drug, device, or biological product where both are required to achieve the intended use, indication, or effect and where, upon approval of the proposed product, the labeling of the approved product would need to be changed, for example, to reflect a change in intended use, dosage form, strength, route of administration, or significant change in dose (a type of cross-labeled combination product)
4. Any investigational drug, device, or biological product packaged separately that according to its proposed labeling is for use only with another individually specified investigational drug, device, or biological product where both are required to achieve the intended use, indication, or effect (another type of cross-labeled combination product)

The constituent parts of a combination product retain their regulatory status (as a drug or device, for example) after they are combined. Accordingly, the CGMP requirements that apply to each of the constituent parts continue to apply when they are combined to make combination products.

1.4.2 The Final Rule

The rule offered two options for demonstrating compliance with the CGMP requirements applicable to a copackaged or single-entity combination product [12]. These options were either (1) to demonstrate compliance with the specifics of all CGMP regulations applicable to each of the constituent parts included in the combination product or (2) to demonstrate compliance with the specifics of either the drug CGMPs or the QS regulation, rather than both, when the combination contains both a drug and a device under certain conditions. These conditions included demonstrating compliance with specified provisions from the other of these two sets of CGMP requirements. In addition, for a combination product that included a biological product, the CGMP's requirements for biological products in 21 CFR Parts 600 through 680 would apply, and, for a combination product that included any human cell, tissue, and cellular and tissue-based products (HCT/Ps), the regulations in 21 CFR Part 1271 would apply.

The rule is organized in the four sections addressing scope (Section 4.1), definitions (Section 4.2), the CGMPs that apply to combination products (Section 4.3), and how to comply with these CGMP requirements for a single-entity or copackaged combination product (Section 4.4).

Section 4.1 states that the rule establishes which CGMP requirements apply to combination products, clarifies the application of these requirements, and provides a regulatory framework for designing and implementing CGMP operating systems at facilities that manufacture copackaged or single-entity combination products.

Section 4.2 provides definitions for terms used in the regulation. Some of these definitions are included for convenience, for example, cross-referencing an existing definition (such as for "combination product") or to establish the meaning for a reference term (such as "drug CGMP"). In addition to cross-referencing the definition for "device," the rule states that a device that is a constituent part of a combination product is considered a finished device within the meaning of the QS regulation and also states that a drug that is a constituent part of a combination product is a drug product within the meaning of the drug CGMPs. The definition for "current good manufacturing practice operating system" states that such a system is the operating system within an establishment that is designed and implemented to address and meet the CGMP requirements for a combination product.

Section 4.3 lists all of the requirements that may apply to a combination product under this rule, depending on the types of constituent parts the combination product includes. The CGMP requirements listed are those found in parts 210 and 211 for drugs, part 820 for devices, and parts 600 through 680 for biological products, and the current good tissue practices found in part 1271 for HCT/Ps.

Section 4.4 addresses how to comply with these CGMP requirements for copackaged and single-entity combination products.

The rule helps ensure that CGMP requirements that apply to single-entity and copackaged combination products are clear and consistent, regardless of which component has lead jurisdiction for the combination product or which type of application is submitted for marketing authorization. The rule also streamlines compliance with CGMP requirements for the combination products and to help ensure appropriate implementation of these requirements while avoiding unnecessary redundancy in CGMP operating systems for these products.

1.4.3 Postapproval Modifications to a Combination Product Approved Under a BLA, NDA, or PMA

A draft guidance is available on the underlying principles to determine the type of marketing submission that may be required for postapproval changes to a combination product as defined in 21 CFR 3.2(e) that is approved under one marketing application, that is, a biologic license application (BLA), a new drug application (NDA), or a device premarket approval application (PMA) [13].

The following section gives examples of significant changes that may be made to constituent parts of a combination product (i.e., changes that may require prior approval from FDA). The types of submissions that such changes may require, depending upon the submission type used to obtain approval of the combination product, are identified.

1. Certain changes in the combination product device constituent part (e.g., those that result in a combination product new indication for use, new clinical effects, or in a modified analyte and indication/patient population for an *in vitro* diagnostic) customarily require new preclinical and clinical data to provide support for safety and effectiveness. For any such changes that do not affect the primary mode of action, select the submission type to match the application type used to obtain approval of the combination product:
   a. PMA Original
   b. NDA Original
   c. BLA Original

2. Changes in the drug constituent part substance, drug constituent part production process, quality controls, equipment, or facilities that affect controlled release or drug particle size or have a substantial potential to have an adverse effect on the identity, strength, quality, purity, or potency of the drug constituent part. Such changes include those that may affect the sterility assurance of the drug constituent part, such as process changes for sterile drug substances and sterile packaging components. For such change, select the submission type to match the application type used to obtain approval of the combination product:
   a. NDA Prior Approval Supplement
   b. BLA Prior Approval Supplement
   c. PMA 180-day Supplement

3. Modified chemical formulation of the device constituent part (not a chemical that would be considered a drug constituent part of the combination product), hardware or software modification of the device constituent part, or other design modification to the device constituent part (without also changing the indication or patient population) for which only new preclinical testing and/or limited confirmatory clinical data are necessary to demonstrate reasonable assurance of safety and effectiveness of the modified device constituent part. For such change, select the submission type to match the application type used to obtain approval for the combination product:

   a. PMA 180-day Supplement

   b. BLA Prior Approval Supplement

   c. NDA Prior Approval Supplement

4. Changes in the biological product constituent part, production process, quality controls, equipment, facilities, or responsible personnel that have a substantial potential to have an adverse effect on the identity, strength, quality, purity, or potency of the product. Generally, for any such change, select the submission type to match the application type used to obtain approval for the combination product:

   a. BLA Prior Approval Supplement

   b. NDA Prior Approval Supplement

   c. PMA 180-day Supplement

5. Changes in indication or in patient population (without any other change to the combination product itself or to any constituent part except for relevant changes to the labeling) that require substantial clinical data to provide reasonable assurance of safety and effectiveness for the change but either no or very limited new preclinical testing. For such change, select the submission type to match the application type used to obtain approval for the combination product:

   a. PMA Panel-Track

   b. NDA Prior Approval Supplement

   c. BLA Prior Approval Supplement

If the applicable submission requirements for each change do not match (e.g., one change requires a Prior Approval supplement and another requires a Changes Being Effected supplement), then the type of submission should be that associated with the most significant change being submitted. For example, a manufacturer of a drug eluting stent approved under a PMA would like to modify the design of the stent and delete a test for the drug to comply with an official compendium that is consistent with FDA statutory and regulatory requirements. In isolation, the change in the design of the stent would generally require the submission of a PMA 180-day supplement, whereas the change in the test to comply with an official compendium for the drug would generally be submitted in an NDA Changes Being Effected-30 day supplement. In this case, when submitted together, the manufacturer should submit the PMA 180-day supplement for both changes.

1.5   CONCLUSION

Recent breakthroughs in science and technology are transforming the ability to treat diseases and related regulatory challenges for its approval. Different therapeutic delivery solutions give different regulatory challenges to its approval whether it is a single-entity pharmaceutical agent, medical device, cellular therapy, or combinations of any three of these therapeutic solutions. The development of regulatory science and new initiatives in FDA is intended to streamline compliance with CGMP and make effective medication to the patients in a timely manner. The common regulatory goal of all three platforms of therapeutic delivery solutions is to comply with CGMP, although the detailed process of achieving this goal is different.

REFERENCES

1. a. Advancing Regulatory Science for Public Health, October 2010; b. Guidance for Industry: CGMP for Phase 1 Investigational Drugs, July 2008; c. Guidance for Industry: INDs for Phase 2 and Phase 3 Studies. Chemistry, Manufacturing, and Controls Information, May 2003.
2. FDA Recall Notice. Shiatsu recall; August 31, 2011.
3. 21 CFR parts 210 and 211.
4. *FDA Draft Guidance: Classification of Products as Drugs and Devices and Additional Product Classification Issues*, June 2011.
5. 21 CFR parts 600, 601 and 610.
6. *FDA Compliance Program Guidance Manual.*
7. Guidance for Industry: Expedited Programs for Serious Conditions – *Drugs and Biologics*, May 2014.
8. a. Generic Drug User Fee Act; b. Pharm Technol 37(8):24–26.
9. The State of the National Vaccine Plan, 2013 Annual Report, U.S. Department of Health and Human Services.
10. a. Draft Guidance for Industry and Food and Drug Administration Staff: *Expedited Access for Premarket Approval Medical Devices Intended for Unmet Medical Need for Life Threatening or Irreversibly Debilitating Diseases or Conditions*, April 23, 2014; b. Draft Guidance for Industry and Food and Drug Administration Staff – *Balancing Premarket and Postmarket Data Collection for Devices Subject to Premarket Approval*, April 23, 2014; c. FDA Medical Device Innovation Initiative White Paper, CDRH Innovation Initiative, February 2011.
11. Draft Guidance for Industry and Food and Drug Administration Staff: *Applying Human Factors and Usability Engineering to Optimize Medical Device Design*, June 22, 2011.
12. Federal Register, Vol. 78, No. 14, January 22, 2013, 4307.
13. Draft Guidance for Industry and Food and Drug Administration Staff: *Submissions for Postapproval Modifications to a Combination Product Approved Under a BLA, NDA, or PMA*, January 2013.

SECTION 2

TRADITIONAL PHARMACEUTICAL DRUG THERAPY DEVELOPMENT

2

DEVELOPMENT OF TABLETS

Kwok Chow

2.1 INTRODUCTION

Compressed tablets are believed to have been first manufactured in 1844 in England [1]. The technology and ingredients have since been adapted and improved for high-speed production. Related theories and practice are written in many textbooks [2, 3] and scientific articles. Despite these advances, development and manufacturing of pharmaceutical tablets is often considered as less sophisticated and more problematic than that of nonpharmaceutical products. This is due to extensive regulations, complexity of drug delivery biology, and diverse material science challenges in formulation development of typical drug substances.

Administering pharmaceutically active substances in tablets is the most common mode of drug delivery. However, development of an orally administered drug product is a long and complex process that takes approximately at 8–10 years from the identification of a new molecular entity to product launch. It is a multidisciplinary program in science, business, and project management. The cost of bringing a new drug to market has been rising steadily to about 1 billion USD [4, 5]. It is expected to be safe and effective. The active ingredient in a tablet should be released upon administration quickly for immediate release tablets or at a specified rate for modified release tablets. The tablets should also have a satisfactory shelf life when stored under recommended storage conditions. Ideally, the formulation should be designed for low-cost and high-speed manufacturing. The design of the market image such as size, shape, surface characteristics, and packaging should promote patient compliance.

Therapeutic Delivery Solutions, First Edition. Edited by Chung Chow Chan,
Kwok Chow, Bill McKay, and Michelle Fung.
© 2014 John Wiley & Sons, Inc. Published 2014 by John Wiley & Sons, Inc.

TABLE 2.1   Content of a typical target product profile for an
immediate-release tablet

Section	Content
1	Development plan and milestones
2	Clinical targets
3	Dosage form and strengths
4	Drug substance characteristics
5	Drug product characteristics
6	Target drug substance and product specifications
7	Product composition
8	Manufacturing process
9	Development budget

A typical pharmaceutical development project involves the execution of a number of critical and noncritical tasks at targeted stages. Failure to successfully complete critical tasks such as product stability studies, choice of delivery system, and selection of product composition may cause significant delays and even project failure. Not finishing a small number of less critical tasks on time may be less problematic. However, accumulation of many less critical issues may create major problems as the development project progresses. In addition, it is not uncommon that major efforts are required at a late stage to correct poor decisions or mistakes made without careful consideration of information available at early development. Hence, the use of a suitable project-tracking/monitoring tool and applying sound strategies are important in order to avoid common development issues as well as to ensure smooth and successful execution of product development programs.

Target Product Profile (TPP) was introduced by the US FDA as part of the Quality-by-Design (QbD) initiative [6] for planning and managing development programs [7]. TPP for pharmaceutical development comprises scientific, development, and business components to facilitate dosage form design and development (Table 2.1). It can be used as a checklist to track quality, cost, time, and therapeutic attributes. In this chapter, TPP concept and development strategies are discussed as holistic user guides for designing and development of pharmaceutical (immediate-release) tablets. Information gathering, product definition, and decision-making through the preparation and use of TPP are included in the discussion.

2.2   DEVELOPMENT PLAN AND MILESTONES IN TPP

Setting and managing goals is a key success factor in formulation development. Including and updating the development plan and milestones using a TPP enable the project team to focus on value-added tasks in a complex and dynamic project. The purpose of the program is to design and prepare suitable pharmaceutical formulation(s) to bring therapeutic successes in target patient populations in a timely and cost-effective manner. The pharmaceutical product should be designed and manufactured for the therapeutic indication, dosage, dosing frequency, and patient compliance. Parallel and interdependent activities including drug substance characterization, formulation development, process development, manufacturing of clinical supplies, analytical testing,

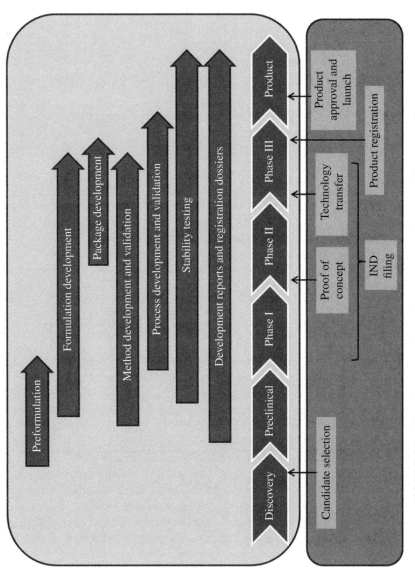

FIGURE 2.1 High-level product development stages/phases, activities, and milestones.

TABLE 2.2   Typical target product profile: project plan and milestones for an immediate-release tablet

Project plan and milestones

Milestone	Start date	Finish date
Preparation of target product profile		
Biopharmaceutics assessments		
Preformulation (solid-state characterization, polymorph screen, excipient compatibility)		
Target product characteristics		
Clinical formulation development (phase I)		
Phase I clinical trial material manufacture		
Phase I IND filing		
Clinical formulation development for phase II		
Phase II IND filing		
Confirmation of market image		
Analytical method validation		
Development of phase III and commercial formulation		
Stability and testing programs		
Phase III IND filing		
Registration batches, preparation, and stability program		
Product registration NDA filing (may register in more countries)		
FDA preapproval inspection if required		
Product approval		
Process validation		
Product launch		

and regulatory filing are expected to be completed on a defined timetable (Fig. 2.1). An abbreviated process plan/milestone in a typical TPP is shown in Table 2.2.

A typical tablet development program comprises complex scientific, regulatory, business, and management tasks. It must be well-managed to produce satisfactory results. It is generally conducted as a gate-staging program [8, 9], where major decisions and major refinements of the TPP/project plan are handled at milestones (Table 2.2). The time and events are controlled by the drug discovery life cycle, clinical phases, and pharmaceutical development stages such as drug synthesis, drug product design, and process development/scale-up. Cross-functional team members from clinical, biopharmaceutics, chemistry (organic, analytical, and physical), material science, production engineering, sales and marketing, and project management are engaged at multiple stages. The degree of success is influenced by leadership, expertise, teamwork, and execution. A high-level drug development program with decision points/milestones is provided in Figure 2.1.

To master and manage a complex, multidisciplinary program, a TPP that comprises a process map (development plan) and target chemistry, manufacturing, and control attributes/strategies is recommended. In the design of a TPP, a front-loading innovation approach [10] to assess the clinical goal, physicochemical characteristics of the drug substance, available delivery technology, development capability, and resource can be applied to clearly identify the goals and strategies to target the essential activities that may bring better results earlier. Assessment of influencing factors before the experiments can help prevent unexpected outcomes and allow proactive problem solving. TPP also serves as a communication tool for the project team to set, track, and revise common goals.

2.3 CLINICAL TARGET PROFILE

The design criteria of a pharmaceutical product are determined by the clinical requirements, manufacturing capability, and physicochemical properties of the drug substance (Fig. 2.2). Clinical Target Profile addresses the clinical aspect of formulation design and would include pharmacology, biopharmaceutics, and dosing strategy information. The selection of a dosage form is influenced by the Clinical Target Profile for drug delivery to the target site given the treatment goal, biopharmaceutics, and patient populations. The Clinical Target Profile also serves as a risk analysis/management and product design tool. A Clinical Target Profile could be created using front-end innovation approach and refined by the stage-gate process throughout a development program to ensure the product development objective is consistent with the treatment goal.

For a target clinical profile of an immediate release tablet, the drug substance is expected to dissolve rapidly, followed by GI absorption at a suitable rate to the systemic circulation in order to produce the desired therapeutic effect. Based on the mechanism of action, pharmacology, and pharmacokinetics, the dosing regimen could be proposed to support patient compliance. In general, once or twice daily administration of an easy-to-swallow tablet would be adequate for most adult and older children populations. The Target Clinical Profile helps identify potential challenges in formulation development such as tablet size for large-dose formulations and drug content uniformity for low-dose products before the experimental work. The Target Clinical Profile and its revisions (revised as the project progresses) also provide the direction for achieving the eventual goal of the program (Table 2.3).

TABLE 2.3 Typical target product profile: clinical targets

Clinical target	Information
Treatment goals and indication	Mechanism of action of the therapeutic agent
	Diseases to be treated
	Patient population
	Treatment goal
Biopharmaceutics— pharmacokinetics and pharmacodynamics	Absorption of drug in the gastrointestinal tract
	Dose response profile
	Pharmacokinetic parameters such as half-life, clearance, and volume of distribution
	Potential drug and food interactions
	Potential toxicity
Effective dose and dosage strategy [Pediatric (age 2–12; to be confirmed) or adult]	Target dosage
	Frequency and route of administration
	Chronic vs. acute therapy, treatment period
Clinical program	List of planned clinical studies
	List of bioequivalence studies for clinical formulations

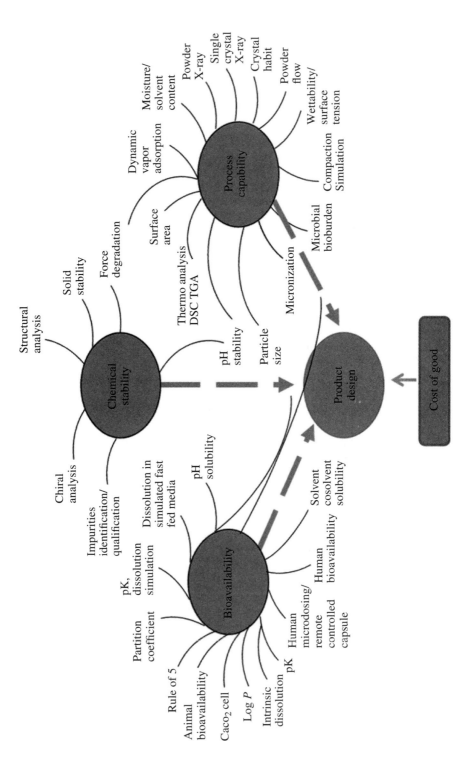

FIGURE 2.2 Elements defining a pharmaceutical dosage form design.

2.4 DRUG SUBSTANCE CHARACTERISTICS IN TPP

The physicochemical and biological properties of a drug substance play an important role in defining the developability and target delivery system. The characteristics of a drug substance in a tablet have direct impacts on the clinical performance, chemical stability, and processing behavior (Fig. 2.2). Therefore, the use of physicochemical data of the drug substance such as those collected during salt selection and preformulation stages can be valuable for establishing formulation strategies [11, 12]. Mapping the existing and target drug substance characteristics in the TPP (Table 2.4) helps create the design concept for the target dosage form.

A highly water-soluble and permeable compound such as Biopharmaceutics Classification System (BCS) class I molecule [13] can be formulated as an immediate release tablet if the drug substance has suitable compaction (physical) behavior and pharmacokinetics. A poorly soluble (drug) compound may also be developed as an immediate release tablet if it can be dissolved quickly *in vivo* for absorption. In this case, the particle size, solid form, pH, and pK_a of the drug substance become important, and these characteristics should be included in the TPP. The chemical structure and solubility may also influence the oral delivery of the drug substance. The Rule of five (proposed by Lipinski in 1997) addresses how solubility and molecular structure could be a useful guide to determine not only whether a new molecular entity should enter drug development from discovery research [14] but may also help estimate the effort required to conduct the

TABLE 2.4 **Typical target product profile: drug substance characteristics for an immediate-release tablet**

Drug substance characteristics	Information
Molecular	Molecular structure, molecular weight, functional groups, potential specific and nonspecific interactions, pK_a, counter ion if applicable
Solid-state	Crystal form, crystallinity, solvates including hydrate, moisture content
Physical	Solubility, particle size distribution, melting point, flow properties, compressibility, viscoelastic properties, color
Chemical	pH-stability, moisture sensitivity, heat sensitivity, light sensitivity, oxidation potential, excipient compatibility, impurity profile (force degradation profile)
Safety	Toxicity category for safe handling explosivity

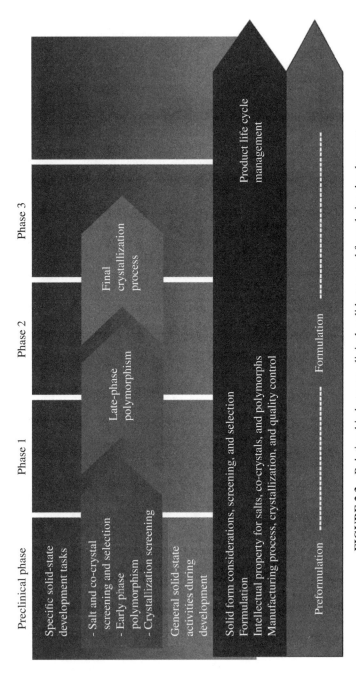

FIGURE 2.3 Relationship between clinical, solid-state, and formulation development.

formulation development program. Other factors such as chemical stability, processing behavior, target dose, and patient population also influence formulation decisions such as the choice of excipients.

Particle size distribution and particle morphology affect the process and clinical performance of a drug substance. These powder properties should be addressed as soon as possible and adjusted as necessary throughout the development program. The use of different particle-sizing techniques such as mechanical milling, jet milling, and modification of final crystallization conditions will impact the process behavior, manufacturing technology (including safe handling procedures), and bioavailability of drug in the product.

In general, there is insufficient time and opportunity to fully characterize the solid-state properties of a drug candidate at the preformulation stage. However, important characteristics such as processing behavior and additional solid forms can often be revealed after the early API characterization study, for example, during chemical development and formulation/process development. Typically, drugs that exhibit biological activities are large organic molecules with multiple functional groups. These molecules may pack in various ways in the crystal lattice without increasing the crystal energy significantly. Hence, it is not uncommon to detect additional polymorphs or solvates, especially hydrates of drugs at different stages of development and even after commercialization. It is generally accepted that "the number of forms known for a given compound is proportional to the time and energy spent in research on that compound" [15]. Collaboration between clinical, chemical, and formulation development (Fig. 2.3) to establish strategies to handle existing and potential solid-state property issues can make a difference in the execution of a successful program.

2.5   DRUG PRODUCT CHARACTERISTICS IN TPP

Setting and refining the target product characteristics from the beginning and throughout the formulation program helps focus the project team to develop a quality product in a time- and cost-effective manner. Ideally, a tablet should be sufficiently easy to manufacture, acceptable in global markets, and easy to be administered by patients. The shelf life, established from data generated in stability programs based on the proposed specifications and testing/storage criteria detailed in ICH guidelines [16], should not be less than 18 months to allow for manufacturing, testing, warehousing, product distribution, and retail inventory management. The drug product characteristics in the TPP (Table 2.5) can be defined and refined during the course of development. It is a useful guide to ensure that the product is developed to meet the clinical, marketing, manufacturing, budget, and patient compliance goals.

TABLE 2.5 Typical target product profile: drug product characteristics of an immediate-release tablet

Drug product characteristics	Information
Shelf life (room temperature)	Typically shelf life of 2 years or more, at least 18 months. The shelf life should be supported by ICH stability testing
Color	A color acceptable to the patient population and markets in difference countries
Size and shape	For example
	Debossed modified capsule-shape tablet of a certain dimension and thickness
	Logo and/or unique product identification number
Coating	Film coating is often required for enhancement of appearance, safety in packaging, ease of swallowing, and improvement of product stability
Dose uniformity	Meet in-house acceptance criteria and compendial requirements
Excipients	Meet USP, PhEur, JPE requirements for global markets
	Below maximum daily intake levels
	BSE free certification
Storage condition	Ideally, room temperature storage
Microbiology quality	Meet harmonized requirements for tablets
Pack	A plastic bottle pack is recommended for transportation and lower cost of packaging materials. For example: white 60 counts HDPE bottle closed with child resistant cap
	Blister calendar (unit dose) pack is a common packaging configuration to enhance patient compliance. Opaque foil-foil blister may be required for light and/or moisture-sensitive drugs
Markets	Global or local. For example, USA, Europe, Asia Pacific, Africa, Middle East, and Japan
Sales volume (units)	An early forecast in early development
	At late development, country-by-country forecast at product launch and at 5–10 years
Price	The cost-of-good changes dramatically from early to late development. The estimates are useful for development and marketing budget as well as rate of return consideration. Estimated price based on cost-of-good, market competition, and reimbursement opportunities
Marketing plan	This could include intelligence of existing and competitor products being developed. High-level strategies for product launch, market penetration, and acquiring market share may be included

2.6 DRUG SUBSTANCE AND DRUG PRODUCT SPECIFICATIONS IN TPP

Drug substance and drug product specifications are test procedures and acceptance criteria established to confirm product quality and consistency. The scope and requirements of drug substance and product specifications are detailed in regulatory guidelines [17]. A market or clinical formulation for treatment of a disease must satisfy the requirements in the specifications provided in approved regulatory submission(s). Test data in compliance with the registered specifications must be available for every lot of the drug substance and product released from manufacturing for distribution. The proposal, preparation, and justification of specifications at different stages of a pharmaceutical development program are fundamental steps to assure that the intended product has been developed. An example of a drug substance and drug product specifications in a typical TPP for an immediate release tablet is illustrated in Table 2.6 and Table 2.7. It should be noted that additional properties such as powder flow, compressibility, crystal habit, and surface tension are normally not included in a drug substance specification. These parameters may become important for dissolution or processing. They should be tracked, if needed, in the development program.

2.6.1 Strategies to Develop Specifications

Establishment of specifications is an evolving process. Proposed specifications are generally introduced early in the development program based on regulatory requirements and experience of the formulation team. They are revised and justified based on stability, manufacturing, and clinical/toxicity data as the program progresses. The proposed specifications are commonly used as a planning tool as well as targets for formulation and analytical development. Formulations are designed, prepared, and manufactured to meet the acceptance criteria of the specifications. Based on the specifications, specific and sensitive analytical methods are developed and validated for input raw materials, drug substances, and finished product testing.

Development of drug substance specifications is an interactive process, where solid-state data, processing results, and specification refinement are closely linked. An example of how the drug substance specifications can be developed for a typical tablet product is illustrated in Figure 2.4. Drug substance and product specifications are also interdependent on each other. Ideally, specifications should be a measure of important product performance, such as safety and clinical efficacy, although this is not achievable in all cases. As part of the FDA Quality-by-Design initiative, attempts are being made to link product specifications such as *in vitro* dissolution performance with product toxicity and/or effectiveness in humans [18, 19].

2.6.2 Analytical Methods for Specifications

Analytical method development and validation is an essential part of a pharmaceutical development program. It is not possible to establish and confirm the compliance of specifications without appropriate analytical methods. Reliable analytical methods

TABLE 2.6   Typical target product profile: drug substance specifications of an immediate-release tablet

Drug substance specifications	Acceptance criteria
Appearance Usually by visual inspection	Define color of drug powder

Identification 1
Identification 2
  Usually, the drug substance is identified
    by two analytical methods such as
    HPLC and infrared spectroscopy
	Retention time or compliance with reference
	  spectrum will be acceptable for release

Assay Usually by HPLC	Defined by regulatory body Stricter than drug product assay

Impurities
  Usually by HPLC
	Specified individual impurities
	Individual unspecified impurities
	Genotoxic impurities
	Total impurities
	Based on regulatory guidelines and toxicity
	  data

Residual on ignition By compendial method	Based on regulatory requirements
Heavy metals By compendial method	Based on regulatory requirements

Residual solvents
  Usually by gas chromatography method
	Based on regulatory requirements and
	  presence of potential residual solvent

Moisture content
  Usually by Karl Fisher, water activity,
    or loss on drying method
	Based on development data and generally
	  accepted limits

Particle size
  Usually by sieve analysis of laser
    diffraction method
	Based on formulation and process experience
	  and accepted practice

Polymorphism
  By melting point (including meting point
    apparatus or hot-stage microscopy),
    IR, X-ray powder diffraction, thermal
    analysis (e.g., DSC, TGA), Raman,
    and/or solid-state NMR
	Match melting or reference spectrum

Chirality (optical purity)
  By optical rotation, HPLC with chiral
    column
	Set limits based on drug substance, clinical,
	  and toxicology data

Microbial limits
  Usually by compendial method
	Based on compendial requirements, site
	  specific information, drug product
	  characteristics, patient population

TABLE 2.7 Typical target product profile: drug product specifications of an immediate-release tablet

Drug product specification	Acceptance criteria
Appearance	Color, shape, product identification attributes such as coating, debossing, and product-unique product identification information on tablet
Identification	By analytical method that is specific to the drug substance
Assay (content)	Meet FDA, EU, and JP requirements such as 95–105% end-of life, tighter at release
Impurities	Specified individual impurities
	Individual unspecified impurities
	Genotoxic impurities
	Total impurities
	Based on drug substance specification, manufacturing process, regulatory guidelines, and toxicity data
Content uniformity	Assay or weight depending on the potency
Usually by compendial method	Based on current compendial requirements for tablets
Dissolution	Acceptance criteria based on formulation data and regulatory/compendial guidances
Usually by compendial method	*In vivo–in vitro* correlation data
Disintegration	Based on formulation data and generally accepted limits
Usually by compendial method	
Crushing strength	Based on experience and generally accepted limits for the size of the formulation
Usually by use of specific hardness determination equipment	
Friability	Based on generally accepted limits
Use compendial method	
Microbial limits	Based on compendial limits, site-specific information, drug product characteristics, patient population
Use compendial method	

are developed rapidly in the preformulation or toxicology formulation development stage to evaluate the quality and performance of drug substance, process, and potential formulations. The precision, accuracy, efficiency, and procedure of these methods are improved as the program advances to reduce cost and meet clinical-phase specific regulatory requirements [20, 21]. Analytical methods are fully validated for testing clinical supplies in late/pivotal studies. The type and the robustness of analytical methods are essentially mandated by the development stage and the content of the specifications. Setting and revising the proposed specifications throughout the development program is a value-added strategy for the formulation and analytical scientists to work closely together to refine the analytical methods and design the experiments to collect valuable data for formulation and process development as well as optimization. Furthermore, the TPP with specification information can be used as a strategic tool to document and track the development of specifications and analytical methods.

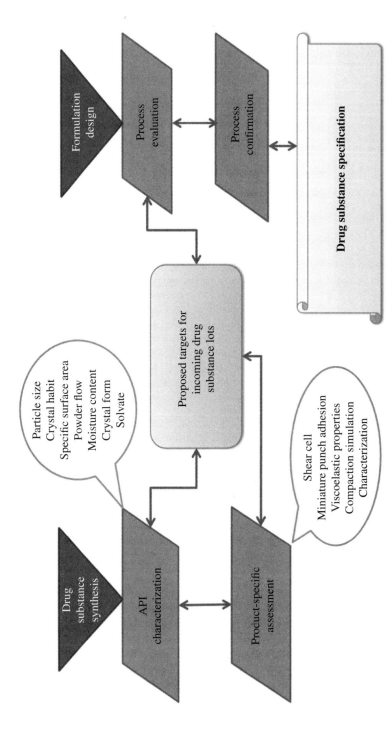

FIGURE 2.4 Example of a decision tree to develop drug substance specifications for an immediate-release tablet.

Recent advances of analytical technologies are playing a significant role in improving product quality and manufacturing efficiency. Process Analytical Technology (PAT) is being adopted to provide real-time feedback in monitoring and/ or adjusting pharmaceutical processes as part of the QbD initiative [22]. This approach enhances product quality and consistency in accordance with set specifications. These analytical tools are also deployed to develop design spaces for formulation compositions and manufacturing processes to allow more flexible limits to produce better products per specifications. With modern technologies such as advanced mass spectroscopic techniques, the analysis of low levels of toxic substances, such as genotoxic impurities and residual solvents, is becoming a reality. The development of specifications for these compounds should be included in formulation development [23].

2.7 PRODUCT COMPOSITION IN TPP

The strategies of putting together the composition of a tablet involve the consideration of intended medical treatment, drug substance properties, cost of good, and excipient quality, toxicity, function, and processing behavior. Application of scientific skills and knowledge related to pharmaceutical products, development experience, manufacturing operations, and financial planning also contributes to successful selection of excipients for the drug product. Typically, excipients are grouped according to their functions, such as processing aid (e.g., glidants, granulating agents), chemical stabilizer (e.g., antioxidant, pH-adjusting agents), and *in vitro* delivery characteristics (disintegrants, surfactants), and selected accordingly. An example of common excipients that may be chosen and included in the TPP for an immediate release tablet is provided in Table 2.8.

2.7.1 Checklist for Excipient Selection

As discussed, the choices of excipients depend on a number of clinical, scientific, and business factors. For example, the presence of reactive impurities in an excipient can adversely affect the stability of a drug substance depending on the degradation mechanism [24]. Establishing the criteria for consideration is crucial because excipients may affect regulatory approval, product release, safety, and effectiveness. A typical checklist for the selection of excipients in an immediate release tablet is provided in Table 2.9. Furthermore, advances in combinatorial chemistry are creating new molecular entities with a larger and more complex chemical structure. These molecules tend to have poor pharmaceutical characteristics such as low solubility, poor GI permeability, and unacceptable powder flow properties. Their formulations are more challenging to design and manufacture. The use of suitable excipients in these products may help create drug delivery and manufacturing solutions and, therefore, is becoming more important in a formulation development project.

The development of new excipients presents new opportunities to solve performance problems for modern drug substances. However, the required toxicity

TABLE 2.8   Typical target product profile: formulation composition of an immediate-release tablet

	Formulation composition	
Excipient function	Typical excipients	Typical use level
Binders/fillers	Microcrystalline cellulose, lactose, mannitol, sorbitol, xylitol, pregelatinized starch, calcium phosphate dihydrate	Up to 90+% (w/w)
Disintegrants	Sodium starch glycolate, Ac-di-sol, Crospovidone	About 2–3% up to 10% (w/w)
Lubricants	Magnesium stearate, stearic acid, glyceryl behenate, sodium stearyl fumarate, glyceryl monostearate	About 1% for magnesium stearate, up to 5% (w/w), depending on type of lubricants
Glidant	Colloidal silicon dioxide, talc	Fraction of 1% (w/w), higher levels used for dissolution enhancement
Granulating agents	PVP, starch, hydroxypropyl cellulose, HPMC	Up to 2–5% (w/w)
Colorants	FD&C dyes, iron oxides, titanium dioxide, natural dyes	Usually fraction of 1% (w/w)
Surfactants	Polysorbate 80, poloxymers	0.5–5% depending on surfactant
Stabilizers	Antioxidants—vitamin C, BHT pH adjusting agents—citric acid, acid or basic salts of different solubilities	About 2–3% for cosmetic coating
Coating materials	Immediate-release polymers: HPMC, PVA, Eudragits, starch Colorants: titanium dioxide, iron oxides, FD&C dyes Plasticizers: PEG, tracetin, glycerin	Coating weight gain of about 2–3% (w/w)

testing [25], development time, and cost are prohibitive even for major excipient suppliers. The cost of introducing a new excipient is estimated at approximately 2 million USD or higher [26]. For most development projects, common excipients that have been used in approved products (such as those listed in USP/NF, PhEur, and JPE) are preferred, and usage levels in the FDA Inactive Ingredient (IIG) list (or equivalent for products in other countries) are taken as upper limits to avoid regulatory hurdles. Qualified vendors with proven quality and supply chain record and up-to-date drug master files are chosen. Cost containment for excipients was not viewed as an important issue in the past because the cost of excipients was regarded as a small portion of the prescription cost, especially for patented medicines. However, cost containment of excipients is gaining popularity with increasing generic competition and interest in healthcare cost reduction.

TABLE 2.9 Example of a checklist to select excipients for an immediate-release tablet

Selection factors	Selection criteria	Excipients for selection
Treatment of disease	Dosage	Excipients for high drug loading for compression
	Rapid onset	Fast disintegrating excipients
	Patient acceptance	Colorant for the disease state or patient population, e.g., geriatric or pediatric. Film coating materials to enhance appearance, taste mask, enhance ease of swallowing
Drug substance properties	Toxicity or side effects	Excipients for slightly delayed absorption to avoid acute side effect
	Bioavailability	Lipid-based excipients may enhance bioavailability
		Disintegrants to enhance dispersion
		Microenvironment to enhance solubility and dissolution for pH-sensitive compounds
		Surfactant(s) to enhance dissolution
		Water-soluble or water-insoluble excipients to modulate dissolution behavior
		Crystallization inhibitors
	Particle size and distribution	Excipients with suitable particle size and flow properties to produce product with suitable content uniformity
	Chemical stability	Compatible excipients to avoid chemical degradation. For example, avoid lactose for amines
	Physical compatibility	Excipients with suitable hygroscopicity for drug substance that is a hydrate or that can form a hydrate
		Avoid excipients that produce physical interaction with drug substance, e.g., magnesium stearate forms a eutectic mixture with ibuprofen
	Light sensitivity	Opaque excipients for film coating
	Oxidation	Avoid excipients with peroxide content
		Antioxidants
		Suitable packs and oxygen scavengers
	Hydrolysis	Excipients with low water activities or moisture content
		Acidic or basic excipients to provide microenvironment to prevent pH catalytic hydrolysis
		Moisture barrier packaging
		Moisture barrier coating to provide in-use stability
	Taste	Film coating excipients
		Flavouring agents include sweeteners and flavors
	Poor compressibility	Compressible ingredients
		Ingredients for unit processes such as roller compaction or wet granulation process to enable compression

(Continued)

TABLE 2.9 **(Cont'd)**

Selection factors	Selection criteria	Excipients for selection
	Cohesive and adhesive	Use of less cohesive excipients
		Use of lubricants or glidants
		Use of bulking agents to dilute the drug substance to reduce punch adhesion
	Color	Film coating materials and colorants to enhance appearance and visual consistency of product
Quality	Excipient toxicity	Use excipients with acceptable toxicity data
		Use excipients at the same or lower levels than those in approved pharmaceutical products
		If possible, use excipients listed in FDA's IGG list
		Avoid lactose if lactose intolerance is an issue
	Quality and quality control	Use vendors that can be qualified and audited successfully
		Define purchase specification
	Supply chain	Where possible, avoid single-source excipients
		Excipients from reliable vendors
	Synthetic source	Use nonanimal-source excipients
		Use certified bovine serum encephalitis excipients
Ease of formulation	Database	Excipients that are well studied by researchers, vendors, or in-house scientists
	Processing	Excipients with in-house expertise
	Equipment and facility	Excipients suitable for process at all scales using equipment in existing facility
Function	Processing technology	Use the right excipients for the intended process and technology for the manufacture of the dosage form
	Processing aid	Excipients to assist granulation, wetting, powder flow, mixing, and compression
	Product performance	Disintegrants to allow rapid disintegration and dissolution
		Surfactant to improve wetting properties and solubility
Manufacturing cost	Purchase cost	Low-cost excipients
	Testing cost	Excipients that do not require extensive testing on release
		Excipients that do not require expensive tests
	Process efficiency	Excipients to reduce processing steps, time, and complexity. For example, use direct compressible excipients such that a granulation process may not be required
Intellectual properties	Patent infringement	Compositions that may infringe existing patents
	Patent filing	Compositions that may provide patent protection
	Trade secret	Choose compositions that allow application of pharmaceutical processes that are difficult to be copied by another organization

2.7.2 Excipient Compatibility Study

Excipient compatibility plays a significant role in defining the composition of the formulation. A good description of the strategies and procedures to conduct an excipient compatibility program can be found in Narang et al. [27]. In practice, excipient compatibility may begin with a paper exercise by:

1. Predicting degradation pathways based on the molecular structure of the drug substance and potential excipients
2. Studying the force degradation data of the drug substance
3. Modeling the required dosage form characteristics given the estimated dosage and route of administration
4. Considering the proposed manufacturing process

Then, a short list of noninteraction (target) excipients and an experimental plan (or protocol) can be prepared. The proposed excipient levels can be based on published use levels such as those listed in the *Handbook of Pharmaceutical Excipients* [28], the content of existing (marketed) products, manufacturing requirements, the weight of the dosage unit, and the properties of the drug substance.

Although binary (one drug and one excipient per sample) excipient compatibility studies are frequently used in the pharmaceutical industry, statistical design of experiment (DOE) provides additional benefits of evaluating multiple factors and interactions in the same experiment [29]. DOE is especially valuable for difficult/complex formulations where multivariate optimization is required to achieve the development objectives. Also, appropriate use of DOE provides better understanding of the impact of excipients on product stability. This helps avoid unexpected results and collect valuable information for troubleshooting downstream in the development program. Traditionally, chromatographic and spectroscopic methods are used to assess the potency and impurity content in excipient compatibility samples after storage under defined conditions including elevated temperature, and high humidity after milling and/or different packaging configurations. Modern high-throughput techniques and nonspecific thermal methods such as differential scanning calorimetry and microcalorimetry have also been applied in compatibility studies. In theory, reaction mechanisms and kinetics should also be studied for a better prediction of drug degradation. However, due to the complexity of formulation compositions, a full understanding of the degradation mechanism is not always possible, and the use of kinetic data to accurately predict product stability (shelf life) remains difficult. In these cases, statistical design of experiment can be a complementary tool in revealing complex excipients and drug substance interactions. Before the excipients and levels are finalized, abbreviated/simulated process trials such as powder flow properties should also be conducted to explore the manufacturing capability.

Drug-related degradation impurities should be measured and assessed against the proposed specifications. Any impurities that are potentially carcinogenic, teratogenic, or genotoxic would be strictly controlled [23]. When a potential impurity contains structural alerts, additional genotoxicity testing or assessment of the impurity, typically in a bacterial reverse mutation assay, or using established software such as Deductive Estimation of Risk from Existing Knowledge (DEREK) or Multiple Computer Automated Structure Evaluation (MCASE), should be considered [30]. The presence of genotoxic impurities may occur under certain processing conditions such as heating of an alcohol in the presence of a mesylate salt. Hence, a molecular-based assessment of the chemical compatibility of a drug substance with excipients is valuable.

2.7.3   Strategies in Defining Formulation Composition and Excipient Levels

Because of the long clinical and regulatory process in drug development, there is a reluctance to change the formulation composition and excipient levels, especially at late clinical stages. Ideally, the composition should be defined as early as possible. A change of excipients impacts product performance such as dissolution and bioavailability as well as processing characteristics. This may significantly delay the development program because additional clinical and bioequivalence testing may be required. Unfortunately, changes in formulations are often required as clinical data are collected and modification of drug substances is observed in scale-up or synthetic route/process changes. When a change of the product composition is required, the FDA SUPAC [31] and biowaiver [32] guidances are useful reference regulatory documents to assist planning. Direct communication with regulatory bodies is recommended for clarification to avoid any costly misinterpretations.

2.8   PRODUCT MANUFACTURING IN TPP

It is not uncommon that most resources are devoted to the manufacture of clinical and registration materials, stability testing for regulatory submission, and process scale-up and validation in a drug development program. A pharmaceutical formulation and its manufacturing process are often decided early in the program on limited data. The time spent on formulation and process design can be limited, for example, as little as a few months for a program of as much as 10 years. This approach often causes expensive and difficult-to-correct issues downstream in the clinic and manufacturing facility. At the end, a less-than-desirable clinical or commercial product may be produced causing program failure, quality problems, or supply issues.

Process design and development are normally conducted in parallel with other activities in the development of new molecular entities. Minimal process work is often performed at early development compared to that in late development because of the high attrition rate of drug candidates. A low budget and a narrow scope are generally set in anticipation of a potential change of formulation/process, clinical failure, adverse toxicological events, limited drug substance availability, and change of drug substance properties including purity, solid-state characteristics, and processing behavior. Less-than-optimal process selection/design and a lack of process understanding, however, often create significant downstream risks in achieving the desired product performance and quality. Changes of manufacturing process,

TABLE 2.10 Typical target product profile: manufacturing process of an immediate-release tablet

Manufacturing process	
Manufacturing process	Process: direct compression, aqueous granulation, solvent granulation, or roller compression
Blending	Lubrication in a V-blender
	Achieve blend content uniformity
	PAT-controlled
Granulation	For example
	Wet granulation using a high-shear mixer followed by wet milling using a Comil
	Optimized process to provide to meet target in-process limited
	PAT
	Fluidbed dry to achieve a moisture content of not more than 1% (w/w) and a defined particle size distribution
	Lubrication
Compression	Compression using a rotator tablet press
	Compression at a defined output and quality attribute
Film coating	Normally, the product is aqueous film coated in a perforated coating pan
Packaging	Tablets are filled in HDPE bottles, and the bottles are closed using child-resistant caps
Site of manufacture	Primary and secondary manufacturing at qualified GM facilities in designed countries
	No supply chain issues
Cost of good	Defined cost per unit
Forecast of manufacturing volume	Launch and yearly forecast

formulation, or drug substance characteristics to improve quality and performance at a late stage can be a major undertaking. These changes often require supporting data from additional clinical studies and pharmaceutical development experiments. To remedy these issues, strategic upfront and continued risk assessment and planning are recommended to avoid the escalating cost and impact of changes as a project advances. TPP with manufacturing information is a useful tool for defining/tracking the strategies, goals, and activities to develop a manufacturing process. An example of a TPP (manufacturing process) for an immediate release tablet formulation is provided in Table 2.10.

As shown in Figure 2.5, a successful process development program for a new molecular entity begins with the collection of background knowledge for risk assessment and planning. Then, a lead process can be defined and selected with feasibility study results as needed. Development trials, validation, and eventual commercial manufacturing are executed following the clinical trial schedule. High-level information for using a risk-based, Quality-by-Design approach can be found in regulatory guidelines such as ICH Q8, ICH Q9, and FDA validation [33–35] for risk assessment, pharmaceutical development, and process validation. Experience shows that incorporation of available information in expert risk assessment, the use

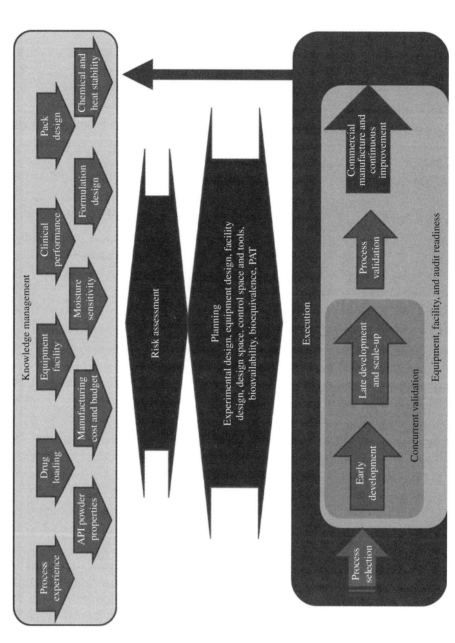

FIGURE 2.5 Process development steps for a typical immediate-release tablet.

of well-designed experiments, and attention to details throughout the development program improve the chance of success.

2.8.1 Selection of a Manufacturing Process

The selection of a manufacturing process is an interactive program where the clinical goal, drug substance properties, product characteristics, cost, and timeline are assessed. For an immediate release tablet, a direct compression process is often preferred for lower cost and faster development and manufacturing. However, it is not always possible to formulate a product with suitable input material consistency and properties using direct compression. If the compressibility of the drug substance is poor, a roller compression process may be selected. Although a wet granulation process is more labor and equipment intensive compared to direct compression and roller compression, it is useful when the flow and wetting properties of the drug substance are poor. Wet granulation can be valuable when the content uniformity, compressibility, and dissolution of the drug product are difficult to control.

Aqueous granulation is frequently used to produce suitable granulation sizes to improve the powder flow properties and compressibility. Depending on the composition and compressibility of the formulation components, wet granulation is often more suitable to produce tablets with a higher drug loading compared to other processing technologies. The drug substance can be better dispersed in modern high-shear granulation mixers or by dissolving the drug substance in the granulating fluids. Suitable surfactants may be incorporated in the wet granulation process to improvement dissolution. A microenvironment (e.g., using an acid or base in granulating fluid) may be introduced at a molecular level for stability and dissolution enhancements. For moisture-sensitive products, a solvent granulation procedure may be selected, but the equipment and facility costs for managing pollution, solvent toxicity, and explosivity are high. Wet granulation methods are also more suitable when the properties of the drug substance are susceptible to significant batch-to-batch variations.

A decision tree diagram for selecting the pharmaceutical process and feasibility trials for an immediate release tablet is provided in Figure 2.6. In addition to technical

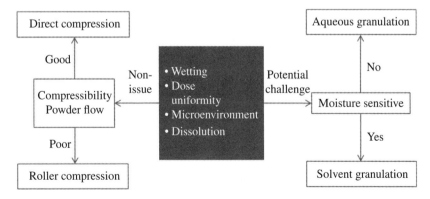

FIGURE 2.6 Decision tree of selecting a manufacturing process for an immediately-release tablet.

and clinical factors, the choice of the manufacturing process also depends on the risk assessment based on the manufacturing knowledge and process experience of the development team as well as the availability of equipment and facility at pilot, development, and commercial scales.

2.8.2 Establishment of a Pharmaceutical Process

Establishing a formulation and process successfully is a balance between innovative ideas, knowledge of products, and team work. At this stage, different companies and research groups employ individually or a combination of experience and scientific or experimental tool-based approaches to achieve their development goals. An experience-based approach alone is often empirical with limited scientific basis to understand a manufacturing process. The process developed is often less optimized. It is difficult to troubleshoot problems following this approach because additional experiments are often not conducted to test the boundary conditions. A scientific approach is based on a theoretical understanding of the properties of material and process technologies to complement an experience-based method. However, material science is often not sufficiently advanced in pharmaceutical sciences to answer many questions. An experimental tool-based approach such as Failure Mode and Effect Analysis, Lean Sigma, and various brainstorming techniques are useful for planning, designing, and execution. Specialty companies are available to train leaders to enable operational excellence, quality excellence, time management, and cost reduction. These tools are a part of Quality-by-Design [34]. However, it should be recognized that even the use of these well-structured programs may not be sufficient at certain development stages or areas of a pharmaceutical development. For example, although a tool-intensive approach encourages system development and team function, this may not always trigger the required creativity, knowledge, and problem-solving skill depending on the team composition and the complexity of the project. Pharmaceutical development is most successful where experience, knowledge, science, creativity, and system/process are applied strategically.

Once the manufacturing process is selected, the establishment of a pharmaceutical process may start with risk assessment and is followed by planning and execution (Fig. 2.5). A fishbone diagram for risk assessment of a wet granulation process for a tablet formulation is provided in Figure 2.7. The robustness of a pharmaceutical process relies on a number of factors that span process engineering, facility engineering, material science, clinical performance, quality control, and cost. Although the available resource often allows the study of a limited number of variables only, a systematic risk assessment of all possible process parameters and factors is cost and time effective. This helps avoid missing key elements that should be controlled or studied. Product experts, scientists, and engineers can be invited to share their experience, knowledge, and innovation ideas in the risk assessment exercise.

A common issue in process risk assessment is the tunnel vision in a specific process or technology. Limited aspects of the manufacture process may have been analyzed in detail, but certain critical factors may have been missed in the onset. For example, a set of elegant process parameters for wet granulation might be considered without an evaluation of the impact of input materials such as particle size and morphology (Fig. 2.7).

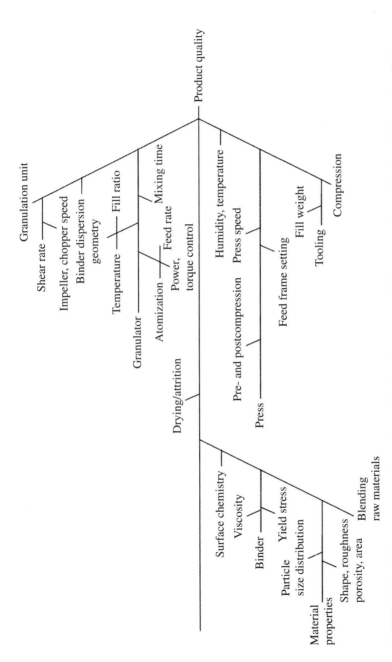

FIGURE 2.7 Risk assessment and experimental design of a wet granulation process for a tablet formulation using a fishbone diagram.

Sometimes, a lack of technical depth in understanding some of the critical risk factors is also problematic. Incorporation of a "big picture" assessment based on input from experienced development scientists and an analysis of "hardcore" technologies by area experts is essential to the success of the program. Because of a large number of factors involved in a process, it is not possible to include all factors for evaluation in a typical program. It is necessary to define factors for control and factors for study, based on the impact of their changes alone and in combination on the manufacturing process. A plan to integrate engineering science to control the processing parameters and technology to monitor processing-related parameters/variables is important. Judicious use of Process Analytical Technologies as a Quality-by-Design is valuable for process monitoring and online/at-line feedback to control and adjust the process.

Applying DOE can make a difference in creating a robust manufacturing process that produces the product consistently. It also helps troubleshoot based on the established design space.

Studying one factor at a time can be resource-intensive and unreliable in understanding the effect of how processing parameters interact with each other. Empirically increasing or decreasing the intensity of one or more factors without a proper design of experiments often leads a program in wrong directions. For example, studying one factor at a time, granulation massing time or addition of water alone in the aqueous granulation of microcrystalline cellulose (MCC) would increase the particle size to enhance particle flow and compressibility in tablet compression. When statistical design of experiments was applied to study the aqueous granulation of MCC and lactose using polyvinyl pyrrolidone as the granulating agent, an improvement of particle size was observed, but compression behavior could only be improved up to a certain level of massing time and water addition because of the interaction of the two factors

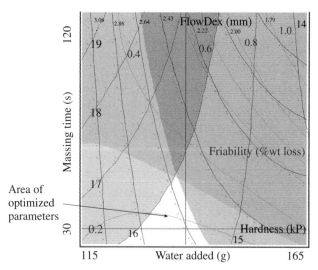

FIGURE 2.8 Process optimization of MCC and lactose granulates by statistical design of experiment.

(Fig. 2.8). Therefore, results showed that studying one factor (e.g., massing time or water addition) and measuring a single response (e.g., compression or particle size) at a time introduces risks and may not produce an optimal process [36]. For a difficult-to-compress drug substance, the product development objective may not be achievable or difficult to attain without statistical design of experiments. Well-designed screening experiments followed by DOE are highly recommended.

2.9 CONCLUSION

This chapter provided an overview of the strategies and the concept of developing a tablet formulation for immediate release. A successful program requires a project team with technical expertise and experience. Discipline and team work together with suitable project management tools such as TPP are success factors for project execution. These approaches are applicable to a variety of formulation programs from the development of immediate release tablets to complex dosage forms or molecules with delivery, stability, or processing challenges.

REFERENCES

1. Wood J. *Tablet Manufacture: Its History, Pharmacy and Practice*. Philadelphia: J.B. Lippincott Company; 1906. p. 10.
2. Qiu Y, Chen Y, Zhang GGZ, Liu L, Porter WR. *Development and Solid Oral Dosage Forms: Pharmaceutical Theory and Practice*. London: Academic Press; 2009.
3. Augsburger LL, Hoag SW, editors. *Pharmaceutical Dosage Forms: Tablets*. Volume 2, Rational Design and Formulation. 3rd ed. New York: Informa Healthcare; 2008.
4. DiMasi JA, Hansen RW, Grabowski HG. The price of innovation: new estimates of drug development costs. J Health Econ 2003;22:151–185.
5. Harper M. The Truly Staggering Cost of Inventing New Drugs [Online]. February 3, 2012. Available at http://www.forbes.com/sites/matthewherper/2012/02/10/the-truly-staggering-cost-of-inventing-new-drugs/. Accessed February 14, 2014.
6. Guidance for Industry and Review Staff. *Target Product Profile—A Strategic Development Process Tool (Draft Guidance)*. Rockville: FDA; 2007.
7. Yelvigi M, Chambliss W. The role of a target product profile in pharmaceutical product development. *Pharma Times* 2010;42(4):27–29.
8. Cooper RG. March/April 2006. The Seven Principles of the Latest Stage-Gate® Method Add up to a Streamlined, New-Product Idea-to-Launch Process [Online]. Available at http://stage-gate.net/downloads/working_papers/wp_23.pdf. Accessed February 14, 2014.
9. Cooper RG. How companies are reinventing their idea-to-launch methodologies. Res Technol Manage 2009;52(2):47–57.
10. Aagaard A. 2010. Idea management in facilitation of pharmaceutical front end innovation. Available at http://www.det-danske-ledelsesakademi.dk/2010/papers_2010/Paper_Aagaard.pdf. Accessed February 14, 2014.

11. Gould PL. Salt selection for basic drugs. Int J Pharm 1986;33:201–217.

12. Huang LF, Tong WQ. Impact of solid state properties on developability assessment of drug candidates. Adv Drug Deliv Rev 2004;56:321–334.

13. Amidon GL, Lennernäs H, Shah VP, Crison JR. A theoretical basis for a biopharmaceutic drug classification: the correlation of in vitro drug product dissolution and in vivo bioavailability. Pharm Res 1995;12(3):413–420.

14. Lipinski CA, Lombardo F, Dominy BW, Feeney PJ. Experimental and computational approaches to estimate solubility and permeability in drug discovery and development settings. Adv Drug Deliv Rev 1997;23:3–25.

15. Mccrone WC. Polymorphism. In: Labes MM, Weissberger A, Fox D, editors. *Physics and Chemistry of the Organic Solid State. 2.* New York: Wiley Interscience; 1965. p. 725–767.

16. International Conference Harmonisation Q1A(R2). Stability testing of new drug substances and products. International Conference on Harmonisation of Technical Requirements for Registration of Pharmaceuticals for Human Use, Geneva, February 2003.

17. International Conference on Harmonisation Guideline Q6A. Specifications: test procedures and acceptance criteria for new drug substances and new drug products: chemical substances. International Conference on Harmonisation of Technical Requirements for Registration of Pharmaceuticals for Human Use, Geneva, October 1999.

18. Short SM, Cogdill RP, D'Amico F, Drennen III JK, Anderson CA. A new definition of pharmaceutical quality: assembly of a risk simulation platform to investigate the impact of manufacturing/product variability on clinical performance. J Pharm Sci 2010; 99(12):5046–5059.

19. Short SM, Cogdill RP, Drennen III JK, Anderson CA. Performance-based quality specifications: the relationship between process critical control parameters, critical quality attributes, and clinical performance. J Pharm Sci 2011;100(4):1566–1575.

20. International Conference on Harmonisation Guideline Q2(R1). Validation of analytical procedures: text and methodology. International Conference on Harmonisation of Technical Requirements for Registration of Pharmaceuticals for Human Use, Geneva, November 2005.

21. Boudreau SP, McElvain JS, Martin LD, Dowling T, Fields SM. Method validation by phase of development an acceptable analytical practice. Pharm Technol 2004;28:54–67.

22. Guidance for Industry. *PAT—A Framework for Innovative Pharmaceutical Development, Manufacturing, and Quality Assurance.* Rockville: FDA; 2004.

23. Draft Guidance for Industry. *Genotoxic and Carcinogenic Impurities in Drug Substances and Products: Recommended Approaches.* Rockville: FDA; 2008.

24. Wu Y, Levons J, Narang AS, Raghavan K, Rao VM. Reactive impurities in excipients: profiling, identification and mitigation of drug–excipient incompatibility. AAPS PharmSciTech 2011;12(4):1248–1263.

25. Guidance for Industry. *Nonclinical Studies for the Safety Evaluation of Pharmaceutical Excipients.* Rockville: FDA; 2005.

26. DeMerlis C, Goldring J, Velagaleti R, Brock W, Osterberg R. Regulatory update: the IPEC novel excipient safety evaluation procedure. Pharm Technol 2009;33(11):72–82.

27. Narang AS, Rao VM, Raghavan KS. Excipient compatibility. In: Chen Y, Liu L, Zhang GGZ, Qiu Y, editors. *Developing Solid Oral Dosage Forms: Pharmaceutical Theory and Practice.* Amsterdam: Academic Press; 2009. p 125–145.

28. Rowe RC, Sheskey PJ, Owen SC. *Handbook of Pharmaceutical Excipients*. 5th ed. Washington, DC: American Pharmacists Association Publications; 2005.

29. Bohanec S, Peterka TR, Blažič P, Jurečič R, Grmaš J, Krivec A, Zakrajšek J. Using different experimental designs in drug-excipient compatibility studies during the preformulation development of a stable solid dosage formulation. Acta Chim Slov 2010;57:895–903.

30. EMA/CHMP/SWP/431994/2007 Rev. 3. *Questions and Answers on the 'Guideline on the Limits of Genotoxic Impurities'*. London: European Medicines Agency; September 2010.

31. Guidance for Industry. *Immediate Release Solid Oral Dosage Forms—Scale-Up and Postapproval Changes: Chemistry, Manufacturing, and Controls, In Vitro Dissolution Testing, and In Vivo Bioequivalence Documentation*. Rockville: FDA; 1995.

32. Guidance for Industry. *Waiver of In Vivo Bioavailability and Bioequivalence Studies for Immediate-Release Solid Oral Dosage Forms Based on a Biopharmaceutics Classification System*. Rockville: FDA; 2000.

33. Guidance for Industry. *Process Validation: General Principles and Practices*. Rockville: FDA; 2011.

34. International Conference on Harmonisation Guideline Q9. Quality risk management. International Conference on Harmonisation of Technical Requirements for Registration of Pharmaceuticals for Human Use, Geneva 2005.

35. International Conference on Harmonisation Guideline Q8. Pharmaceutical development. International Conference on Harmonisation of Technical Requirements for Registration of Pharmaceuticals for Human Use, Geneva 2005.

36. Huang J, Yan A, Lum S, Chow K. Defining scaling criteria in high shear granulation. American Institute of Chemical Engineers (AChIE) Annual Meeting, San Francisco; November 12–17, 2006.

3

FORMULATION OF POORLY SOLUBLE DRUGS FOR ORAL ADMINISTRATION

Kwok Chow

3.1 INTRODUCTION

With the application of combinatorial chemistry and high-throughput screening in drug discovery, many new molecular entities with high molecular weight, complex structure, and poor aqueous solubility are created. These molecules often do not possess "drug-like" properties for formulation development. It has been estimated that between 40% and 70% of all new molecular entities have insufficient aqueous solubility to produce adequate and consistent absorption after oral administration [1]. Because larger molecules tend to dissolve slowly *in vitro/in vitro* and exhibit poor intestinal permeability, the absorption after oral administration of these compounds is often rate-limited by dissolution and transportation across the gastrointestinal tract. Consequently, the impact of solid-state properties and gastrointestinal (GI) physiology such as food effect and stomach emptying on the absolute and variation of oral bioavailability is often more pronounced than that expressed from traditional, small, highly water-soluble drugs.

To date, the formulation development strategies of poorly soluble compounds are generally focused on improvement and control of dissolution as well as permeability. In practice, the set of variables to improve/control is unique for each molecule although the elements are generally known to formulation scientists. The success of a development program has heavily relied upon the identification of the critical elements and suitable application of scientific tools to address the bioavailability

Therapeutic Delivery Solutions, First Edition. Edited by Chung Chow Chan, Kwok Chow, Bill McKay, and Michelle Fung.
© 2014 John Wiley & Sons, Inc. Published 2014 by John Wiley & Sons, Inc.

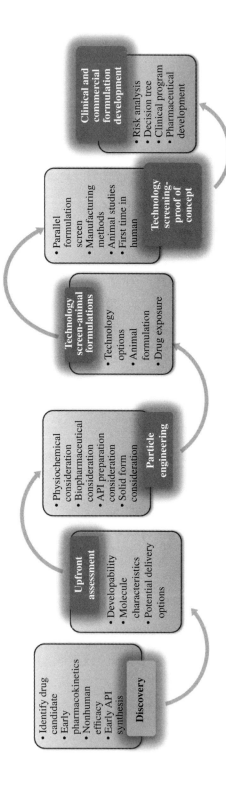

FIGURE 3.1 Typical approach of formulation screening to improve the bioavailability of poorly soluble compounds.

challenges. Upfront and continuous scientific assessment of the drug molecule/ candidate to create multidisciplinary solutions will enhance the chance of success. Then, the application of material engineering solutions for the active pharmaceutical ingredient, formulation options to influence the biopharmaceutics, suitable delivery systems to improve the rate of release, and processing technologies to produce more desirable dosage forms can be considered and applied. A typical approach from drug discovery to first-in-human clinical study is shown in Figure 3.1. It is important to recognize that a suitable formulation for most poorly soluble compounds can be developed using conventional or standard techniques. Applying complex technologies unnecessarily may introduce additional risks, cost, and time to the project. For more difficult compounds, the choice of delivery technologies should be based on scientific and therapeutic reasons and not on the expertise in a specific laboratory alone.

3.2 UPFRONT ASSESSMENTS

The bioavailability of an orally administered drug is dependent on the concentration of the dissolved drug (dissolution) at the site of absorption, GI permeability, absorption window, metabolism, elimination, and dose administered. These physical and physiological events are influenced by the biopharmaceutics, drug substance physicochemical properties, and delivery technology applied. Unfortunately, it is not uncommon that drug delivery development programs are conducted haphazardly in an attempt to improve the bioavailability of poorly soluble molecules. Technical programs are often designed around proximity and availability of the testing laboratory and/or preference of the lead formulation scientists. A poorly designed and unfocused program often causes budget and time overruns on unsuitable technologies or inefficiency experiments (Fig. 3.2). The chance of program failure was increased because the root causes of bioavailability problems were not adequately addressed. Therefore, an upfront consideration of the developability and risk assessment is highly recommended. It is often possible to identify promising delivery solutions and approaches for screening although a paper exercise alone is unlikely for the entire program to succeed given the complexity and integrated nature of the physical and biological systems. An integrated assessment, however, should help develop the scope and timeline, for example, of a parallel formulation screening program. Assessment of a poorly soluble compound before and during its development program also helps define value-added strategies and create cost-effective experimental plans.

3.2.1 Initial Assessment of Developability

A number of high-level approaches such as Biopharmaceutical Classification System (BCS) [3], Biopharmaceutics Drug Disposition Classification System (BDDCS) [4], Lipinski's Rule of Five [5], and Developability Classification System [6] were proposed and employed to predict the developability of new molecular entities from a bioavailability perspective. These classification systems attempt to predict the absorption of a drug substance/candidate based on the solubility, molecular mass,

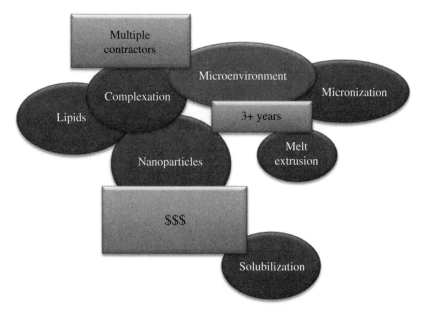

FIGURE 3.2   An example of an uncertain path to improve bioavailability. Reproduced from Ref. 2, with permission of Patheon, Inc.

molecular structure, permeability, and dissolution. In addition to these popular classification approaches, more developability assessments based on the influence of physiochemical properties not only on bioavailability but also on formulation performance, manufacturing capability, and product stability can also be used. Properties such as pK_a, pH-solubility, melting point, crystallinity, particle size, surface area, crystal habit, residual solvent content (including water), and hygroscopicity can be included. For the entire drug development program, a more holistic developability assessment that also includes business, synthesis, and therapeutic aspects of a compound is valuable [7], but beyond the scope of this chapter.

A molecule with exceedingly high molecular weight, poor solubility, and excessive hydrophobicity or hydrophilicity tends to exhibit less-than-optimal bio-availability. Based on the Lipinski Rule of Five, a drug candidate with a molecular weight of not more than 5 hydrogen bond donors, not more than 10 hydrogen bond acceptors, a molecular mass of less than 500 Da, an octanol-water partition coefficient (log P) of not greater than 5 will likely be more developable [5]. The Rule of Five is particularly useful as a high-level assessment tool in drug discovery and early pharmaceutical development when limited physiochemical and bio-pharmaceutics data are not available.

Biopharmaceutical Classification Systems (BCS) was introduced by Amidon et al. [3] in an attempt to correlate drug dissolution and bioavailability, based on solubility and permeability of pharmaceutical compounds (Fig. 3.3). Based on the BCS concept, correlation between dissolution rate and bioavailability is expected for poorly soluble and highly permeable compounds, unless the dosage is very high. Three dimensionless numbers were used to explain oral drug absorption, and case

High solubility	Low solubility
Class I High solubility High permeability	**Class II** Low solubility High permeability
Class III High solubility Low permeability	**Class IV** Low solubility Low permeability

FIGURE 3.3 Biopharmaceutical classification system.

studies, where permeability (related to Absorption Number), dissolution (related to Dissolution Number), dose (related to Dose Number), and solubility (related to Dose Number) became rate-limiting for bioavailability [3], were presented. BCS is a useful tool to predict potential bioavailability issues and define strategies to meet regulatory requirements. It has been applied frequently not only for biowaivers [8], but also for making early development decisions in formulation development. For example, polymorphism of BCS Class I (highly soluble and highly permeable) and III (highly soluble and poorly permeable) compounds is less likely to impact the bioavailability, provided that immediate dissolution is achieved. Therefore, extensive polymorphism screening may be delayed until more efficacy data are collected. On the other hand, polymorphism and salt selection could be prioritized for BCS Class II (poorly soluble and highly permeable) and IV (poorly soluble and poorly permeable) candidates to reduce development risks. BCS can also be used to provide direction to design animal and human formulations earlier and faster and avoids bridging BA/BE (bioavailability/bioequivalence) studies [9].

The BCS classification is based on intestinal permeability and dose-corrected solubility. Different methods have been used to estimate the permeability:

- Extension of absorption intravenous reference dose, humans determined by mass balance [8, 10]
- Partition coefficient [10]
- Metabolism, BDDCS [4]
- Epithelial cell culture model, for example, Caco2 cells
- Perfusion studies

The FDA guideline on *Waiver of In Vivo Bioavailability and Bioequivalence Studies for Immediate-Release Solid Oral Dosage Forms Based on a Biopharmaceutics Classification System* indicated that a drug substance is considered to be highly permeable when there is an absence of evidence suggesting instability in the gastrointestinal tract and the extent of absorption in humans is determined to be 90% or more of an administered dose based on a mass balance determination or in comparison to an intravenous reference dose. In this definition, first-pass metabolism has not been taken into consideration. In addition, a drug substance is considered highly soluble if the highest strength is soluble in 250 ml or less of aqueous media throughout the pH range of 1.2–7.5 [8]. A lower–upper range of 1.2 (0.1N HCl) to 6.8 has been proposed by European Medicines Agency (EMEA) [11]. BDDCS has been proposed to include metabolism in the classification of drug molecules and prediction of permeability of drug molecules [4]. This approach may provide mechanistic information for formulation scientists to find bioavailability solutions or make development decisions earlier in the program.

In most cases, BCS gives a good prediction of bioavailability. However, BCS classification should be treated with some caution as a first approximation in planning a development program. Relevant physicochemical factors that can be rate-limiting in processing, dissolution, and solid-state transformation could also be valuable for assessing the amount of drug deliverable *in vivo*. Formulation variables can also come into play in affecting the delivery of poorly soluble compounds. Factors such as dose volume, aqueous equilibrium solubility, and solid-state characteristics should also be reviewed for formulation (drug delivery) development. In the case of digoxin, although it is poorly soluble (low aqueous equilibrium solubility), it is classified as a highly soluble compound based on the BCS system (BCS Class I or III), because a low dosage is required for oral administration [6, 12]. Because of its low equilibrium solubility and high melting point (248°C), the dissolution of the drug substance can be slow and affected by the solid-state properties (such as particle size and crystallinity), formulation composition, and manufacturing technologies. Source-to-source and batch-to-batch variations of digoxin products, a narrow therapeutic index drug, are well documented [13].

United States Pharmacopeia (USP) categorized drug solubility into very soluble (<1 part per solute), freely soluble (1–10 parts per solute), soluble (10–30 parts per solute), sparingly soluble (30–100 parts per solute), slightly soluble (100–1,000 parts per solute), very slightly soluble (1,000–10,000 parts per solute), and practically insoluble (>10,000 parts per solute) [14]. The USP solubility is based on thermodynamics (equilibrium solubility) and is useful for predicting dissolution and processing behaviors as well as developing analytical techniques for product testing. It is an invaluable tool for bioavailability predictions in early development, especially when accurate permeability and dosage information is not readily available. Using both BCS and USP solubilities to assess the developability of a molecule on bioavailability will provide a better understanding of molecular and solid-state variables that affect the performance of the intended product.

The Developability Classification System was proposed by Butler and Dressman [6]. This classification system further divided BCS II drugs into dissolution-controlled (BCS IIa) and solubility-controlled (BCS IIb) classes for absorption predictions. The concept of using biorelevant dissolution media containing solubilizers such as bile acids for assessment of drug developability and prediction of formulation performance of poorly soluble drugs was also introduced. The use of fasted-state simulated intestinal fluid (FaSSIF), fed-state simulated intestinal fluid (FeSSIF), fasted-state simulated gastric fluid (FaSGF), and several fed-state simulated gastric fluids (FeSSGF) were recommended [6, 15]. To date, these simulated fluids are widely used in the pharmaceutical industry for solubility and dissolution studies.

This approach may provide a better prediction of the *in vivo* release of the drug than the use of USP-simulated fluids, provided other physiological factors such as GI motility and physical factors such as recrystallization are not dominating the drug release process. However, the time and cost of preparing biorelevant dissolution media are relatively high. Variations of a biorelevant dissolution medium have been proposed [16]. At this stage, the benefits of these compositions are not easy to establish because of a lack of significant body of *in vitro–in vivo* correlation data.

3.2.2 Molecule-Specific Considerations

The delivery of an orally administered drug to the systemic circulation is a series of events in the lumen of the GI, enterocytes, and liver (Fig. 3.4). The occurrence of these events is affected by the dose administered, drug dissolution, GI permeation, and metabolism/elimination and is also governed by a number of physiological, physicochemical, and delivery technology factors (Fig. 3.4). More detailed examination of these key factors in addition to an initial assessment of developability or biopharmaceutical classification often further identifies and improves delivery approaches and solutions. For example, using Lipinski's Rule of Five together with metabolism, permeability, and solubility information, Thomas et al. [17] have proposed a road map for risk assessment of bioavailability in drug discovery. This approach enables efficient and effective design of experiments to reach the objectives. Due to the diversity of the chemical structure of drug substances, each poorly soluble molecule is expected to exhibit unique chemical, physical, and biological properties. Molecule-specific assessment of drug release, permeability, and pharmacokinetics is an integral part of the formulation development of a poorly soluble drug with potential bioavailability issues.

3.2.2.1 Assessment of Drug Release The drug release of a poorly soluble compound often relies heavily on the solid-state characteristics of the active and excipients. An orally administered formulation becomes available for absorption when it is disintegrated, followed by dissolution *in vivo*. The disintegration process is influenced by the physical properties of the drug substance and excipients, manufacturing process (e.g., compression force affecting tablet-crushing strength), formulation characteristics (e.g., moisture content), and GI physiology. Similarly, the dissolution process is also affected by the solid-state characteristics of the drug substance/excipients and environment in the gastrointestinal tract. Therefore, modeling drug release based on material science and biopharmaceutical information is beneficial for designing formulations and selecting excipients for poorly soluble drugs.

Physical properties of the drug substance such as particle size can modulate the dissolution rate of by increasing the surface area and solubility of poorly soluble drugs. Other factors such as hygroscopicity, melting point, and lipophilicity may also influence drug release and (physical and chemical) stability. Therefore, the rate and extent of drug release from a dosage form can also be affected. An ionizable, poorly soluble molecule will present a different release profile at different pHs, for example, a different dissolution rate between the stomach and the intestine. In some cases, the dissolved drug substance precipitates in the gut lumen due to supersaturation as a result of changes in counterion concentration, pH, and composition of the GI fluid (e.g., presence of food), as well as solid-state transformation such as crystallization of a thermodynamically more stable form. An assessment of available solid-state data and the potential impact of a GI environment/physiology on the bioavailability may help set the direction and expectation of the development program.

3.2.2.2 *Assessment of Transport Across the Gut Wall*

The movement of drug molecules through the gastrointestinal barrier involves a combination of physical and physiological events that may include passive diffusion, active transport, facilitated (carrier-mediated) passive diffusion, and/or pinocytosis [18], depending on the chemical structure of the drug molecule. Common factors that affect drug transport across the gut wall are listed in Figure 3.4. Defining the rate-determining steps of gastrointestinal transportation is valuable in constructing the development strategy and delivery options. For example, poor permeability may be caused by slow passive diffusion if the molecule is relatively large, for example, greater than 500 Da. The rate-limiting step may also be related to lipophilicity as expressed by the octanol – water partition coefficient, for example, log $P > 5$ based on Lipinski's Rule of Five [5]. Another rate-limiting factor for passive diffusion is ionization of dissolved drugs, and pK_a can provide valuable information on whether a drug will be ionized (poorly absorbed) at physiological pH. A zwitterion drug with a large molecular weight will likely be poorly absorbed, unless the transport is supported by active or facilitated mechanisms such as ion pairing with another substance.

Active transport is a rate-limiting step for the absorption of many drugs. Recent advances in the understanding of both intestinal uptake and efflux transporters have raised the awareness of the role of transporters on pharmacokinetics, drug delivery [4], and interactions between drugs and their excipients in affecting bioavailability. Drug molecules may also be transported by pinocytosis with lipid-based substances. The active molecules may then reach the systemic circulation through the hepatic and/or lymphatic routes. In this case, the presence of suitable lipids (food- or lipid-based excipients) at the site of absorption will facilitate the absorption process. Hence, excipients are expected to have a larger impact on the performance of this type of products, and a better understanding of their properties is required to develop a successful formulation.

The increasing usage of cell culture and the increased knowledge of transporters and metabolic enzyme functions allow a better understanding of drug absorption and the prediction of bioavailability. Caco-2 monolayer cell culture is widely used for the assessment of drug permeability [19] and the mechanism of absorption, including passive diffusion and active transport such as transporter-mediated uptake and P-glycoprotein efflux. Despite some noted deficiencies in the expression of certain transporters and metabolic enzymes, using Caco-2 cell culture model is a popular/standard method for predicting drug absorption [20] to support formulation development. Other cell lines are available; each has advantages and disadvantages

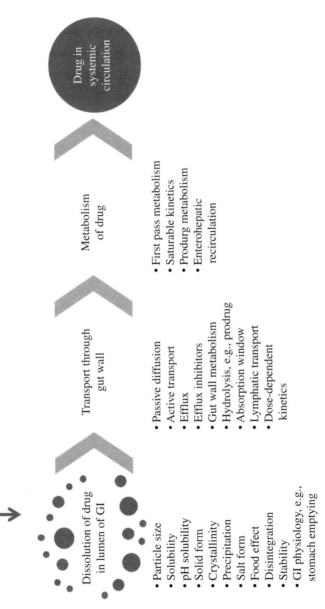

FIGURE 3.4 From dosing to systematic circulation—factors affecting bioavailability of poorly soluble compounds.

Dose administration

Dissolution of drug in lumen of GI

• Particle size
• Solubility
• pH solubility
• Solid form
• Crystallinity
• Precipitation
• Salt form
• Food effect
• Disintegration
• Stability
• GI physiology, e.g., stomach emptying

Transport through gut wall

• Passive diffusion
• Active transport
• Efflux
• Efflux inhibitors
• Gut wall metabolism
• Hydrolysis, e.g., prodrug
• Absorption window
• Lymphatic transport
• Dose-dependent kinetics

Metabolism of drug

• First pass metabolism
• Saturable kinetics
• Produrg metabolism
• Enterohepatic recirculation

Drug in systemic circulation

over each other [21]. Accurate prediction of drug transport through a human GI wall may still be difficult using cell culture methods alone. Animal perfusion studies may provide better answers [12]. Nevertheless, cellular *in vitro* methods remain valuable for hypothesis building and formulation design in preparation of animal studies as well as a proof-of-concept or validation of the intended delivery system in humans.

3.2.2.3 Assessment of Pharmacokinetics Identification of critical pharmacokinetic issues is an important aspect of scientific assessment for the development of poorly soluble drugs. For example, first-pass metabolism or degradation (e.g., hydrolysis) in the intestine or liver will diminish the oral bioavailability of a drug substance or introduce the active metabolite(s). It is a common recommendation that a poorly soluble substance with significant first-pass metabolism will not be further progressed for oral administration after accounting for dissolution, permeability, and first-pass metabolism issues, unless the dosage is very low.

For drugs that eliminate rapidly from the body, a very large dose or a modified release dosage form may be required. Oral modified release dosage forms often need an acceptable absorption throughout the GI track to be effective. This may not be possible for poorly soluble compounds with a narrow absorption window. Other metabolic processes for formulation development considerations include saturable kinetics and the metabolism and hydrolysis of prodrugs.

When sufficient information about the physicochemical properties, permeability, and dosage is available, in silico simulation or estimation may be performed for the planning of formulation screening. In general, a lower bioavailability target (compared to that of most highly soluble compounds) is set for poorly soluble drugs. A 25% target [22] could be used for simulation work, but approximately 10% was used for low-dosage or potent molecules.

3.2.3 Defining Potential Delivery Options

Based on the assessment of the molecule, drug delivery options may be evaluated for both preclinical (discovery) and clinical programs. The available techniques are primarily based on solubility enhancement, particle size reduction, and permeability enhancement. A summary of techniques that can be used in both development and commercial settings is given in Figure 3.5. Solid dispersion may be considered as a special class of solubility enhancement technology because of the theory and techniques applied. Many bioavailability enhancement techniques offer synergistic effects to improve bioavailability. For example, both particle size reduction and the use of certain permeation enhancers may increase the solubility of the drug substance for bioavailability enhancement.

Using griseofulvin (a popular model drug in research) as an example, the solid-state properties, biopharmaceutics, and available delivery solutions can be considered in a scientific assessment. Griseofulvin is an antifungal agent with a strong crystal lattice that is reflected by the high melting point and poor solubility in both

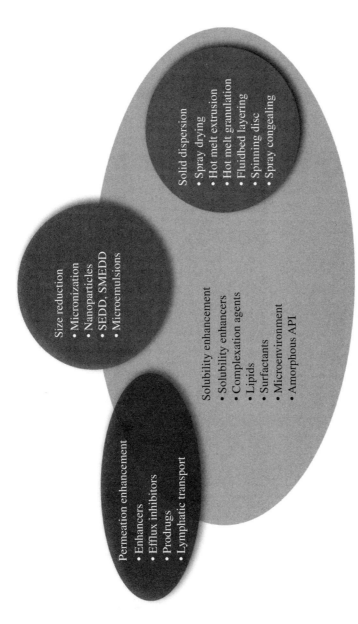

FIGURE 3.5 Summary of bioavailability enhancement technologies available in both development and commercial settings. Courtesy of Patheon Inc., with modification.

polar and nonpolar solvents. The molecule has several polar functional groups although it is not ionizable (log $P \sim 2$). It is a BCS II compound with an orthorhombic well-packed crystal lattice (no polymorphs are reported to date). Griseofulvin is administered at 1 g/day. In view of the solid-state properties and biopharmaceutics, micronization can be included in the screening because of the high melting point and well-formed crystal structure. Other delivery systems such as high-energy milling to produce submicron particles and solubility enhancement by amorphous solid dispersion may also be considered to improve bioavailability. However, the strong crystal lattice suggests that griseofulvin is susceptible to Ostwald ripening in suspension or crystallization in amorphous solid dispersion. A suitable crystallization inhibitor(s) is recommended. For drug delivery by solubilization, both highly polar and nonpolar solvents should be avoided. The solubility will likely be improved in solvent systems with a combination of Debye interactions, dispersion forces, and specific interactions. The relatively high dosage and low solubility suggest that the development of a liquid-filled capsule for griseofulvin will be challenging.

3.3   PARTICLE ENGINEERING OF DRUG SUBSTANCE

Particle engineering plays an important downstream role in achieving the target product profile that satisfies the physicochemical, bioavailability, and manufacturing requirements for many poorly soluble molecules (Fig. 3.6). The solid-state properties of an active pharmaceutical ingredient (API) can significantly affect the biological and chemical performance of the finished product. For example, solid form, solubility, and particle size can drastically affect dissolution and processing behaviors that impact bioavailability. A poorly soluble drug normally possesses a relatively complex chemical structure with multiple functional groups that favors the formation of polymorphs and solvates including hydrates. Moreover, poorly soluble compounds are often more susceptible to processing, wetting, and tablet compression problems as a result of hydrophobic surface properties, larger specific surface area (due to smaller particle size required for dissolution), and high drug loading (lower excipient content available to enhance compression characteristics) in dosage forms to compensate for the lower bioavailability. Consequently, more demanding/vigorous, upfront particle engineering and specifications development, for example, solid form and particle size specifications of the input drug substance and excipients, are expected for the development program.

Once a drug candidate is identified for development, the physicochemical properties, bioavailability, and processing capability of the synthesized material can be assessed to determine whether the pharmaceutical solid is suitable for toxicology, formulation, and/or clinical studies (Fig. 3.6). The evaluation can be abbreviated and/or miniaturized using a small quantity of drug substance [23]. A salt or cocrystal form may be developed, should the base, acid, or unionized form of a molecule not be desirable for delivering the intended toxicology exposure and potential human doses. For more difficult cases, such as molecules with very poor solubility (BCS Class IIb or IV), amorphous drug substances, chemical structure changes (e.g., use of

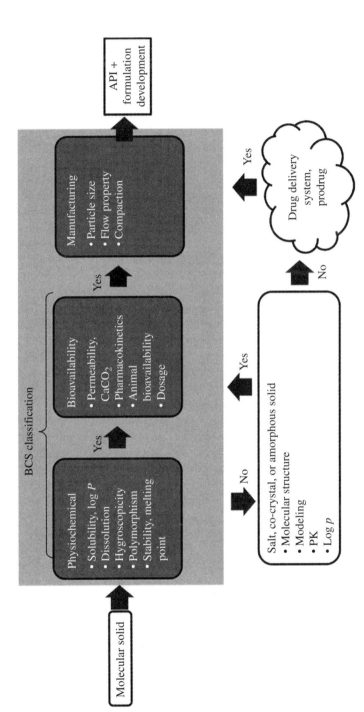

FIGURE 3.6 Particle engineering of drug candidate for formulation development.

a prodrug), or special delivery technologies may be considered (Fig. 3.6), after conventional particle engineering options are exhausted.

Salt selection is a critical step to define the physicochemical properties, product release behavior, and mechanical properties (manufacturability) of a poorly soluble drug candidate. If the molecule is easily ionizable (e.g., pK_a sufficiently close to the physiological pH), a salt may be produced to increase the solubility such that the dissolution and absorption rate may also be enhanced. A list of FDA-accepted counterions for basic drugs was provided by Gould in 1986. The selection of counterions is generally based on the solubility, melting point, stability, and wettability of the salt [24] and the toxicity of the counterion [25]. Hydroscopicity and processing behavior also play a strong role in salt selection, in addition to the improvement of dissolution and bioavailability. Preformulation tests and the choice of approaches for salt selection will further define whether a suitable salt can be employed for formulation development [26]. Aspects of salt application, selection, evaluation, and development strategies are summarized in Stahl and Wermuth [27].

A number of very poorly soluble drugs cannot be ionized easily and are quite hydrophobic, such as those in the BCS IIb classification. Producing a physically stable salt of these molecules with desirable solubility is not possible or quite difficult without using a strong acid or base. Even if a salt can be produced, it may be hygroscopic, irritate the GI tract, corrode pharmaceutical equipment, and disproportionate easily [28]. Polymorphs, solvates, and amorphous forms may also be produced easily as a pure solid or a mixture of forms in a batch, due to the complex, multipolar nature of the chemical structure of the drug molecule. Such salts may not be suitable for further development, and other means to deliver the drug substance will be required. Despite the potential challenges in salt selection and preparation for poorly soluble drugs, successful bioavailability enhancement employing this approach is often very rewarding. Applying salt selection is often an easier solution compared to the use of complex and more risky delivery technologies. Modern high-throughput equipment coupled with state-of-the-art solid-state characterization techniques allows extensive salt searching and selection. This approach is being applied in large pharmaceutical organizations and specialty pharmaceutical/contract companies [29]. A rational approach also enhances the chance of success. Prudent and scientific evaluation of the solid-state properties to determine the suitability of salts [30] is a key step in formulation development of poorly soluble drugs.

Because of the poor solubility of the unionized form of the drug substance, a dissolved salt may "crash out" or precipitate *in vivo* easily. For example, the salt of a basic drug may fully dissolve in the acidic medium of the stomach. However, the drug substance may become supersaturated at a higher pH in the intestine, causing precipitation and aggregation as the unionized form. The drug will not be available for absorption until it is dissolved again in the GI tract. The second dissolution process can be variable or incomplete. It is influenced by the effective specific surface area of the precipitate and the composition of the GI fluid. The use of crystallization inhibitors or local pH modifiers in the formulation may be useful to allow the drug substance to remain dissolved to enhance the bioavailability.

3.4 ANIMAL FORMULATIONS AND SCREENING OF DELIVERY TECHNOLOGIES

The screening and selection of delivery systems for poorly soluble compounds often starts in parallel with the animal toxicology program early at the preformulation stage and refines as needed until the commercial formulations are defined at the Phase II or III clinical stage. Formulations for toxicology studies present significant challenges, due to the fact that a very high concentration of drugs in formulation is needed to deliver an *in vivo* exposure equivalent to multiple folds greater than that of the efficacious level [31]. For poorly soluble molecules, the delivery of sufficient quantities of the drug substance to support the exposure of the drug substance in different animal species for the human program is often more difficult than that of other classes of compounds. It is advisable to improve the bioavailability of poorly soluble drugs as early as possible in the drug discovery and development program. Designing toxicology formulations can be a crucial and rate-limiting process that affects downstream product performance, manufacturing capability, cost, and chance of success in the clinic. To achieve a sufficiently high no-observable-adverse-effect level (NOAEL) is strategically important because it helps define the toxicology profile and allows flexibility for dose ranging later in the clinic.

To achieve the exposure level to determine NOAEL for a poorly soluble molecule requires dissolutions and dispersions of a sufficiently large dosage from an animal formulation *in vivo*, based on the target systemic exposure level and the permeability. This target dosage is limited by the dose volume and is inversely proportional to its solubility, which is also a function of the molecular structure, solid-state properties, and formulation employed. Hence, a more intense solid-state characterization effort to define the delivery options and particle engineering at early development for poorly soluble and poorly permeable compounds is often required.

Advances in screening approaches and automation have produced more poorly soluble compounds, putting increasing demands on the formulation capabilities in drug development. An early focus on formulation strategy including the collection of information and an understanding of the compound than is routinely/traditionally required in the drug discovery stage is becoming a new paradigm [32, 33]. Depending on the properties of the drug candidate, special attention to dissolution for reaching the required concentration *in vivo* and permeability enhancement is required. Solubility enhancement by pH adjustment, solvents/surfactants, and cyclodextrins is commonly used to increase the concentration of a drug substance in a formulation. The use of BCS classification and solubility information in defining the animal formulation can be valuable [9]. Based on the *in vitro* permeation and the dosing (amount and volume) information, the decision to apply advanced drug delivery technologies such as suspensions, lipid-based excipients, nanoparticles, and amorphous solid dispersion may be proposed [33]. Although the animal formulations may not be intended for clinical use due to toxicology and administration reasons, early formulation experience supporting the toxicology and pharmacokinetic programs is valuable for setting directions for human formulation development.

Use of early exploratory (animal) formulations to define the drug delivery system and the manufacture process is advantageous and has a significant impact on subsequent development. Upfront effort in qualifying delivery technologies and lead formulations is recommended for poorly soluble molecules to avoid downstream technical problems and project delays. Preclinical formulations for poorly soluble drugs are often prepared in small quantities quickly. The method of preparation can be less expensive and less time-consuming due to the smaller scale of manufacture, except that the availability of drug substances is limited. Procedures to ensure safety, consistent drug administration, and manufacturing standards are less vigorous than those employed for clinical and commercial formulations. Changes of delivery technology and composition are permitted with reasons and data. Several drug delivery technologies for animal administration can normally be explored in a short period of time. A well-designed and executed program starting at an early development stage will likely save time and money and avoid costly changes as the project progresses.

3.5 PARALLEL TECHNOLOGY SCREENING PROGRAM FOR HUMAN FORMULATIONS

Improving solubility and bioavailability is a multifaceted challenge in the drug development process. Many technologies exist with varied degrees of success. Selecting and employing the most appropriate technology is often required to achieve the development goals. Evaluating any set of technologies in sequence can be very expensive, time-consuming, and a serendipitous exercise. A parallel screening approach with an upfront scientific assessment is a powerful tool that provides fast, cost-effective results. Integration of animal bioavailability evaluation provides quick feedback to decide a path forward. Accelerated development through parallel formulation screening could lead to a faster and more successful path to first-in-human (FIH) trials.

Although many technologies are available to improve the bioavailability of poorly soluble drugs, appropriate application should be tailored to the physicochemical properties, biopharmaceutics, and the dosage requirements of the drug substance to enhance the chance of success. The choice of the technology is also influenced by time and the cost of development. Complex formulations, such as amorphous solid dispersions that are thermodynamically unstable and more difficult to manufacture, may be reserved for difficult cases. Due to the strategic nature of developing poorly soluble drugs, a decision tree approach adds value, especially at proof-of-concept or first-time-in-human stages, for the planning and selection of a drug delivery technology to achieve the bioavailability goal (Fig. 3.7).

A typical parallel screening program begins with an upfront scientific assessment to define potential technology solutions. Then, a series of screening tools is applied to fine-tune the strategy and define prototype compositions. This is followed by the preparation and evaluation of pilot formulations using miniaturized equipment and predictive methods before animal pharmacokinetic studies. The results are

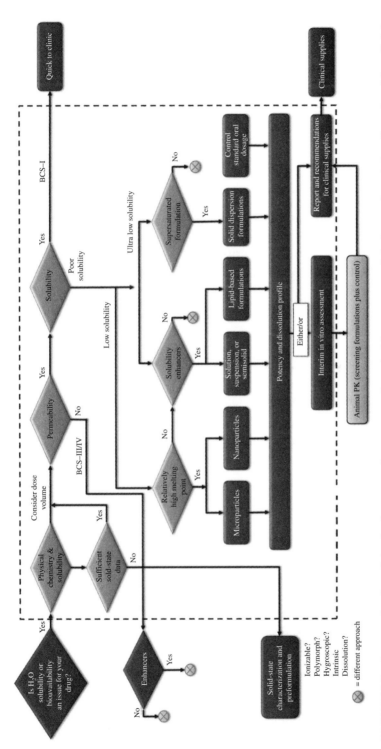

FIGURE 3.7 An example of parallel technology screening program using a decision tree approach for human formulation. Reproduced from Ref. 2, with permission of Patheon, Inc.

expected to enhance the probability of success of developing suitable clinical formulations for BCS II and IV compounds in first-in-human (FIH) and proof-of-concept studies (Fig. 3.7).

As discussed, the delivery technologies for bioavailability improvement primarily involved particle size reduction, solubility enhancement, and permeation enhancement (Fig. 3.5). Parallel screening programs can be conducted quickly and efficiently with minimal resources. It is well-suited for early development, where financial support is limited until promising drug delivery and clinical results are available. A number of established screening studies including thermal analysis, optical microscopy and spectroscopic methods, targeted solid-state characterization, solubility studies, and permeation studies (e.g., bi-directional Caco-2 assay) can be applied to collect data, fine-tune delivery strategies, and construct prototype compositions. For example, based on the molecular structure and properties of the API and polymers, experiments can be designed to produce solid dispersions using milligram quantities of API and excipients in a differential scanning calorimeter (DSC). In general, an excipient-screening program using DSC can be completed in a few days to identify lead solid dispersion formulations. Other excipient-screening studies such as the preparation of simple quaternary phase diagrams followed by dilution tests may be employed to define compositions and to predict the *in vivo* behavior of Self-Emulsifying Drug Delivery Systems (SEDDS). Effective use of screening tools reduces or eliminates trials-and-errors in prototype preparations. It also creates improved design spaces to improve the probability of success downstream *in vitro* and *in vivo*.

3.5.1 Particle Size Reduction

Certain bioavailability-enhancing approaches are technically easier, less time-consuming, and less costly to achieve. One such approach is particle size reduction, if the absorption is rate-limited by dissolution. Considering the Noyes–Whitney equation below:

$$\frac{dW}{dt} = \frac{DA(C_s - C)}{L}$$

where:

dW/dt is the rate of dissolution

A is the surface area of the solid

C is the concentration of the dissolved material in the dissolution medium

C_s is the concentration of the material at the diffusion layer surrounding the solid

D is the diffusion coefficient

L is the diffusion layer thickness

An increase of specific surface area (A) by particle size reduction will increase the rate of dissolution. An increase of solubility and reduction of the thickness of the boundary (diffusion) layer resulting from particle size reduction also improve the dissolution rate, especially at nanosize scale. The solubility increase can be explained by the Kelvin and the Ostwald–Freundlich equations, and boundary layer reduction is described by Prandtl's equation [34].

Pharmaceutical micronization of poorly soluble solids is well-established and is frequently performed by air jet milling for brittle solids with a melting point well above room temperature. To avoid dust explosion, an explosivity and safety assessment is performed, and the process may be conducted under an inert atmosphere (e.g., nitrogen). Micron-size drug substances may also be obtainable at the final stage of API synthesis under suitable crystallization conditions. This approach is getting popular to obtain crystals of more desirable physical properties and processing behavior than those produced by jet milling.

To achieve satisfactory bioavailability improvements by micronization, the particle size should be adequately reduced (e.g., <10 μm). In most cases, micronized particles of poorly soluble drugs are cohesive because of increased surface area and dispersion forces. Use of a suitable formulation composition and process technology to assist powder flow, disintegration, and dissolution will likely be required for tablet and capsule products. The drug substance could be blended with water-soluble carriers/excipients at a suitable level to allow good dispersion for dissolution and successful compaction if a tablet product is required [35]. A larger quantity of superdisintegrants and processing aids such as glidants and lubricants are often incorporated in the formulation. One or multiple surfactants at suitable levels will likely be required to achieve the dissolution profile of a tablet formulation [36].

The creation of a suitable microenvironment using local pH modifiers can complement micronization to increase the rate of dissolution and the concentration of pH-solubility-sensitive drugs. The presence of a crystallization inhibitor to prevent the precipitation of poorly (pH-sensitive) soluble drugs at supersaturated levels upon pH changes in the gastrointestinal tract may be needed to ensure the drug is available for absorption [37, 38].

Drug particles are generally more sensitive to chemical degradation and moisture after micronization. Also, the reduction of crystallinity and poor powder properties are often encountered upon micronization. Processing of micronized materials is expected to be more difficult. The physical size of a dosage unit may be quite large to address the low bioavailability, especially for molecules with low potency, low permeability, and ultralow solubility (e.g., <5 μg/ml in water). The size of solid dosage units (dose volume in the case of a liquid) may need to be increased until it is not practical to swallow, given the patient population. Nevertheless, the benefit of applying micronization to enhance bioavailability may still outweigh the extra risks, time, and effort to develop a product using more complex options.

As predicted by the Noyes Whitney Equation, compounds with exceedingly low aqueous solubility require further size reduction than that provided by micronization to achieve the desired dissolution rate. For spherical drug particles, a size reduction from 10 μm to 100 nm generates a 100-fold increase in surface area-to-volume ratio. Hence, a nanodelivery system is a viable delivery option to improve the bioavailability of highly permeable molecules with very low solubility, provided that a suitable manufacturing process is available and a physically stable formulation can be developed for the molecule. Use of nanodrug particles may also reduce the impact of food on bioavailability [39].

High-energy dry and high-energy wet millings are established methods for producing particles with a mass-median diameter ranging from approximately 100 to 500 nm in the pharmaceutical industry. However, formulation of these products can be challenging because submicron particles have a tendency to grow in suspension on storage and *in vivo* due to Ostwald ripening in order to minimize the crystal energy. Particle aggregation as a result of an increase of contact surface area (due to particle size decrease) or Ostwald ripening also causes poor dissolution and processing problems. To suppress these problems, a significant quantity of crystallization inhibitors, stabilizers (steric or ionic), processing aids, and surfactants will likely be required for the formulations [40]. Reduction of solubility by controlling ionization (pH), use of common ions, or suitable excipients (e.g., surfactants) also reduces Ostwald ripening and aggregation. Some of these ingredients present safety issues when they are used in large quantities. Many surfactants have a poor taste, and a suspension formulation with these ingredients may require taste masking. To stop Ostwald ripening, converting the high-energy milled suspension to a powder, for example, by fluid bed layering or spray drying, is desirable, although the manufacturing process is time- and energy-consuming. Because of the potential physical instability and more complex manufacturing process, only a limited number of pharmaceutical products containing nanoparticles have been commercialized. Nevertheless, at least four oral products (Rapamune®; Emend®, TriCor 145®, and MegaceES®) incorporating the NanoCrystal technology are marketed in the United States and other countries [39]. Nanoparticles are frequently used in preclinical and toxicology settings to achieve drug exposure and for proof-of-concept studies. These formulations can be used within a short period of time to avoid physical and chemical stability issues. Pilot-scale equipment to generate nanoparticles is less capital-intensive, and its operation is less time consuming than that of commercial equipment.

Submicron-size particles can also be delivered as a microemulsion or self-emulsified nanoemulsion. Microemulsions are thermodynamically stable but usually contain significant quantities of surfactants to maintain the physical stability on storage. The presence of a high concentration of surfactants may cause toxicity, irritation, taste, and stability issues. These factors should be considered carefully in selecting the technology and defining the composition for a formulation. A number of surfactants such as poloxamers and crystallization inhibitors such as polyvinylpyrrolidones (PVPs) contain peroxides or generate peroxides on storage. They may not be suitable for use with drug substances that are highly sensitive to oxidation. When the drug in a microemulsion is prone to crystallization upon dilution *in vivo*, the risk of decrease or variation in bioavailability is also increased. The use of an *in vitro* dilution test may help predict such *in vivo* events.

To assess whether a simple particle size reduction method will likely achieve the bioavailability target for poorly soluble molecules, a Maximum Absorbable Dose (MAD) approach may be applied [41]. MAD is defined as

$$MAD = k_a C_s V_t$$

where:

k_a is the absorption constant

C_s is the solubility

V_t is the volume of fluid in the gastrointestinal tract

The MAD is reached if the rate of GI wall transport (k_a) and the volume of GI fluid are unchanged and a saturation is attained. Therefore, if the MAD is high relative to the therapeutic dose, particle size reduction may be useful to enhance the bioavailability. If the MAD is significantly lower that the target dosage, using the micronization method to improve the bioavailability may not be suitable. Less-than-optimal absorption may occur. Moreover, further reduction of particle size such as the use of nanotechnologies will also unlikely be meaningful if the MAD is much smaller than the dosage requirement, because the absorption may not be rate-limited by dissolution alone. The MAD will need to be increased, for example, by solubility (increase C_s) and/or permeability enhancements (increase k_a) in addition to size reduction.

Extension of MAD to include intestine transit time was proposed by Curatolo [42] for use in drug discovery. Butler and Dressman [6] also proposed to include the rate of dissolution to calculate the maximum particle size of a drug substance to achieve the target bioavailability given a jejunal permeability value. In the absence of permeability data, a dose volume estimate alone may still be suitable to consider whether a more advanced technology should be applied. For a highly permeable compound, a dose volume higher than 250 ml may still provide good absorption. A dose volume as high as 5000 ml has previously been used as the breakpoint for the selection of a solubilized dosage form for BCS Class II compounds [9].

3.5.2 Solubility Improvement

In theory, the concentration of a dissolved drug available for absorption can be increased by several orders of magnitude by administrating a drug substance dissolved in a solvent or solvent mixture, provided that precipitation does not occur on dilution with the GI fluid. The drug vehicle can be formulated as a liquid or liquid-filled capsule. The choice of solvent or solvent systems should be based on the solubility of the drug, toxicity of the vehicle, impact of the preparation on GI physiology, potential of drug precipitation *in vivo*, and, where applicable, compatibility of the solubilization system with a capsule shell [43–45]. A typical list of common solvents in oral and injectable formulations can be found in the literature [46], and toxicity information on common excipients and solubilization agents is available in the WHO website [47].

The amount of dissolved drug substance that can be administered is defined by the dosing volume and solubility of the API in the vehicle. For example, if the dosing volume of a dosage unit is up to 1 ml and the solubility is 25% (w/w), respectively, the maximum quantity of drug that can be administered per dosage unit would be approximately 250 mg, assuming the density is 1. Multiple units of a liquid would be required for large doses (>250 mg). This may not be acceptable for some patients or indications and could also be particularly problematic for multiple drug therapies such as those employed to treat HIV.

The solubility of a drug in various solvents is related to the molecular structure, crystal packing, and the property of the solvent systems. Understanding solubility theory [48, 49] may help identify the possible solvent systems to maximize drug loading in a liquid or liquid-filled capsule. Instead of picking the solvent system and measuring the solubility at random, solubility data for common or structurally similar molecules may be retrievable from the literature as initial estimates. Crude estimates may also be obtained from molecular structures using thermodynamic theories, for example, using Hildebrand or Hanson solubility parameters. Yalkowsky et al. [50] also used physical parameters such as melting point and water–octanol partition coefficient to predict solubility with the assumption that the presence of water does not alter the properties of the crystals. In silico software is available commercially for solubility prediction [51]. Kinetic (DMSO) solubility data from high-throughput studies generated at discovery stage, if available, can also be used for initial estimates. Despite recent advances in solubility theory and computer technology, an accurate prediction of the solubility of complex drug molecules remains difficult because of an insufficient understanding of molecular interactions, for example, polymorphism in the solid state [52] and intermolecular forces. Nevertheless, some solubility prediction early in the development program is a valuable starting point for conducting solubility experiments, finding solubilization agents, and developing formulation strategies.

Accurate equilibrium solubility measurement is useful for the formulation design of a poorly soluble compound to estimate the physical size/volume of a dosage unit. An overestimation of the solubility may result in supersaturation in the dosage unit, causing precipitation on storage, especially when the temperature fluctuates. The traditional shake-flask method involves equilibrating an excess quantity of representative solids in the solvent and determining the concentration from time to time until a constant value is obtained. This assumes that the properties of the solid in the solvent system have not been changed during the equilibration period. Solid-state characterization of the excess solid is useful for detecting thermodynamically more stable polymorphs or solvates. For aqueous solubility measurements, the pH and ionic strength should be known and kept constant. Accurate solubility measurement is often limited by the stability of the molecule, the time to reach equilibrium, temperature control, resources to conduct all the experiments, and the amount of drug substance available, especially in early development, where drug substance supply is limited because the synthesis method is being developed and the cost of API is high. Use of miniaturized techniques, statistical design of experiments, high-throughput methods, and API sparing methods can be valuable for achieving the product design goal more easily and effectively [53, 54].

Oral lipid-based formulations are often used to deliver drugs that require solubility enhancement. Lipid-based excipients may act as a drug carrier, enable fast dispersion in the GI, reduce the food effect, and improve permeability. About 2–4% of marketed oral formulations contain lipid-based ingredients [1]. Using lipid-based formulations to solve drug delivery issues has recently attracted considerable attention in the scientific community. A large number of lipid-based excipients are available for formulation development, although experience with each of these excipients varies significantly. The safety, toxicity, and physical and chemical properties of some of these ingredients are not fully understood. Lipid-based excipients represent a wide range of compounds such as fatty acids; natural oils and fats; semisynthetic mono-, di-, and triglycerides; semisynthetic polyethylene glycol (PEG) derivatives of glycerides and fatty acids; polyglyceryl fatty acid esters; and cholesterol and phospholipids [55]. To fully explain the bioavailability improvement by lipid-based excipients, one or more of the following mechanisms may be involved:

- Solubility enhancement
- Carrier to the site of absorption, for example, a drug in a self-emulsified preparation
- Improve wetting
- Efflux inhibition
- Inhibit crystallization
- Lymphatic absorption
- Prolong GI transit, increase stomach-emptying time, reduce food effect
- Protect from GI degradation

A Lipid Formulation Classification System (LFCS) based on the dispersion of a lipid formulation was proposed by Pouton in 2000 [56, 57]. Lipid formulations were divided into five types (Types I, II, IIIa, IIIb, and IV). Type I formulations will be dispersed into coarse dispersions. Type II, IIIa, and IIIb are self-emulsifying formulations that are classified by particle size on dispersion and emulsion stability on dilution. Type IV is oil-free or mixed micelle formulations that may undergo phase changes upon dilution. LFCS is a valuable design tool for lipid-based formulation development because it is useful for matching the physiochemical/biopharmaceutic properties of the drug molecule and behavior with the type of lipid formulations. LFCS also helps the formulation scientists to define the class (e.g., surfactants, oil, cosolvents), digestibility, and hydrophilic-lipophilic balance (HLB) value of excipients as well as the experiments required for the target (type of) formulations.

Lipid-based formulation is particularly useful for the delivery of certain poorly water-soluble drugs, for example, those with high lipophilicity, food sensitivity, and permeability issues. The dosage of these drugs cannot be too high so that API can be dissolved (if required) in the dosage unit for administration, unless a liquid formulation is used. Many lipid-based excipients contain multiple components and are also chemically less stable than solid excipients. The quality control of these ingredients to assure stability and performance can be challenging.

A solid oral dosage form is probably the most preferred choice of drug administration for business and compliance reasons. For this reason, liquid lipid-based preparations are often filled in soft or hard gelatin capsules, provided that the fill ingredients are compatible with the capsule shell and potential leakage issues are addressed. A liquid matrix may also be adsorbed on porous excipients such as Neusilin® (a synthetic amorphous form of Magnesium Aluminometasilicate), such that the resulting ingredients are compressible to form tablets. However, the adsorbed materials tend to be more cohesive and difficult to flow compared to conventional solid excipients. A number of investigators have recently proposed to use solid self-emulsifying agents for tablets and capsules of poorly water-soluble compounds with bioavailability issues [58, 59]. In this case, the drug substance should be completely dissolved by melting and cooling a mixture of the vehicle and the drug substance at a concentration below the solubility limit at the storage temperature. Unfortunately, measuring the equilibrium solubility of drugs accurately in solids or semisolids is quite difficult. An overestimation of the solubility may lead to the preparation of a supersaturated matrix, which (in theory) has a tendency to precipitate on storage, resulting in dissolution failure. Most lipid-based excipients have low glass-transition temperatures and melting points relative to the room temperature; they will unlikely inhibit API crystallization by slowing down the molecular mobility of the drug substance if the system is supersaturated, that is, the respective formulation will be thermodynamically unstable. In general, formulation development using lipid excipients alone for amorphous solid dispersions (supersaturated formulation) is discouraged. Consideration of the benefits and limitations of lipid-based formulations is an important aspect of selecting the appropriate technologies for bioavailability improvement of poorly soluble drugs.

The development of an amorphous solid dispersion delivery system (SDD) is reserved as a last resort in achieving the bioavailability when less risky methods such as particle size reduction and solvent solubility enhancement effort are exhausted. An amorphous drug in SDD has a higher free energy than its crystal form, causing a significant increase of solubility, for example, by as much as 1600 times [60]. Such an increase of solubility creates a large driving force to transport the drug across the intestinal wall to an extent that is difficult to be matched by the use of other delivery systems. The measured solubility, dissolution, and bioavailability benefits of SDDs, however, are often less than the estimated solubility increase [61] due to solid-state transformation during dissolution and reduction of the effective surface area available for dissolution as a result of large primary particles and/or aggregates. More importantly, developing amorphous SDD is fairly risky, because it is prepared as a metastable structure (thermodynamically unstable). The drug in an amorphous SDD can eventually be converted to a more physically stable form, and, therefore, the product shelf life and performance may be affected on storage and *in vivo*. Although the concept of SDD to improve the bioavailability was introduced for many years [62], the kinetics of the solid-form transition was considered, until recently, to be too unpredictable and difficult to control because it is associated with nucleation and growth in a solid solution.

There is a renewed interest in amorphous SDD for BCS IIb and IV drugs for early development, clinical studies, and commercialization. Recent improvements of manufacturing technologies and better use of crystallization-inhibiting polymers added confidence in developing this type of formulations. Advances in spray drying and twin-screw hot-melt extrusion techniques allow a more reliable and more consistent preparation of amorphous structures from gram quantities to 100 s of kg scales. The use of solid-state characterization tools including thermoanalysis, X-ray diffraction (including high-resolution powder X-ray diffraction), solid-state nuclear magnetic resonance (NMR), microcalorimetry, and Raman spectroscopy provides valuable molecular mobility and solid-state structure information to help assess, understand, and predict the physical stability and performance of the preparations [63]. To date, important physicochemical and physiological factors that affect the therapeutic performance, stability, dissolution, and bioavailability of amorphous dispersions are better understood for formulation development [64]. A significant body of literature demonstrating the benefits and risks of amorphous solid dispersion is available, and the following is a typical list of amorphous SDD products developed:

- Cesamet® (Nabilone) Lilly
- Gris-PEG® (Griseofulvin) Novartis
- Nimotop® (Nimodipine) Bayer
- IsoptinSR-E 240® (Verapamil) Abbott
- Sporanox® (Itraconazole) Janssen/J&J
- Crestor® (rosuvastincalcium) AstraZeneca
- Kaletra® (lopinavir/ritonavir) Abbott
- Intelence® (etravvirine) Tibotec/J&J

Despite the recent advances, the development of amorphous SDD remains fairly difficult because of the incomplete understanding of the material science behind this technology, the molecular interaction of molecules of diverse chemical structures, and the absorption/metabolism of drugs. It is suspected that amorphous materials may still exist in certain molecular structures (i.e., not totally amorphous). Its performance and physical stability is influenced by the method of preparation. The industry is still gaining experience in designing and developing dosage forms for this emerging delivery system. In view of the physical, chemical, and physiological challenges in developing an amorphous SDD, a systematic approach could be considered and may involve:

- Excipients and formulation screening
- Pilot-scale processing
- Evaluation of physical and chemical stability
- Animal evaluation
- Human test
- Formulation optimization
- Process development
- Stability programs

A formulation-screening study for an amorphous SDD begins with the collection and review of solid-state characterization and biopharmaceutics information for the API. For example, the chance of the successful preparation of an amorphous SDD may be increased if an amorphous form was previously observed even for a brief period of time by the discovery or development team. The ratio between the melting point and the glass-transition temperature may also be used to predict whether the development of an amorphous SDD would be successful. A high ratio suggests that the propensity for the compound to crystallize in a solid dispersion is high [65]. An unionized form of the drug substance, if available, is generally more (physically) stable in certain polymeric systems.

The selection of excipients for screening could be based on the ability of the excipient(s) to prevent or inhibit crystal growth. A polymer with a sufficiently high T_g above the storage temperature of the product will likely present a slower molecular mobility to inhibit the movement of drug molecules to trigger crystallization. In fact, polymers that are commonly used in solid dispersions have a relatively high T_g above the ambient temperature (Table 3.1). Examples are HPMCP, Soluplus, Kollidon VA-64, PVP (12, 25, 30, or 90), hydroxypropyl methylcellulose acetate succinate (HPMCAS) LF, L-HPC, and Eudragit L100-55. The "$T-50$" rule for glass transition [66] is often applied as a starting point to predict whether a drug substance, polymer, and dispersion will likely be physically stable. The Gordon Taylor Equation [67] can also be used to calculate the dispersion glass-transition temperature for prediction and comparison. In addition, the molecular mobility of a basic drug may be reduced in the presence of an interacting polymer with acidic functional groups. To prevent rapid dissolution and precipitation in the stomach before the drug is absorbed and transported across the small intestine, a pH-sensitive polymer with enteric properties may be utilized to target the dissolution near the site of absorption. HPMCAS is pH-solubility-sensitive in water and forms a colloidal solution that may help release the drug molecules without significant precipitation *in vivo*. Another selection criterion for SDD excipients is whether the drug and polymer system will form a stable continuous phase. A quick paper assessment may be performed based on their solubility parameters [68], followed by confirmatory (screening) tests.

Once the lead polymers are selected, a variety of techniques can be conducted for formulation screening. Differential scanning calorimeters using physical mixtures can be used to evaluate whether an amorphous solid can be produced (Fig. 3.8). For example, the glass-transition temperatures and the number of glass transitions (number of phases) can provide a useful indication of physical stability. Unlike older equipment, modern DSCs are very sensitive in measuring T_g quickly, and high-throughput-like studies can be conducted with automation accessories. Film casting by solvent evaporation [69] with or without high-throughput detection of an amorphous structure may also be used to determine the suitability of the polymer systems. Because film forming is a slow process, possible false negative results may affect formulation decisions. Optical microscopy and powder X-ray diffraction are common methods to evaluate lead solid dispersion formulations.

Spray drying is frequently used in early amorphous solid dispersion development programs because of the ease of process control and low drug substance requirements. In this process, the drug substance should be sufficiently soluble in the solvent system to allow efficient processing. This can be difficult for some very poorly soluble molecules. The mixture of solvents and the order of addition of solvents will need to be evaluated in solubility studies. The residual solvent content can influence the physical stability, release behavior, chemical stability, and product toxicity. Therefore, their levels should be well-controlled to maintain product consistency. Depending on

TABLE 3.1 **Common polymers used in amorphous solid dispersions and their glass transition temperature**

Polymer types	Polymers	Glass transition temperature, T_g (°C)	Comments
Neutral polymers that may inhibit crystallization	Kollidon VA-64	~107	PVP-based
	PVP K 90	180	PVP-based
	PVP K 30	~164 (175 (BASF))	
	PVP K 25	~160	Hygroscopic
	PVP K 12		
	Hypromellose 2910 (3 cps)	~175 (170–180)	Cellulosic
	Hypromellose 2910 (5 cps)	~178 (170–180)	
	Soluplus	70	PVP-based, peroxide-free
pH-sensitive polymers with acidic functional groups	Hypromellose phthalate (HP-50)	~137	Cellulosic
	Eudragit L100-55	~110	Acrylic-based
	HPMCAS (various grades)		Cellulosic Inhibit crystallization Form colloids

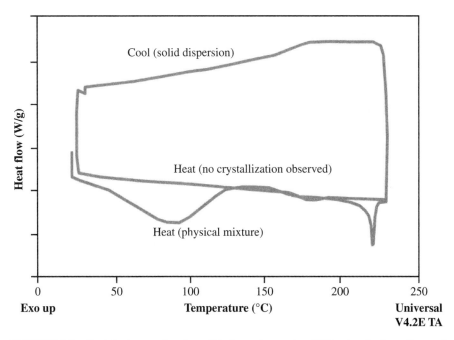

FIGURE 3.8 Excipient screening for solid dispersions using DSC—*in situ* formation and analysis of griseofulvin and PVP K 30 (10% (w/w)) solid dispersion. Courtesy of Patheon Inc.

the molecular structure, residual solvents can be too toxic for use in pharmaceutical products or should be controlled at very low levels in order to meet regulatory requirements. It should be noted that the property of amorphous solid dispersion is sensitive to operational conditions and batch size. Process control, scale-up procedure, and equipment are some of the factors to consider in pilot-scale, scale-up, and commercial manufacture. Because of the differences in equipment design between development- and commercial-scale spray driers, the solid-state properties (such as particle size) often change on scale-up.

Recently, the use of twin screw equipment for hot-melt extrusion (HME) of amorphous solids has attracted considerable interest [70, 71]. A number of pharmaceutical operations including mixing, heating, and formation of solid dispersion can take place in a twin-screw extruder. It is a process that does not require the use of solvents. Therefore, intensive effort in controlling solvent content and costly explosion-proof facility/equipment are not required. HME is a well-established technology in the plastic industry. It allows continuous processing and effective quality control. Although the product is only exposed to higher temperatures for a short period of time, it may still not be suitable for highly temperature-sensitive activities. The extrusion process requires processing materials of suitable viscosity and T_g to produce amorphous SDDs. The minimum batch size for the preparation of amorphous SDDs by twin-screw extruders is generally larger than that required by spray dryers at pilot scale. Therefore, spray drying is more frequently used in early development when drug availability is limited. The SDD properties produced by spray drying often differ from those by HME. HME is a more cost-effective tool at or after proof-of-concept, whereas spray drying is more API-sparing (a critical requirement in some cases) and economical at early development.

An array of predictive methods can be utilized to test and select pilot formulations for bioavailability studies. Optical microscopy is a simple technique to study particle size and the habit of powders. It is also a quick, complimentary technique to X-ray diffraction to reveal the presence of crystalline vs. amorphous materials (Fig. 3.9). Modified dissolution testing (paddle) and microcentrifuge dissolution will provide

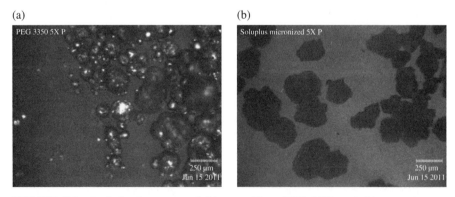

FIGURE 3.9 Optical microscopy of (a) crystalline—PEG 3350 and (b) amorphous—SoluPlus® materials under cross-polarized light. Courtesy of Patheon Inc.

the API release characteristics at normal and supersaturated conditions in a bio-relevant medium as required for a wide range of prototype formulations. Thermoanalysis provides solid-state, solid form, and physical stability information, for example, based on the glass-transition temperature of solid dispersions, for formulation assessment and animal pharmacokinetic formulation selection.

3.5.3   Other Development Options

Improvement of dissolution and solubility may not be adequate to improve the bioavailability of certain molecules that exhibit or require:

- Ultralow solubility
- High efficacious dosage
- Fed/fasted variability
- Lack of dose proportionality
- Extremely poor permeability
- Significant first-pass due to metabolism in the intestine or liver

Specialty methods or delivery systems that carry more inherent risks and development time may need to be applied for therapeutically promising molecules. It is hoped that a drug can be administered by the mouth instead of injections if the permeability is increased. A number of excipients affect transporter functions, such as those influencing efflux, and judicious use of excipients may enhance bioavailability [72]. For example, the bioavailability is improved by increasing the permeability of amprenavir across the gut wall using D-α-tocopheryl polyethylene glycol succinate (vitamin E TPGS) to reduce efflux [73]. Oral absorption enhancers such as bile salts, fatty acids (e.g., oleic acids, surfactants, and chitosan) were also studied and reported. These compounds improved membrane permeation through different mechanisms including transient opening of tight junctions, disruption of lipid bilayer packing and complexation/carrier/ion pairing [74]. Permeation enhancers, however, may present regulatory and toxicity hurdles because they may enhance the absorption of both drugs and other materials. GI physiology such as transit time may influence the effect of absorption enhancers for poorly soluble compounds. It is suspected that the permeation enhancement function of excipients exists in a number of approved products already. This may have not been clearly identified and reported to date. Classifying these excipients as enhancers will likely require more scientific/clinical evidence and justifications for safety in regulatory submissions.

Numerous prodrugs were designed to overcome formulation, delivery, and toxicity barriers to drug utilization. Approximately 20% of all small molecular drugs approved during the period 2000–2008 were prodrugs. A number of prodrugs have been developed successfully to improve bioavailability. For example, the bioavailability of Amprenvair, an HIV medication that was used in high dose in combination with other medicines, can be improved using a prodrug. In this case, Fosamprenavir is rapidly hydrolyzed to amprenavir by cellular phosphatases in the gut epithelium as it is absorbed [75]. Amprenvair was replaced by Fosamprenavir after the original product was commercialized. Other examples of successful bioavailability enhancing prodrugs are oseltamivir, dabigatran etexilate, midodrine, and valacyclovir [76]. The prodrug approach should be considered early in the development of poorly soluble molecules to avoid extensive and unnecessary drug delivery exercise, delay in development, and risk of failure.

Prodrug in combination with lipid-based vehicles may also bypass first-pass metabolism to significantly enhance the bioavailability, for example, lymphatic absorption of testosterone undecanoate (a regulatory approved product). Lymphatic transport has shown to be a contributor to oral bioavailability of a number of highly lipophilic drugs (typically, log $P > 5$, long-chain triglyceride (TG) solubility >50 mg/g) [77], including two lipophilic cannabinoids, halofantrine, moxidectin, mepitiostane, testosterone derivatives, MK-386 (a 5α-reductase inhibitor), penclomedine, naftifine, probucol, cyclosporine, ontazolast, CI-976, fat-soluble vitamins and their derivatives, retinoids, lycopene, dichlorodiphenyl trichloroethane (DDT) and analogs, benzopyrene, PCBs (polychlorinated biphenyls), and a number of lipophilic prodrugs [78]. The delivery is dependent on the transportation of the drug with lipids into the enterocyte, association of the drug with colloidal lipoproteins to form chylomicron, and their transfer into the lymphatic system. A good understanding of the properties of lipid-based excipients is required to design a formulation of this delivery system. Because the flow of lymph is 500 times less than that of blood in the intestine, the quantity of drug deliverable through the lymphatic route is low. Hence, lymphatic delivery is reserved for relatively low-dose molecules.

3.6   ANIMAL PHARMACOKINETIC STUDIES AND CLINICAL FORMULATION DEVELOPMENT

The performance of animal pharmacokinetic studies is a significant milestone of formulation screening to improve the bioavailability of poorly soluble drug candidates. It is important to conduct a risk analysis before choosing the animal for study. The choice of animal model and the bioavailability study design depend on the physical properties and pharmacokinetics of the API, the animal GI physiology and metabolism, and the prototype formulations to be evaluated. Animal pharmacokinetic data should be evaluated in consideration of the biopharmaceutics and metabolism of the molecule as well as GI physiology of the animal species used in the bioavailability study. With the animal data, a technical recommendation for the development of the clinical formulation(s) is expected after the *in vitro/in vivo* performance, biopharmaceutics of the molecule, clinical indication, and complexity of formulation/manufacturing/scale-up of the delivery technology are reviewed.

Many formulation-screening studies are conducted using mice or rats because of the low cost, API sparing, speed, and data quality, especially for proof-of-concept and initial assessment of drug delivery systems. Since rodents are commonly used in toxicology and pharmacokinetic studies, a rapid protocol and analytical method can be developed to allow a quick assessment of a large number of prototype formulations.

However, the metabolism of mice and rats is faster than that of humans. The bioavailability achieved may be lower. Since the doses and dosage forms administered are adjusted for the body mass and metabolism, the formulations administered may not be representative of those intended for humans. For small animals, a dosage unit is often administered as a solution, a suspension, or a very small capsule. For example, an amorphous solid dispersion formulation may dissolve prematurely if it is prepared as a suspension and precipitated in the GI tract. Therefore, the expected increase of bioavailability by solubility enhancement cannot be observed. Assessment of a pH-sensitive delivery system may also be difficult to study if the formulation is dispersed in a liquid to assist drug administration. Beagle dog is also a common species used in bioavailability studies. In terms of metabolism, it is a more representative species than rodents, and dogs can swallow large pills. It is easier to conduct crossover studies using dogs than rodents as they are not normally sacrificed after dosing. The stomach of dogs has a different pH and emptying time than that of humans. The GI motility also differs. These differences are particularly important for testing pH-sensitive drugs and formulations containing pH-sensitive polymers. The dog stomach is often acidified with citric acid or predosed with pentagastrin [79, 80] to simulate the human stomach environment in a fasted state for bioavailability studies. Although a monkey model can be more predictive, it is normally reserved for confirmation studies because of the high cost to conduct the experiments.

3.7   DECISION TREE FOR COMMERCIALLY VIABLE DELIVERY SYSTEMS AND FORMULATIONS

After a formulation screening or proof-of-concept program has provided satisfactory animal and human data, a definitive formulation of the appropriate delivery technology may be selected. This decision can be critical to the progress and eventual success of the development program. The formulation will likely be used in a significant number of clinical studies. Ideally, little further modifications will be needed toward the end of the clinical program to construct a final commercial image product.

Bioavailability, technical difficulties, development time, cost, patient compliance, and commercial potential are some of the factors that are common considerations for picking clinical and commercial formulations for poorly soluble drugs. Conventional tablets and capsules are regarded as a low-risk option in the pharmaceutical development of poorly soluble drugs if the bioavailability requirement is fulfilled (Fig. 3.10; options in white boxes). They are also preferred because of their acceptance by patients and are less expensive and relatively easy to develop. They can be formulated with simple excipients and prepared by directly mixing wet granulation or dry granulation. In some cases, a slightly more complex tablet or capsule containing a higher level of glidants, disintegrants, and surfactants may be needed to overcome processing and wetting difficulties. A liquid-filled hard-gelatin capsule or soft-gelatin capsule may be needed if the drug is delivered as a solution or suspension. In this case, compatible ingredients should be selected, and potential leakage of content from the capsules on storage should also be addressed. A simple modified release delivery system may be used to provide the release profile. For example, an enteric coating may allow a pH-sensitive drug substance to dissolve in the intestine instead of precipitating in the stomach.

A next level of technology (Fig. 3.10; options in grey boxes) can be considered when conventional dissolution and solubility improvement methods alone cannot achieve the bioavailability solution. Delivery systems that are technically more complex and specialized may be applied. However, applying this level of technology will be scientifically and financially more challenging. For example, the development of a tablet or capsule containing nanoparticles will likely require special processing equipment and scientific know-how to establish the processing procedure. Taste masking for a liquid or suspension is a specialized skill and a technology that may not be readily available in a typical formulation laboratory. Certain patented technologies such as the stomach retentive delivery system increase the resident time and may improve the bioavailability. However, such technologies often require royalty/ milestone payments. Potential patent infringement may also prohibit an investigator to pursue the technology. More demanding time management will likely be needed for special technologies, so that the formulation activities are not becoming the bottleneck of the development program, although the time available for complex formulation is often possible given the usual gaps between long clinical studies.

Very difficult molecules may require drastic measures to deliver (Fig. 3.10; options in black boxes). Last-resort methods that present a higher risk of failure in formulation development, at product registration, and after product launch may be considered for molecules with promising commercial and therapeutic potentials. Amorphous solid dispersion is a typical high-risk option with known formulation/ processing challenges [81]. Another example is the use of absorption or permeation enhancers, where their usage may create unwanted toxicity or regulatory issues. These powerful delivery techniques should be reserved for high-value molecules, when less risky options are no longer available. If possible, backup options such as prodrugs, alternate administration route, and follow-up molecules with better drug properties should be explored in parallel.

3.8 CONCLUSION

The development of poorly soluble drugs with bioavailability issues is a stimulating experience that often involves multidisciplinary contributions from experts in material science, biopharmaceutics, drug delivery technologies, regulatory, and drug development processes. Conducting an upfront assessment to establish the development goal and a strategic plan quickly to evaluate suitable drug delivery technologies (as needed) will improve the chance of success. Early strategic screening of delivery systems using animal formulations will not only support the pharmacokinetics and toxicology programs, but also streamline the effort to introduce the first-in-human and future clinical formulations. After formulation screening, the clinical and market formulation development of poorly soluble compounds should be a science and risk management exercise.

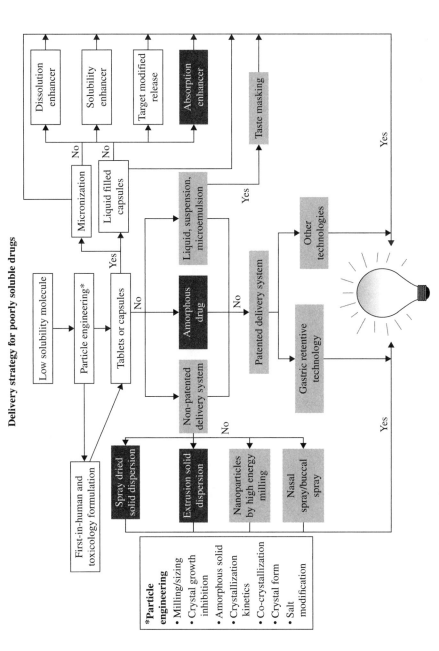

Delivery strategy for poorly soluble drugs

FIGURE 3.10 Decision tree for selection of technology and formulation to improve the bioavailability of poorly soluble drugs.

REFERENCES

1. Hauss DJ, editor. *Oral Lipid-Based Formulations: Enhancing the Bioavailability of Poorly Water-Soluble Drugs.* New York: Informa Healthcare; 2007. p vii.

2. Chow K, Kane A. Drug development, bioavailability and solubility: parallel screening. *Eur Pharm Contract* September 2011:100–104.

3. Amidon GL, Lennernäs H, Shah VP, Crison JR. A theoretical basis for a biopharmaceutic drug classification: the correlation of in vitro drug product dissolution and in vivo bioavailability. *Pharm Res* 1995;12(3):413–420.

4. Shugarts S, Benet LZ. The role of transporters in the pharmacokinetics of orally administered drugs. *Pharm Res* 2009;26(9):2039–2054.

5. Lipinski CA, Lombardo F, Dominy BW, Feeney PJ. Experimental and computational approaches to estimate solubility and permeability in drug discovery and development settings. *Adv Drug Deliv Rev* 1997;23(1–3):3–25.

6. Butler JM, Dressman JB. The developability classification system: application of biopharmaceutics concepts to formulation development. *J Pharm Sci* 2010;99(12): 4940–4954.

7. Han C, Wang B. Factors that impact the developability of drug candidates: an overview. In: Siahaan TJ, Soltero R, Wang B, editors. *Drug Delivery: Principles and Applications.* Hoboken: John Wiley & Sons, Inc; 2005. p 1–14.

8. U.S. Department of Health and Human Services. *Guidance for Industry: Waiver of In Vivo Bioavailability and Bioequivalence Studies for Immediate-Release Solid Oral Dosage Forms Based on a Biopharmaceutics Classification System.* Rockville: Food and Drug Administration, Center for Drug Evaluation and Research (CDER); 2000.

9. Ku MS. Use of the biopharmaceutical classification system in early drug development. *AAPS J* 2008;10:208–212.

10. Dahan A, Miller JM, Amidon GL. Prediction of solubility and permeability class membership: provisional BCS classification of the world's top oral drugs. *AAPS J* 2009;11(4): 740–746.

11. European Medicines Agency (EMEA). *Guideline on the Investigation of Bioequivalence (CPMP/EWP/QWP/1401/98 Rev. 1/Corr).* London: Committee for Medicinal Product for Human Use. European Medicines Agency; January 2010.

12. Lindenberg M, Kopp S, Dressman JB. Classification of orally administered drugs on the World Health Organization model list of essential medicines according to the biopharmaceutics classification system. *Eur J Pharm Biopharm* 2004;58:265–278.

13. Fraser EJ, Leach RH, Poston JW, Bold AM, Culank LS, Lipede AB. Dissolution and bioavailability of digoxin tablets. *J Pharm Pharmacol* 1973;25(12):968–973.

14. *The United States Pharmacopeia (USP 26, NF 21).* Rockville: United States Pharmacopeial Convention; 2003.

15. Galia E, Nicolaides E, Hörter D, Löbenberg R, Reppas C, Dressman JB. Evaluation of various dissolution media for predicting in vivo performance of class I and II drugs. *Pharm Res* 1998;15(5):698–705.

16. Fotaki N, Vertzoni M. Biorelevant dissolution methods and their applications in in vitro–in vivo correlations for oral formulations. *Open Drug Deliv J* 2010;4:2–13.

17. Thomas VH, Bhattachar S, Hitchingham L, Zocharski P, Naath M, Surendran N, Stoner CL & El-Kattan A. The road map to oral bioavailability: an industrial perspective. *Expert Opin Drug Metab Toxicol* 2006;2(4):591–608.

18. Merck. September 17, 2012. Drug absorption. *The Merck Manual for HealthCare Professionals* [Online]. Whitehouse Station: Merck Sharp & Dohme Corp. Available at http://www.merckmanuals.com/professional/clinical_pharmacology/pharmacokinetics/drug_absorption.html. Accessed February 13, 2014.

19. Hidalgo IJ, Raub TJ, Borchardt RT. Characterization of the human colon carcinoma cell line (Caco-2) as a model system for intestinal epithelial permeability. *Gastroenterology* 1989;96:736–749; (a) The Backstory. *AAPS J* 2011; 13(3):323–327.

20. Hu M, Kulkarni K. Caco-2 cell culture model for oral drug absorption. In: Hu M, Li X, editors. *Oral Bioavailability: Basic Principles, Advanced Concepts, and Applications.* Hoboken: John Wiley & Sons; 2011. p 431–442.

21. Kwatra D, Boddu SHS, Mitra AK. MDCK cells and other cell-culture models. In: Hu M, Li X, editors. *Oral Bioavailability: Basic Principles, Advanced Concepts, and Applications.* Hoboken: John Wiley & Sons; 2011. p 443–461.

22. Hendriksen BA, Felix MV, Bolger MB. The composite solubility versus pH profile and its role in intestinal absorption prediction. *AAPS J* 2003;5(1):35–49.

23. Balbach S, Korn C. Pharmaceutical evaluation of early development candidates "the 100 mgapproach". *Int J Pharm* 2004;275:1–12.

24. Gould PL. Salt selection for basic drugs. *Int J Pharm* 1986;33:201–217.

25. Thackaberry EA. Non-clinical toxicological considerations for pharmaceutical salt selection. *Expert Opin Drug Metab Toxicol* 2012;8(11):1419–1433.

26. Bastin RJ, Bowker MJ, Slater BJ. Salt selection and optimization procedures for pharmaceutical new chemical entities. *Org Proc Res Develop* 2000;4:427–435.

27. Stahl PH, Wermuth CG, editors. *Handbook of Pharmaceutical Salts: Properties, Selection, and Use.* 2nd rev. ed. Zürich: Verlag Helvetica Chimica Acta; Weinheim: Wiley-VCH; 2011.

28. Guerrieri P, Taylor LS. Role of salt and excipient properties on disproportionation in the solid-state. *Pharm Res* 2009;26(8):2015–2026.

29. Morissette SL, Almarsson O, Peterson ML, Remenar JF, Read MJ, Lemmo AV, Ellis S, Cima MJ, Gardner CR. High-throughput crystallization: polymorphs, salts, co-crystals and solvates of pharmaceutical solids. *Adv Drug Deliv Rev* 2004;56:275–300.

30. Guerrieri P, Rumondor ACF, Li T, Taylor LS. Analysis of relationships between solid-state properties, counterion, and developability of pharmaceutical salts. *AAPS PharmSciTech* 2010;11(3):1212–1222.

31. U.S. Department of Health and Human Services. *Guidance for Industry: Estimating the Maximum Safe. Starting Dose in Initial Clinical Trials for Therapeutics in Adult Healthy Volunteers.* Rockville: Food and Drug Administration, Center for Drug Evaluation and Research (CDER); 2005.

32. Wilson AG, Nouraldeen A, Gopinathan S. A new paradigm for improving oral absorption of drugs in discovery: role of physicochemical properties, different excipients and the pharmaceutical scientist. *Fut Med Chem* 2010;2(1):1–5.

33. Li P, Zhaob L. Developing early formulations: practice and perspective. *Int J Pharm* 2007;341:1–19.

34. Muller RH, Jacobs C, Kayser O. Nanosuspensions as particulate drug formulations in therapy rationale for development and what we can expect for the future. *Adv Drug Deliv Rev* 2001;47:3–19.

35. Nystrom C. How to formulate a tablet for instant dissolution of practically insoluble drugs. *Eur J Pharm Sci* 1996;4 (Suppl 1):64.

36. Chowdari S, Huang J, Minchom C, Chow K, Aurora J. Case study: formulation development of high dose tablets of a poorly soluble and poorly compressible drug substance. AAPS Annual Meeting & Exposition, Toronto, November 8–14, 2002.

37. Bi M, Kyad A, Kiang YH, Alvarez-Nunez F, Alvarez F. Enhancing and sustaining AMG 009 dissolution from a matrix tablet via microenvironmental pH modulation and supersaturation. *AAPS PharmSciTech* 2011;12(4):1157–1162.

38. Ramani C, Wang J, Kane A, Chow K, Lambing J. Solid composition for controlled release of ionizable active agents with poor aqueous solubility at low pH and methods of use thereof. US patent. US 2010/0151019 A1. 2009.

39. Merisko-Liversidge EM, Liversidge GG. Drug nanoparticles: formulating poorly water-soluble compounds. *Toxicol Pathol* 2008;36:43–48.

40. Liversidge GG, Cundy KC, Bishop JF, Czekai DA. Surface modified drug nanoparticles. US patent 5,145,684. 1992.

41. Johnson KC, Swindell AC. Guidance in the setting of drug particle size specifications to minimize variability in absorption. *Pharm Res* 1996;12:1795–1798.

42. Curatolo W. Physical chemical properties of oral drug candidates in the discovery and exploratory development settings. *Pharm Sci Technol Today* 1998;1(9):387–393.

43. Cole ET. *Liquid Filled and Sealed Hard Gelatin Capsules*. Arlesheim/Basel: Capsugel Division,Warner-Lambert Co; 2000.

44. Rowley G. Filling of liquids and semi-solids into hard two-piece capsules. In: Jones BE, Podczeck F, editors. *Pharmaceutical Capsules*. London: Pharmaceutical Press; 2004. p 169–194.

45. Gullapalli RP. Soft gelatin capsules (softgels). *J Pharm Sci* 2010;99(10):4107–4148.

46. Strickley RG. Solubilizing excipients in oral and injectable formulations. *Pharm Res* 2004;21(2):201–230.

47. World Health Organization (WHO). September 18, 2012. Joint FAO/WHO Expert Committee on Food Additives (JECFA) [Online]. Available at http://apps.who.int/ipsc/database/evaluations/search.aspx?fc=39. Accessed on February 13, 2014.

48. Grant DJW, Higuchi T. *Solubility Behavior of Organic Compounds*. Volume 21, New York: Wiley-Interscience; 1990. Techniques of Chemistry.

49. Yalkowsky SH. *Solubility and Solubilization in Aqueous Media*. Washington, DC: American Chemical Society; New York: Oxford University Press; 1999.

50. Yalkowsky SH, Valvani SC, Roseman TJ. Solubility and partitioning VI: octanol solubility and octanol–water partition coefficients. *J Pharm Sci* 1983;72(8):866–870.

51. Abbott S, Hansen CM, Yamamoto H, Valpey III RS. *Hansen Solubility Parameters in Practice*. 3rd ed. 2008. Available at Hansen-Solubility.com. Accessed on February 13, 2014.

52. Lipinski CA, Lombardo F, Dominy BW, Feeney PJ. Experimental and computational approaches to estimate solubility and permeability in drug discovery and development settings. *Adv Drug Deliv Rev* 2001;46:3–26.

53. Zhou L, Yang L, Tilton S, Wang J. Development of a high throughput equilibrium solubility assay using miniaturized shake-flask method in early drug discovery. *J Pharm Sci* 2007;96(11):3052–3071.

54. Alsenz J, Kansy M. High throughput solubility measurement in drug discovery and development. *Adv Drug Deliv Rev* 2007;59(7):546–567.

55. Gibson L. Lipid-based excipients for oral drug delivery. In: Hauss DJ, editor. *Oral Lipid-Based Formulations: Enhancing the Bioavailability of Poorly Water-Soluble Drugs.* New York: Informa Healthcare; 2007. p 33–59.

56. Pouton CW. Lipid formulations for oral administration of drugs: nonemulsifying, self emulsifying and 'self microemulsifying' drug delivery systems. *Eur J Pharm Sci* 2000;11 (Suppl 2):S93–S98.

57. Pouton CW, Porter CJH. Formulation of lipid-based delivery systems for oral administration: materials, methods and strategies. *Adv Drug Deliv Rev* 2008;60: 625–637.

58. Tulieu C, Newton M, Rose J, Euler D, Saklativala R, Clarke A, Booth S. Comparative bioavailability study in dogs of a self-emulsifying formulation of progesterone presented in a pellet and liquid form compared with an aqueous suspension of progesterone. *J Pharm Sci* 2004;93(6):1495–1502.

59. Serajuddin ATM. Solid dispersion of poorly water-soluble drugs: early promises. *J Pharm Sci* 1999;88(10):1058–1066.

60. Hancock BC, Parks M. What is the true solubility advantage for amorphous pharmaceuticals? *Pharm Res* 2000;17(4):397–404.

61. Murdande SB, Pikal MJ, Shanker RM, Bogner RH. Solubility advantage of amorphous pharmaceuticals: II. Application of quantitative thermodynamic relationships for prediction of solubility enhancement in structurally diverse insoluble pharmaceuticals. *Pharm Res* 2010;27(12):2704–2714.

62. Chiou WL, Riegelman S. Pharmaceutical applications of solid dispersion systems. *J Pharm Sci* 1971;60(9):1281–1302.

63. Chow K, Tong HH, Lum S, Chow AH. Engineering of pharmaceutical materials: an industrial perspective. *J Pharm Sci* 2008;8:2855–2877.

64. Newman A, Knipp G, Zografi G. Assessing the performance of amorphous solid dispersions. *J Pharm Sci* 2012;101(4):1355–1377.

65. Friesen DT, Shanker R, Crew M, Smithey DT, Curatolo WJ, Nightingale JAS. Hydroxypropyl methylcellulose acetate succinate-based spray-dried dispersions: an overview. *Mol Pharm* 2008;5(6):1003–1019.

66. Hancock BC, Shamblin SL, Zografi G. Molecular mobility of amorphous pharmaceutical solids below their glass transition temperatures. *Pharm Res* 1995;12:799–806.

67. Gordon M, Taylor JS. Ideal copolymers and the second-order transitions of synthetic rubbers. I. Non-crystalline copolymers. *J Appl Chem* 1952;2(9):493–500.

68. Rambaud E, Patwardhan K, Asgarzadeh F. Prediction of API solubility in solid solutions using modified solubility parameters, film casting and melt extrusion. AAPS Annual Meeting & Exposition, Washington, DC, October 23–27, 2011.

69. Weuts I, Van Dycke F, Voorspoels J, De Cort S, Stikbroekx S, Leemans R, Brewster ME, Xu D, Segmuller B, Turner YT, Roberts CJ, Davies MC, Qi S, Craig DQ, Reading M. Physicochemical properties of the amorphous drug, cast films, and spray dried powders to predict formulation probability of success for solid dispersions: etravirine. *J Pharm Sci* 2011;100(1):260–274.

70. Crowley MM, Zhang F, Repka MA, Thumma S, Upadhye SB, Battu SK, McGinity JW, Martin C. Pharmaceutical applications of hot-melt extrusion: part I. *Drug Dev Ind Pharm* 2007;33:909–926.

71. Repka MA, Battu SK, Upadhye SB, Thumma S, Crowley MM, Zhang F, Martin C, McGinity JW. Pharmaceutical applications of hot-melt extrusion: part II. *Drug Dev Ind Pharm.* 2007;33:1043–1057.

72. Johnson BM, Charman WN, Porter CJH. An in vitro examination of the impact of polyethylene glycol 400, pluronic P85, and vitamin E D-α-tocopheryl polyethylene glycol 1000 succinate on P-glycoprotein efflux and enterocyte-based metabolism in excised rat intestine. *AAPS PharmSci* 2002;4(4):193–205.

73. Yu L, Bridgers A, Polli J, Vickers A, Long S, Roy A, Winnike R, Coffin M. Vitamin E-TPGS increases absorption flux of an HIV protease inhibitor by enhancing its solubility and permeability. *Pharm Res* 1999;16(12):1812–1817.

74. Aungst BJ. Absorption enhancers: applications and advances. *AAPS J* 2012;14(1): 10–18.

75. Furfine ES, Baker CT, Hale MR, Reynolds DJ, Salisbury JA, Searle AD, Studenberg SD, Todd D, Tung RD, Spaltenstein A. Preclinical pharmacology and pharmacokinetics of GW433908, a water-soluble prodrug of the human immunodeficiency virus protease inhibitor amprenavir. *Antimicrob Agents Chemother* 2004;48(3):791–798.

76. Huttunen KM, Raunio H, Rautio J. Prodrugs—from serendipity to rational design. *Pharmacol Rev* 2011;63:750–771.

77. Charman WN, Stella VJ. Estimating the maximum potential for intestinal lymphatic transport of lipophilic drug molecules. *Int J Pharm* 1986;34:175–178.

78. Trevaskis NL, Charman WN, Porter CJ. Lipid-based delivery systems and intestinal lymphatic drug transport: a mechanistic update. *Adv Drug Deliv Rev* 2008;60:702–716.

79. Lentz KA, Quitko M, Morgan DG, Grace JE, Gleason C, Marathe PH. Development and validation of a preclinical food effect model. *J Pharm Sci* 2007;96(2):459–472.

80. Akimoto M, Nagahata N, Furuya A, Fukushima K, Higuchi S, Suwa T. Gastric pH profiles of beagle dogs and their use as an alternative to human testing. *Eur J Pharm Biopharm* 2000;49:99–102.

81. Dobry DE, Settell DM, Baumann JM, Ray RJ, Graham LJ, Beyerinck RA. A model-based methodology for spray-drying process. *J Pharm Innov* 2009;4:133–142.

SECTION 3

OVERVIEW, CURRENT TRENDS AND STRATEGIES OF SPECIAL MEDICAL DEVICE DEVELOPMENT

4

OVERVIEW OF DRUG DELIVERY DEVICES

Iain Simpson, James Blakemore, and Chung Chow Chan

4.1 INTRODUCTION

Medical devices, covering a wide range of products, from simple bandages to the most sophisticated life-supporting products, play a crucial role in the diagnosis, prevention, monitoring, and treatment of diseases and the improvement of the quality of life of people suffering from disabilities.

Medical devices have definition that differs in details in different parts of the world and hence may have different regulatory requirements between countries. In the United States, FDA defines a device as being[1]:

- an instrument, apparatus, implement, machine, contrivance, implant, *in vitro* reagent, or other similar or related article, including a component part, or accessory which is:
  - recognized in the official National Formulary, or the United States Pharmacopoeia, or any supplement to them,
  - intended for use in the diagnosis of disease or other conditions, or in the cure, mitigation, treatment, or prevention of disease, in man or other animals, or

---

[1] http://www.fda.gov/medicaldevices/deviceregulationandguidance/overview/classifyyourdevice/ucm051512.htm

Therapeutic Delivery Solutions, First Edition. Edited by Chung Chow Chan, Kwok Chow, Bill McKay, and Michelle Fung.
© 2014 John Wiley & Sons, Inc. Published 2014 by John Wiley & Sons, Inc.

- intended to affect the structure or any function of the body of man or other animals, and which does not achieve any of its primary intended purposes through chemical action within or on the body of man or other animals and which is not dependent upon being metabolized for the achievement of any of its primary intended purposes.

In Europe, medical devices are defined as articles that are intended to be used for a medical purpose. The medical purpose is assigned to a product by the manufacturer. The manufacturer determines through the label, the instruction for use, and the promotional material related to a given device its specific medical purpose. The medical purpose relates in general to finished products regardless of whether they are intended to be used alone or in combination.

In Canada, Health Canada defines the term "device" as any article, instrument, apparatus, or contrivance, including any component, part, or accessory thereof, manufactured, sold, or represented for use in the following:

1. The diagnosis, treatment, mitigation, or prevention of a disease, disorder, or abnormal physical state, or its symptoms, in human beings or animals
2. Restoring, correcting, or modifying a body function or the body structure of human beings or animals
3. The diagnosis of pregnancy in human beings or animals or
4. The care of human beings or animals during pregnancy and at and after birth of the offspring, including care of the offspring, and includes a contraceptive device but does not include a drug

Medical devices are classified into different classes in different parts of the world. Table 4.1 summarizes the different classes in the United States, Europe, and Canada. The general classification hierarchy is similar based on the risk the medical devices present to the patient.

4.2 TRENDS AND DRIVERS FOR DRUG DELIVERY DEVICES

The recent advent of personalized medicine and wellness healthcare, coupled with innovations in miniaturization technology and software development, has forced industry analysts to reconsider the untapped potential of device and diagnostic companies. Attitudes toward device companies are rapidly changing, and many consider these companies to be the new source of innovation in the life sciences industry.

The continuing convergence of biotechnology, electronics, and medical devices is primarily responsible for increased interest in combination products. While combination products such as drug-eluting cardiovascular stents have received a lot of attention, they are probably the least complex (i.e., mostly placing a small molecule onto a polymer). Some of most exciting yet technical and regulatory challenging

TABLE 4.1   Classes of medical instruments in the United States, Europe, and Canada

United States [1]	Europe [2]	Canada [3]
Class I	Class I (including Is and Im)	Class I devices present the lowest potential risk (e.g., thermometers). Class I devices do not require licenses, but manufacturers must ensure that devices are designed and manufactured to be safe
Class II	Class IIa	Class II, III, and IV devices receive increasingly rigorous reviews and must be licensed before being sold in Canada
Class III	Class IIb	Class IV devices present the greatest potential risk (e.g., pacemakers)
Most class I devices do not require premarket notification or approval and so are just subjected to general controls as defined in 21 CFR 800-814	Class III	
Most class II devices require premarket notification through a 510k process		
Most class III devices require premarket approval (PMA), for example, through the PMA process		

combination products are those that contain biologics and biotechnology component parts. For example, Neurologix is developing a combination product that will deliver a gene product into the brain of patients with Parkinson's disease with the intent that direct delivery of the gene product into the brain should help alleviate many symptoms of the disease. There are many efforts of developing advanced combination products (with cardiovascular) and implantable pacing electrodes (to electrically stimulate the brain and CNS) and catheters (hooked to infusion pumps) to mediate controlled local delivery of monoclonal antibodies (mAbs) and silencing RNA (siRNA) across the blood–brain barrier.

4.3   REGULATION OF MEDICAL DEVICES

A discussion of the regulatory requirements and regulatory initiatives is detailed in Chapter 1. Please refer to that chapter for more details.

4.4   THE DEVELOPMENT OF DRUG/DEVICE COMBINATION PRODUCTS

4.4.1   The Medical Device Development Process

The product design process involves defining the requirements that a product needs to achieve and then developing the design and manufacturing process that ensure the product meets these requirements. Although the development process can be described in many different ways, it will normally progress through a series of phases in which:

- The requirements (design inputs) are identified and documented.
- Design concepts are generated and evaluated to see if they meet the requirements.
- Proof of principle of the design is tested using prototypes and engineering analysis to check that the right level of performance can be achieved.
- Detailed design of the product is created that describes the design (design outputs).
- Verification testing of the design is conducted to confirm that actual design meets the requirements.
- Transfer into commercial manufacturer (industrialization) is completed.

In reality, the process is rarely linear but involves interactions and changes as follows:

- The initial design does not meet the requirements or the requirements change.
- If the process is not well controlled, then there is a risk that requirements will not be properly identified and that the design may not meet them or the design description (design outputs) is inadequate for the product to be manufactured reliably.

It is beyond the scope of this book to provide details on the device development process, but it is appropriate to discuss briefly the design control process for medical devices. A more detailed overview of combination product development has been given by V&S Gopalaswamy [1].

Figure 4.1 summarizes the activities in the development of medical devices from design to validation and verification.

4.4.1.1   *CFR21 Section 820.30: The Required Design Controls for the Development of Class II and III (and Some Class I) Medical Devices for the U.S. Market*

Design and development planning   Each manufacturer shall establish and maintain plans that describe or reference the design and development activities and define responsibility for implementation. The plans shall identify and describe the interface with different groups or activities that provide, or result in, input to the design and development process. The plans shall be reviewed, updated, and approved as design and development evolve.

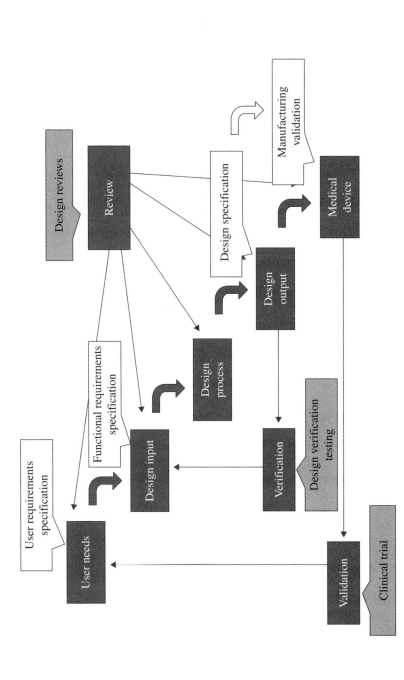

FIGURE 4.1 Verification and validation of medical devices. Based on FDA Design Control Guidance (http://www.fda.gov/downloads/medicalDevices/.../ucm070642.pdf).

Design input   Each manufacturer shall establish and maintain procedures to ensure that the design requirements relating to a device are appropriate and address the intended use of the device, including the needs of the user and patient. The design input requirements shall be documented and shall be reviewed and approved by a designated individual(s). The approval, including the date and signature of the individual(s) approving the requirements, shall be documented.

Design output   Each manufacturer shall establish and maintain procedures for defining and documenting design output in terms that allow an adequate evaluation of conformance to design input requirements. Design output shall be documented, reviewed, and approved before release. The approval, including the date and signature of the individual(s) approving the output, shall be documented.

Design review   Each manufacturer shall establish and maintain procedures to ensure that formal documented reviews of the design results are planned and conducted at appropriate stages of the device's design development. The results of a design review, including identification of the design, the date, and the individual(s) performing the review, shall be documented in the device history file (DHF).

Design transfer   Each manufacturer shall establish and maintain procedures to ensure that the device design is correctly translated into production specifications.

Design changes   Each manufacturer shall establish and maintain procedures for the identification, documentation, validation, or, where appropriate, verification, review, and approval of design changes before their implementation.

Design verification   Design verification [820.30(f)] shall confirm that the design output meets the design input requirements. Verification means confirmation by examination and provision of objective evidence that specified requirements have been fulfilled. The results of the design verification, including identification of the design, the method(s), the date, and the individual(s) performing the verification, shall be documented in the DHF.
   Design verification is always done versus specifications. Therefore, to control the specifications and increase the probability of achieving desired safety and performance characteristics, device, software, labeling, packaging, and any other specifications should be complete and thoroughly reviewed before development commences. As the hardware and software designs evolve, they should be evaluated versus their current specifications.

Validation   Validation [820.30(g)] means confirmation by examination and provision of objective evidence that the particular requirements for a specific intended use can be consistently fulfilled.

Process validation   Process validation means establishing by objective evidence that a process consistently produces a result or product meeting its predetermined specifications.

Design validation   Design validation means establishing by objective evidence that device specifications conform with user needs and intended use(s).

Each manufacturer shall establish and maintain procedures for validating the device design. Design validation shall be performed under defined operating conditions on initial production units, lots, or batches, or their equivalents. Design validation shall ensure that devices conform to defined user needs and intended uses and shall include testing of production units under actual or simulated use conditions. Design validation shall include software validation and risk analysis, where appropriate. The results of the design validation, including identification of the design, the method(s), the date, and the individual(s) performing the validation, shall be documented in the DHF.

Verification and validation should be done with test equipment calibrated and controlled according to quality system requirements. Otherwise, there is limited confidence in the data.

Verification and validation should also be done according to a written protocol(s). The protocol(s) should include defined conditions for the testing. The protocol(s) should be approved before being used. Test protocol(s) is not perfect for a design, particularly a new design. Therefore, the designers and other verification personnel carefully annotate any ongoing changes to a protocol. Likewise, the verification personnel should record technical comments about any deviations or other events that occurred during the testing. The slightest problem should not be ignored. During design reviews, the comments, notes, and deviations may be as important as test data from the formal protocol(s).

4.5   COMMON THERAPEUTIC ROUTES OF ADMINISTRATION THAT REQUIRE A MEDICAL DEVICE

Oral (enteral) drug delivery is generally considered the preferred delivery route if available as it is considered to be preferred by patients and also avoids the complexities of developing and getting approval for a drug/device combination that is required for other routes such as nasal, parenteral, or oral inhalation. Pharmaceutical companies will go to great lengths to make or try to make the oral route available, but there still remain many situations where a different delivery route is required. The main problem with oral delivery of some drugs is first-pass metabolism in the gut and liver that limits their uptake into systemic circulation. Another issue for the treatment of severe pain or conditions such as migraine is the relatively slow rate of onset of the drug effect due to the time it takes the drug to pass into systematic circulation via the digestive system and the liver. A third issue is that taking oral drugs to treat a localized medical condition can result in unacceptably high concentrations of the drug in systemic circulation where no therapeutic benefits are derived. An example of this is the use of corticosteroids to treat respiratory disease. While this treatment might be achievable using oral steroids, the dose would have to be high, and it is safer and more effective to use an inhaler to deliver the drug directly to the lung.

Table 4.2 summarizes the main routes of administration that require a medical device to facilitate drug delivery.

TABLE 4.2 Summary of the main routes of administration that require a medical device to facilitate drug delivery

Route of administration	Description	Example			Rationale for use of drug delivery device
		Generic name (brand; manufacturer)	Device	Indication	
Injection					
Intravenous	Within a vein	Factor VIII (ReFacto® AF; Pfizer)	*Vial and syringe*	Hemophilia	Recombinant factor protein requires direct administration to vein to avoid gastric degradation; requires reconstitution by HCP to ensure stability and shelf life
Intramuscular	Within a muscle	Epinephrine (Meda)	EpiPen®	Acute anaphylaxis	Delivery of epinephrine requires IM injection to achieve required blood serum concentrations[i]
Intrathecal	Within the thecal membrane of the spinal cord	Ziconotide (Prialt®; Eisai)	SynchroMed® II infusion system (Medtronic) CADD-Micro ambulatory infusion pump	Chronic severe pain	Requires direct administration to thecal cavity to bypass blood–brain barrier Requires specialist HCP to administer Delivered using a programmable implanted variable-rate microinfusion device or an external microinfusion device and catheter
Subcutaneous	Beneath the skin	Adalimumab (Humira®; Abbott)	Autoinjector ("Humira pen")	Rheumatoid arthritis	Mw of API, depot injection, patient-centric medication
Intravitreal	Within the vitreal region of the eye	Ranibizumab (Lucentis®; Novartis)	1cc tuberculin syringe	Neovascular (wet) AMD	Drug solution requires specialist preparation and administration by HCP

Route	Location	Drug	Device	Indication	Comments
Transdermal	To the dermal layer of the skin	Fentanyl (Durogesic DTrans®; Janssen)	Transdermal patch	Severe chronic pain	Multiple; convenience (patient-centric medication); pharmacodynamics (sustained release from patch); lipophilic nature of drug passes through dermal space; systemic effects
		Lidocaine (Versatis®; Grunenthal)	Transdermal patch	Postherpetic neuralgia	As aforementioned; presentation results in local analgesic effects only
Ophthalmic	To the eye	Sodium hyaluronate (Clinitas®; Altacor)	Eyedropper	Dry eye conditions	Device provides measured dose to the surface of the eye; little more to be said; good example of device requiring minimal regulatory controls (i.e., class 1)
Transmucosal					
Pulmonary	To the lungs	Tiotropium (Spiriva®; Boehringer)	HandiHaler® (MDI) Respimat® (soft mist inhaler)	COPD	Requires direct administration of drug to the lung (bronchial region of the lungs) to provide local drug delivery and lower systemic levels of the drug that might result in side effects
	To the lungs	Insulin (Affreza®; MannKind)	Inhaler	Type II diabetes	Development stage; example of pulmonary route for systemic delivery of macromolecule
Sublingual	Beneath the tongue	Glyceryl trinitrate (Glytrin Spray®; Sanofi)	Metered pump spray	Acute angina	Requires direct administration to sl region to aid rapid absorption and onset of action; delivery routes utilize lipophilic nature of small-molecular-weight drug

(Continued)

TABLE 4.2 (Cont'd)

Route of Administration	Description	Example		Rationale for Use of Drug Delivery Device
Intranasal	To the nose	Azelastine hydrochloride (Rhinolast®; Meda)	Metered pump spray	Drug presentation facilitates local drug effects in order to alleviate symptoms
	"	Fentanyl (Instanyl)	Metered pump spray	Drug presentation facilitates rapid transmucosal delivery of drug within nasal cavity to facilitate rapid onset of systemic analgesic effects
Other				
Intra-arterial	Within an artery	Fluorouracil	Infusion	Certain oncological conditions require intra-arterial administration via infusion

4.5.1 Parenteral Drug Delivery

In its broadest definition, parenteral drug delivery includes all routes of systemic drug administration other than enteral delivery (i.e., delivery via the gastrointestinal tract). However, it is often taken to cover all forms of injectable drug delivery but excludes inhalation and other forms of mucosal drug delivery.

4.5.2 Injectable Routes of Administration

Injection is the most common nonoral delivery route and has undergone considerable evolution in terms of technical complexity and user benefit since the days of "syringe and vial" approaches to injected medicines.

A main driver for the increase in demand for injectable delivery systems has been the increase in biologic drugs that currently account for nearly 50% of drugs currently under development. In general, biologic drugs are destroyed by first-pass metabolism in the gut or liver and hence cannot easily be delivered orally and so need to be delivered by other routes. While other routes of administration such as pulmonary are suitable for the delivery of biologics and other drugs destroyed in the gut, injectable delivery generally offers better bioavailability and ease of administration by a healthcare personnel (HCP) than other delivery routes.

4.5.2.1 *Intravenous (IV) Administration*   This mode of delivery requires a device to access a vein and historically has been the most common form of parenteral drug administration. This mode of administration should not to be confused with intra-arterial administration. The relatively high blood pressure associated with arterial circulation, compared with venous flow, requires specialist HCP administration. This limits use of intra-arterial administration to niche therapeutic applications such as the dosing of fluorouracil for the treatment of some forms of cancer.

Intravenous (IV) administration utilizes a broad spectrum of device modalities to deliver the therapeutic, for example, implantable, infusion, or bolus. Infusion enables a steady PK state to maintain a steady state of drug concentration. IV administration typically requires HCP expertise and therefore is limited to hospital setting or physician clinic. It is not normally considered a viable route of self-administration because of the difficulties and risks of getting venous access apart from for some specific patient groups such as hemophiliacs for whom the risks of administration outweigh the life-threatening situation of failing to administer a drug rapidly.

IV administration by HCP is a low-cost way to administer drugs in a controlled environment. Safety considerations, for example, risk to HCPs of cytotoxic drug, have brought about new standards for delivery of cytotoxic drugs (reduced exposure of cytotoxic drugs in health workers).

4.5.2.2 *Subcutaneous (SC) Route of Administration*   The drug is injected by syringe and needle into the tissues just beneath the skin. The injection must be a sterile liquid capable of complete absorption. Subcutaneous (SC) medications are usually given into the SC tissue of the upper triceps area (the area in the back of the upper arm) or to the

thigh or abdomen. Most people have an adequate amount of fat in these regions, readily accessed by gently pinching the skin at the site of injection.

SC administration offers potential for self-administration because it is more "patient-friendly" than IV administration as a process of delivery with appropriate training. Training for SC administration is minimal and less complicated.

An example of the evolution of the benefit of SC route of administration is that for the tumor necrosis factor (TNF) blockers. The therapies began as IV formulations for administration at site of clinic that is time consuming requiring cyclic administration and reconstitution prior to administration together with HCP supervision. It was reformulated as an SC depot formulation using a prefilled syringe (PFS) in order to facilitate patient self-administration. The reformulation provides benefits to a range of stakeholders including patients, physicians, and payers.

SC self-administration provides opportunities for drug companies to add value to existing and pipeline drug candidates, critical for life cycle management, in a time where there is reduced innovation at drug R&D level. An example of innovation in self-administration space is incorporation of electronics in devices to monitor concordance and adherence and maintain ideal critical delivery activities (e.g., rate of delivery).

A key driver for use of SC devices is the emergence of recombinant therapies, particularly mAbs, which require parenteral delivery in order to avoid gastric degradation incurred via the oral route. Of particular importance is in the delivery of oncology drugs where larger drug volumes are required in order to carry sufficient active pharmaceutical ingredient (API) "payload." Volume challenges may be mitigated to a degree by concentrating the formulation, but this leads to viscosity challenges. SC bolus injections are limited by volume constraints. Formulation volumes above approximately 1 ml are painful for the patient and increase a risk of drainage from the administration site. Microinfusion (or "patch pump") technologies have emerged to deal with this volume-challenged formulation, for example, MyDose (Roche). However, the new technology comes with higher level of technical and manufacturing complexity and cost.

SC drug delivery devices to facilitate delivery of drug volumes up to 1 ml with high viscosity have been developed and include the ASI AutoInjector Platform from Bespak and a combination product from Glide Pharma, which uses a solid dose of drug for SC administration, administered using a simple, easy-to-use device.

Development of other novel delivery systems, for example, needle-free delivery systems, also tends to be subcutaneous, for example, Zogenix's DosePro device platform.

Widespread adoption of SC devices is dependent upon the device sector's ability to demonstrate material rather than incremental clinical utility. Otherwise, payers will overlook them in favor of lower-cost alternatives. Device companies may not have sufficient funds to do this alone and will require pharma partnership to demonstrate the benefit.

4.5.2.3 Intramuscular (IM) Injection   With intramuscular (IM) injection, the drug is injected directly into a muscle. The commonly used site in adults is the deltoid muscle (the muscle in the top part of the arm, about three fingerbreadths below the shoulder). The ventrogluteal muscle of the buttocks may be used sometimes but is never to be used for vaccines.

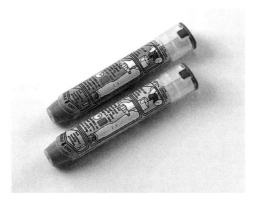

FIGURE 4.2 EpiPen® injector device. Reproduced with permission of Mylan Specialty L.P.

A leading device for IM injection is the EpiPen device (Fig. 4.2), which can administer epinephrine by self-administration for the emergency treatment of severe allergic reaction, including anaphylaxis associated with foods (e.g., peanuts, eggs, etc.), stinging insects (e.g., bees, wasps, and hornets), and other allergens.

EpiPen and EpiPen Jr contain 2 ml of solution but deliver a single dose of 0.3 ml only, with 1.7 ml remaining in the unit after use. EpiPen delivers a dose of 0.3 mg in 0.3 ml of 1:1000 dilution of epinephrine injection, USP, and EpiPen Jr delivers a dose of 0.15 mg in 0.3 ml of 1:2000 dilution of epinephrine injection, USP. Both EpiPen and EpiPen Jr are intended for IM use in the anterolateral aspect of the thigh and should not be injected into the buttock. To use the EpiPen and EpiPen Jr, the cap is removed and the drug injected into the muscle of the thigh through clothing if necessary during emergency.

4.5.3 Intravitreal Route of Administration

Certain ocular diseases, specifically those affecting the rear of the eye, require a relatively invasive route of administration in order to reach the back of the eye. Posterior eye diseases are now a major R&D focus for pharma because they are increasingly associated with major unmet medical needs and large and growing patient potential, for example, age-related macular degeneration (AMD), diabetic retinopathy, and macular edema. These disease conditions are all being driven by an aging patient population. By comparison to front of the eye disease, back of the eye disease is less accessible for noninvasive drug treatment to be effective.

Lucentis® requires direct administration of the drug to the vitreal space. It works as an inhibitor of angiogenesis, which is the main cause of blindness in AMD. To be effective, very high doses are required to be given systemically to maintain therapeutic window in the vitreal region of the eye. The high dose required would cause serious adverse events rendering the drug's risk–benefit profile nonviable. Other follow-on products include Regeneron/Sanofi drug EYLEA® (aflibercept).

4.5.4 Other Injectable Administration Routes

4.5.4.1 Intrathecal Injection   Intrathecal injection is an injection into the spinal canal and more specifically into the subarachnoid space. This means the drug is delivered into the cerebrospinal fluid and is a route to deliver drugs into the brain and neurological system. Examples of drugs delivered by this route include analgesics such as bupivacaine, methotrexate for chemotherapy, and baclofen for spastic cerebral palsy.

4.5.4.2 Intraperitoneal Injection   This delivery route involves administration by injection directly into the intraperitoneal cavity, which is the area of the body that surrounds the abdominal organs. Although more commonly used to treat animals, it does have human applications in the delivery of chemotherapy drugs for conditions such as ovarian cancer. Its benefit is that it can treat the whole of the abdominal cavity eliminating metastases that may be too small to treat by other means.

4.5.4.3 Epidural Injection   Epidural injection involves administration to the region around the dura, the membranes that cover the spinal nerves. Although its main application is for analgesia such as in childbirth and major surgery, it is also utilized in chemotherapy.

4.5.5 Injection Devices

As mentioned in Section 4.5.5.2, "advanced" injection systems have been developed that reduce the burden and risk for HCPs to prepare and administer injectable drugs. More recently, devices such as injection pens and autoinjectors have begun to allow patients to self-administer an increasing wide range of drugs used to treat chronic diseases such as diabetes, rheumatoid arthritis, and multiple sclerosis. The main drivers for self-administration have been improved compliance and cost savings through the avoidance of the need to visit or be visited by an HCP.

4.5.5.1 Prefilled Syringes (PFSs)   PFSs improve convenience and dose preparation errors as the drug is factory metered into the syringe. Although glass PFSs are most common as they ensure good drug stability for many compounds over an acceptable shelf life, there is a move toward more use of polymer syringes, particularly in Japan. The main benefit is a reduced risk of breakage in manufacture or use, as well as eliminating tungsten (used in forming glass syringes) and silicones (used to lubricate glass syringes), both of which can affect drug stability. However, cyclic olefin polymers (COPs) and cyclic olefin copolymers (COCs) have higher permeation rates for oxygen and moisture compared to glass that can limit their suitability for some drugs, although the development of coating technology might address this. In the longer term, as polymer costs potentially reduce due to economies of scale, they may even become price competitive with glass devices.

4.5.5.2   Autoinjectors   Autoinjectors are injection devices that can automate the insertion of the needle into the patient and delivery of a drug and thus are suited to self-administration by patients. They were originally developed for emergency applications such as the delivery of epinephrine for the treatment of anaphylaxis but more recently have gained popularity with a wider range of drugs used to treat chronic diseases such as multiple sclerosis and rheumatoid arthritis. These devices normally use a PFS to contain the drug and a spring-based delivery mechanism, although they can also use electromechanical drives (e.g., the Merck Serono Easypod®). Because the primary drug packing can be unchanged in switching from a PFS to an autoinjector, this reduces the burden of approval, and so this switch has often been used as a life cycle management opportunity for some drugs. However, increased competition in some disease areas such as rheumatoid arthritis has resulted in some products being launched in an autoinjector.

4.5.5.3   Injection Pens   Injection pens provide a convenient way of administering multiple injections from the same device. Some devices such as the Haselmeier Penlet administer fixed doses of drugs, whereas others such as the Sanofi SoloSTAR® allow a variable dose of insulin to be dialed up and administered. Although the most common application for variable dose pens is in the delivery of insulin, they are also used for administering other drugs including growth hormone and fertility treatment. Fixed-dose pens are used for administration of antidiabetic drugs such as glucagon-like peptides (GLPs) and for drugs used to treat osteoporosis. The use of pens to administer insulin has shown rapid growth over the past 20 years, particularly in Europe. The market in Europe has moved toward disposable devices because of their greater convenience although in other more cost-sensitive markets such as in India, reusable devices are common. These allow a new drug cartridge to be inserted into the pen.

4.5.5.4   Needle-Free Delivery Systems   Needle-free delivery systems operate by using a high-pressure jet of drug or vaccine to penetrate the skin without use of a needle. They have been around since the late 1940s when they were developed as a means for mass vaccination. However, it was discovered that if the vaccine was administered from the same reservoir and injector nozzle to successive patients, there was a risk of transfer of a virus due to fluid from one patient contaminating the device and then being injected into a successive patient, which prompted the withdrawal of these devices. More recently, unit-dose and multidose devices were developed for personal use by several companies including Bioject, Antares, and Weston Medical. Success of these devices was limited to some niche markets such as human growth hormone and insulin. Zogenix acquired the Weston Medical Technology and developed Sumavel® DosePro® for the treatment of migraine and achieved some good market growth. The main benefit of the needle-free delivery system was rapid onset of drug effect compared to oral triptans. The relatively pain-free delivery and convenience of the device were considered major factors in encouraging patients to make the switch from oral to injectable forms of the drug. Needle-free devices avoid issues associated with needlestick and needle reuse and have potential for use in mass inoculation programs if the device cost can be reduced, although at current costs they are only likely to continue to be used in high-value niche applications.

While most needle-free systems have been developed to deliver liquid formulations, PowderJect developed a powder-based needle-free delivery system and targeted applications in genetic engineering and pain management. The technology was later acquired by Pfizer, although it has yet to come to market.

4.5.6 Pulmonary Route of Administration

4.5.6.1 Introduction   Inhaled drug delivery represents a significant pharmaceutical market. In 2010, the global pulmonary drug delivery technology market was US\$19 billion, and BCC Research estimated it would grow to US\$44 billion by 2016 at a compound annual growth rate (CAGR) of 14.3%. At present, the inhaled drug delivery market is dominated by treatments for respiratory disease such as asthma and chronic obstructive pulmonary disease (COPD) (where they account for significantly more than half of that market). It is estimated that 300 million people worldwide have asthma, and this is expected to increase by a further 100 million by 2025.

However, over the next few years, inhaled drugs for nonrespiratory conditions are likely to come to market including Alexza's Adasuve (for the treatment of agitation in bipolar disorder and schizophrenia), Mapp's Levadex (for the treatment of acute migraine), and MannKind's Afrezza® (an inhaled insulin for the treatment of diabetes).

There are five main drivers for the use of inhaled drug delivery:

1. Topical administration (i.e., where the target site is the lung itself), for example, respiratory and (some) infectious diseases.
2. Speed of action, for example, CNS applications such as the treatment of migraine and for pain relief.
3. Convenience over other delivery routes, for example, the development of inhaled insulin to avoid patient concerns/stigma over the use of needles.
4. Dry powder formulations can provide better drug stability than for many liquid formulations (e.g., insulin and vaccines).
5. Research suggests administration of a vaccine to the lungs can induce a local immune response more effectively than conventional types of vaccine delivery (opening up the possibility of inhaled vaccines).

4.5.6.2 Topical Drug Delivery to the Lung   As mentioned previously, the most common use of inhaled drug delivery is to treat respiratory disease as this approach allows the drug to target the disease site directly and reduce the adverse effects associated with systemic circulation of the drug. With inhaled delivery of drugs such as corticosteroids, much of the drug remains in the lung and does not get into systemic circulation (Table 4.3).

TABLE 4.3   Examples of inhalers for topical delivery to the lung

Inhaler type	Medication delivered	Remarks
CFC inhaler	Ventolin® (salbutamol) 100 µg Flovent® (fluticasone) 50, 125, and 250 µg	Discontinued and replaced by HFA for environment protection
HFA inhaler	Advair® (salmeterol/fluticasone) 25/125 and 25/250 µg Atrovent® HFA (ipratropium) 20 µg Flovent HFA (fluticasone) 50, 125, and 250 µg Ventolin HFA (salbutamol) 100 µg	HFA is the environmentally friendly solvent
DPI	Turbuhaler Bricanyl® (terbutaline) 0.5 mg Pulmicort® (budesonide) 100, 200, and 400 µg Symbicort® (budesonide/formoterol) 100/6 and 200/6 µg Oxeze® (formoterol) 6 and 12 µg	Dry powder drug substance metered from reservoir
	Diskus Advair (salmeterol/fluticasone) 50/100, 50/250, and 50/500 µg Flovent (fluticasone) 50, 100, 250, and 500 µg Serevent® (salmeterol) 50 µg Ventolin (salbutamol) 200 µg	Dry powder drug substance factory premetered into blister strip
	HandiHaler Spiriva (tiotropium) 18 µg	Dry powder drug substance in capsules used in unit-dose inhaler
Soft mist inhaler	Respimat Spiriva (tiotropium) 2.5 µg	Liquid drug substance

4.5.6.3   Systemic Drug Delivery via the Lung   It has long been known that inhaled substances can be adsorbed from the lung into systemic circulation—nicotine being one example (Table 4.4).

4.5.6.4   Inhaler Device Categories   A wide variety of inhalers have been developed to meet the needs of the market and the specific requirements for a drug in terms of dose size, drug stability, and physical and chemical properties. The main categories are:

- Pressurized metered-dose inhalers (pMDIs)
- Dry powder inhalers (DPIs)
- Soft mist inhalers
- Nebulizers

TABLE 4.4   Examples of inhalers for systemic delivery via the lung

Inhaler type	Medication delivered	Remarks
Exubera®	Insulin	Regulatory approved; withdrawn from market by manufacturer
Afrezza®	Lyophilized human insulin microencapsulated within fumaryl diketopiperazine	Awaiting regulatory approval
Adasuve™	Staccato® Loxapine for the treatment of agitation associated with schizophrenia or bipolar I disorder in adults; 10 mg dry powder	Approved in the United States and EU in late 2012/early 2013; launch in the United States and EU anticipated in Q3 2013
Levadex™	Dihydroergotamine for the treatment of migraine	Awaiting approval

Pressurized metered-dose inhalers (pMDIs)   Although the history of inhalers can be traced back several hundred years (see http://www.inhalatorium.com/), the first effective modern inhalers were pressurized metered-dose devices that were developed in the 1960s. These devices use drug dispersed in a propellant such as chlorinated fluorocarbons (CFCs). When the device is actuated, a metered dose of drug and propellant is released and is drawn into the lung by the user's inspiratory flow. A major challenge with standard pMDIs is the need for the user to coordinate the actuation of the device with their inhalation. Although breath-actuated devices have been developed that automatically release the dose of drug when the user achieves a threshold flow rate, the majority of devices in the market today are the standard pMDIs. Environmental concerns over CFCs that are believed to deplete the earth's ozone layer have resulted in a change to less environmentally harmful HCFCs in pMDIs.

Albuterol/salbutamol is a short-acting beta-agonist that is commonly administered using a pMDI such as the Ventolin inhaler. Research shows that when 100 µg of titrated salbutamol aerosol is administered, plasma levels of radioactively labeled drug were insignificant at 10, 20, and 30 min following inhalation. Approximately 10% of an inhaled salbutamol dose is deposited in the lungs. Eighty-five percent of the remaining dose from the metered-dose inhaler (MDI) is swallowed. Since the dose is low (100 µg), the absolute amount swallowed is too small to be of clinical significance.

Dry powder inhalers   DPIs began to be developed around 30 years ago as they are considered to offer a number of benefits over pMDIs including more stable drug formulations, ease of use (as they do not require coordination of device actuation with inhalation), as well as avoiding the need to use environmentally harmful propellants. Most dry powder formulations consist of an active drug deposited on the surface of larger carrier particles such as lactose. This formulation approach is beneficial as it allows a small active dose to be "bulked up" by the carrier and also allows the drug to be more easily aerosolized during inhalation so that it can penetrate into the lung. During inhalation particle–particle and particle–device wall collisions, coupled with shear forces generated in flows in the device deagglomerate the particles so that they are small enough to be carried into the lung. In reality, this process is not that efficient and typically only between 10% and 40% of the drug actually gets into the lung.

For the majority of DPIs, the drug is deagglomerated solely by the airflow from the patient's breath—these are termed passive devices. If the patient inhales too slowly, then there may be insufficient energy to aerosolize the drug, whereas at higher flow rates, the drug might be deagglomerated, but then a larger proportion deposited in the throat due to higher inhalation velocities. Inhalation rates can be controlled by patient training and by optimizing the airflow resistance of the device for the patient. However, patient-to-patient variability remains an issue.

One route to reducing this issue is to develop more sophisticated formulations that require lower deagglomeration forces and whose transport is more flow independent. Another approach is to develop active devices in which stored energy is used to deagglomerate the drug. These devices tend to be more complex, more costly, and also larger in size, which explains why few devices of this type have come to the market, although some continue to be under development.

Early dry powder devices tended to be unit-dose devices. To use these devices, a capsule of drug is loaded then usually pierced by pins that are actuated by the user pressing button(s) on the device. On inhalation, the capsule then spins in the airflow releasing the drug to be inhaled. While simple in design, capsule inhalers tend to have low efficiency and high drug retention in the capsule. There have also been cases where patients have swallowed the capsules rather than use them in the inhaler perhaps because they have mistaken them as pills. Despite these issues, capsule inhalers are still commonly used with new drugs such as Onbrez® marketed by Novartis.

Spiriva® HandiHaler (tiotropium bromide monohydrate bronchodilator) is another example of a capsule inhaler. The HandiHaler inhaler is a reusable plastic device for use in administration of the Spiriva capsule. The HandiHaler operates with flow rates as low as 20 l/min. To use the device, a Spiriva capsule is placed in the center chamber of the green inhalation device. The capsule is pierced by a pointed pin by pressing and releasing the piercing button on the side of the device. The tiotropium formulation is dispersed into the airstream when the patient inhales slowly and deeply through the mouthpiece. Each Spiriva capsule contains a dry powder blend of 18 μg tiotropium with lactose monohydrate as the carrier.

Multidose dry powder devices have been developed as they offer the patient the benefit of having multiple doses stored in the device. Reservoir devices store the drug in a single chamber from which individual doses are metered (e.g., Turbuhaler), whereas other multidose devices (such as Diskus®) contain a strip or ring of individual doses that are individually filled during manufacture. Such devices are often termed premetered multidose inhalers and generally have the advantage of better drug protection (so improved stability) and more accurate dosing than reservoir devices. On the other hand, reservoir devices have the benefit of smaller size (for a given number of doses) and lower manufacturing cost.

As mentioned earlier, Turbuhaler® (Fig. 4.3) is an example of a reservoir multidose device. The preparation of each dose is achieved by removing the cap from the device and then turning the base in one direction and then back until a click is heard. The device is then ready for inhalation.

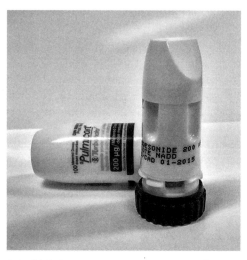

FIGURE 4.3   Turbuhaler® inhaler. Reproduction of Pulmicort® device provided courtesy of AstraZeneca Canada Inc.

The Diskus® inhaler is an example of a premetered multidose device. Diskus is a disposable colored plastic inhaler device containing a foil strip with blisters. Each blister contains the active drug and use lactose as the carrier. The blisters protect the powder for inhalation from effects of the atmosphere. When a dose is required, the patient follows the four simple steps: open, slide, inhale, and close. Sliding the lever of the Diskus inhalation device opens a small hole in the mouthpiece and unwraps a dose ready for inhalation. When the Diskus inhalation device is closed by sliding the thumb grip as far as it will go, the lever automatically moves back to its original position ready for your next dose when it is needed. The outer case protects the Diskus inhalation device when it is not in use. Diskus inhalation device is available for Ventolin® (salbutamol sulfate) and Advair® (salmeterol xinafoate/fluticasone propionate).

Although simple to use, both Diskus and Turbuhaler have the risk of double dosing in that they can be actuated more than once without inhaling the drug. Newer devices such as Sun Pharma's Starhaler and Aptar's Prohaler ensure that the user cannot access multiple doses in one inhalation. These devices are also breath actuated, so the dose is only released when the patient inhalation flow rate is sufficient to deagglomerate the drug and transport it to the lung.

A challenge with passive devices is that they depend on the user's inhalation to aerosolize the drug and transport it to the lung. If the patient inhales too slowly, then there may be insufficient energy to break the drug apart and create aerosolizable particles. If the patient inhales too fast, then the drug may strike the back of the throat and hence not reach the lung. Active inhalation devices can address the former issue and have mainly been developed for systemic drug delivery via the lung. In this situation, it is important for the drug to penetrate deep into the lung where it will be absolved through the alveoli into the bloodstream. Deep lung penetration is best achieved using a long slow inhalation maneuver, which may mean there is insufficient energy to aerosolize the drug unless an active device is used.

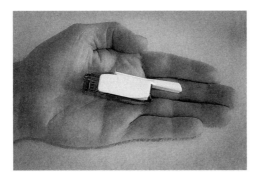

FIGURE 4.4 Afrezza® inhaler (palm size). Reproduction of Afrezza device provided courtesy of MannKind Corporation.

Pfizer worked with Nektar to develop the Exubera device that entered the market for use with inhaled insulin. This device uses a pressure pulse to aerosolize the drug into a large chamber, similar to a spacer used with MDIs, from which the drug can then be inhaled by the user with slow deep breaths. The device is large, is complicated (around 60 parts), and received considerable negative feedback from a usability perspective. Presumably at the time the insulin program was being developed, the device complexity was considered necessary to make the product work, although there have been several claims of much simpler devices achieving the same performance with similar formulations. The product entered the market in September 2006 and following slow sales and some concern about a possible link with lung cancer, Pfizer announced its withdrawal in October 2007.

While most companies terminated product development after Exubera was removed from the market, MannKind Corporation continued developing inhaled insulin and came up with thumb-sized second-generation DreamBoat Afrezza Technosphere insulin (TI) (Fig. 4.4) breath-actuated inhaler (TI is formed when regular human insulin (RHI) is microencapsulated within fumaryl diketopiperazine and lyophilized for inhalation). Upon inhalation, the microspheres dissolve in the neutral pH of the lungs.

MicroDose Therapeutx has developed device technology based on the use of a piezoelectric actuator to disperse the drug. The device is breath actuated—through the use of a flow sensing microphone to trigger the piezo actuator once the correct flow has been established. The technology has been partnered for several programs including Novartis for a number of proprietary respiratory compounds. MicroDose Therapeutx has also initiated its own programs for the treatment of virus, nerve agent, and COPD, some of which have progressed into early clinical trials. More recently, MicroDose Therapeutx has been acquired by Teva who is seeking to exploit the benefits of the device and technology platform.

Alexza has developed the Staccato device platform that operates by heating a thin-layer drug deposited onto a metal substrate in order to vaporize it into the inhalation airflow. Both unit-dose and multidose variants have been in development. Their lead product Adasuve™ received approval in the United States and Europe in late 2012/early 2013 for the treatment of agitation associated with schizophrenia or bipolar I disorder in adults. Other programs for breakthrough cancer pain and some other indications are in early clinical development.

Another approach to developing less flow rate-dependent devices is to develop formulation technologies that require low energy to aerosolize the drug and use particles that have aerodynamic properties that limit their velocity during transport from the device to the patient's lungs. Examples include Alkermes Air (now being developed by Civitas Therapeutics), Prosonix, and Pulmospheres (owned by Novartis following their acquisition of the Nektar's respiratory business).

Nebulizers   The latest devices from Pari and Respironics have to some extent addressed the main disadvantages of nebulizers, portability, and time it takes to deliver the drug.

Nebulizers requiring multiple inhalations to deliver the required dose are not really direct competitors with DPIs. Although devices have the ability to match drug aerosolization to breathing rate and also to provide training and so might achieve better dosing accuracy for systemic drugs, there is not much public data to support this. Need for multiple breaths is a disadvantage from a user's point of view.

i-neb is a portable device providing around 40 doses from one battery charge. Adaptive aerosol delivery (AAD) means that the drug is only aerosolized when patient inhalation flow is correct. Throughout the treatment, the I-neb provides feedback to the patient via the liquid crystal display and audible and tactile signals when dosing is complete.

E-flow appears quite similar, being battery powered, using a vibrating mesh and has AAD. It claims 90% delivered dose.

The main applications for nebulizers have been for rescue therapies for child asthmatics and for the treatment of cystic fibrosis. Although in some situations DPIs are replacing nebulizers (e.g., Novartis launched TOBI Podhaler, which is a unit-dose capsule inhaler for the delivery of tobramycin as an alternative to nebulized delivery for the treatment of lung infection in cystic fibrosis), nebulizers are still seen to have an advantage where the patient inhalation is either weak or variable (e.g., young children) as at least some of the drug will be delivered by even minimal breathing. Furthermore, in the case of rescue asthma therapies, it is possible to "titrate" by observing a reduction in wheezing as the pathways in the lung begin to be dilated by the effect of the drug.

Nebulizers can also be used for the systemic delivery of drugs via the lung. For example, YM BioSciences has used nebulization to deliver fentanyl, and Aerogen used nebulizer-like technology to deliver insulin in clinical trials.

Soft mist inhalers   Although stability may prove to be an issue with many liquid formulations compared to dry powders, they have the advantage of being more likely to produce homogeneous aerosols. Aqueous formulations also require sterile filling lines that add cost. All soft mist devices are active as the inspiratory flow from a patient is unlikely to be sufficient to aerosolize the drug. Fine particle fractions quoted can be quite high (up to 80% or more). Maximum doses are up to 50 μl or so equating to a few milligrams of drug depending on formulation.

Aradigm developed the AERx device to have ability to titrate. The first-generation device also had the ability to monitor breathing to control the release of the drug although the device was relatively large and heavy. A second-generation mechanical version is being developed, which is claimed to offer similar performance to that of the electronic device. Emitted dose and fine particle fraction from AERx are understood to be quite high.

The Respimat® soft mist device has been developed by Boehringer Ingelheim to deliver the drug Spiriva for the treatment of respiratory disease. The device was originally developed as an alternative to pMDIs but with a drug-release time reducing the need for breath control and thus avoiding the need for breath actuation. The device has a good fine particle fraction of between 60% and 80% and a 50% respirable fraction that is comparable with the better active DPI devices.

SHL has been developing aqueous drop inhaler although there is limited public information on this. The device appears to use a mechanical means to create respirable droplets of the liquid drug.

4.5.7 Intranasal (IN) Route of Administration

Development of intranasal (IN) route of administration parallels with pulmonary administration as they share some common diseased states, for example, allergy. IN provides effective means for local and systemic drug delivery in allergic rhinitis and avoids systemic use of antihistamine and corticosteroids. Utility of this route of administration is limited by volume, and therefore, payload is limited by volume.

IN also requires a higher degree of patient involvement, which may lead to issues in terms of concordance unless they are dealt with at the design level.

IN route of administration is successfully utilized in the delivery of biologics, for example, Avastin® (bevacizumab, Roche), to treat congenital nose bleed condition (niche indication, nonetheless, establishes a proof of concept to delivery biologics via IN route) and vaccines, for example, FluMist®.

Historically, it has been difficult to establish dosing of therapeutics by the IN route, due to drain back from the nasal cavity, and therefore, formulations have become increasingly important to ensure adhesion of the formulation to the mucosal membrane of nasal cavity in order to facilitate drug transport.

The IN route provides a notable means to facilitate rapid onset of action for pain management indications. The route has been successfully utilized in the delivery of fentanyl, for example, Nycomed Instanyl®.

4.5.8 Transdermal Route of Administration

It is important to distinguish between topical and transdermal: topical means that the administration of drug exerts its effects at the dermal layer; transdermal is a route of administration employed to transport the drug beyond the dermal layer so that the drug effects are systemic. For example, lidocaine patch operates at a topical level (i.e., pharmacology acts on pain receptors in the dermal region only although there is some absorption of drug leading to some secondary pharmacology), while fentanyl patch facilitates drug transport beyond the dermal layer and into the bloodstream so that it goes into systemic circulation.

Transdermal drug delivery is reliant on the drug's physicochemical properties (i.e., small molecular weight, lipophilic), and the device component largely acts as a safe reservoir to hold the drug and provide adhesion to the skin, for example, Duragesic®.

The exception to this is active transdermal systems that utilize iontophoresis to maintain control of drug transport, for example, Ionsys patch. There are other active devices that enhance the delivery of larger drug molecules, for example, TransPharma and Altea technologies, the latter technology having been recently acquired by Nitto Denko.

Transdermal drug delivery is historically limited to small, lipophilic molecules. However, the advent of microneedles is now opening up this space to molecules with larger molecular weight up to recombinant proteins, for example, 3M Drug Delivery Systems' Solid Microstructured Transdermal System (sMTS) and others (e.g., Kimberly Clark). Strictly speaking, this route is intradermal drug delivery as the device—either by microneedles or other intervention—is creating a space in the dermal region to allow drug directly into this region.

Transdermal system are volume limited, and therefore, they are often linked with vaccine where small dose volume is effective.

4.5.9 Sublingual Route of Administration

Sublingual (sl) drug administration utilizes the highly vascularized region directly beneath the tongue; this permits transport of small, lipophilic molecules through the mucosal membrane to facilitate rapid onset of action. This route of administration has limitations and a need to ensure acceptable taste after drug dissolution and accidental swallowing of the dosage form. Table 4.5 summarizes some successful commercial sl formulations.

An example of the small-molecule (nitroglycerin) formulation is the Nitrolingual® spray and nitroglycerin sl tablets for the relief of angina pectoris. It delivers a dose of 0.4 mg per spray under tongue. One or two metered doses (0.4 or 0.8 mg nitroglycerin) should be administered onto or under the tongue, without inhaling. The mouth must be closed immediately after each dose to avoid loss of medication. Onset of action is within 2–4 min and lasts for 10–30 min.

TABLE 4.5 Examples of sl dosage forms

Dosage form	Onset of action	Duration of action
Isosorbide dinitrate		
Isordil® sl tablets	2–5 min	1–3 h
Isordil oral tablets	20–40 min	4–6 h
Nitroglycerin		
sl tablets	1–3 min	10–30 min
Translingual spray	2–4 min	10–30 min
Lorazepam		
Ativan® sl tablets	dissolve in 20 s	60 min (peak plasma level)

For administration of nitroglycerin, one tablet should be dissolved under the tongue or in the buccal pouch immediately upon indication of an acute anginal attack. Due to the volatility (leading to loss of potency) and chemical reactivity of nitroglycerin, the nitroglycerin sl tablets should be dispensed and kept in tight amber glass containers. The container should be closed tightly immediately after each use, and no more than 100 tablets should be packaged in a container.

4.6   FUTURE DIRECTION FOR DRUG DELIVERY DEVICES

One of the key drivers for innovation in the drug delivery device field is the emergence of advanced therapy medicinal products (ATMPs). ATMPs represent groups of therapeutic interventions for diseases that currently have limited or no effective treatment options. ATMPs comprise gene therapy medicinal products, somatic cell therapy medicinal products, and tissue-engineered products.[2]

Combined ATMPs are products that incorporate, as an integral part, a medical device and viable cells or tissues. It is likely that combined ATMPs will emerge as basic and regulatory sciences evolve. Examples of combined ATMPs include tissue-engineered products incorporated onto an artificial matrix or scaffold for implantation or living cells inserted into a special encapsulation and/or implantation device.

Table 4.6 summarizes definitions of these three categories, according to European pharmaceutical legislation, and provides an example of a combined ATMP approach for each category.

One of the major distinctions between combined ATMPs and conventional drug/device combinations described in previous sections of this book is the requirement for a medical device component early in development to establish clinical proof of principle. Typically, in pharmaceutical development, a drug will be administered to a patient or volunteer via infusion; device considerations will be made post-proof of principle once a favorable pharmacodynamic profile is established. However, for combined ATMP development, the intended medical device component is often required in order to facilitate administration of the genetic or cellular therapeutic agent.

An example of this is the London Project to Cure Blindness, which aims to replace damaged and diseased retinal cells that lead to AMD with healthy cells that are grown from stem cells.[3] Retinal pigment epithelium (RPE) cells form a thin sheet lining the inside of the eye under the retina. A healthy RPE layer is critical to normal sight. When these RPE cells are damaged or lost, they are thought to lead to AMD. The project aims to replace the damaged RPE cells with a sheet of RPE cells created from stem cells. In phase I studies, the stem cells used are human embryonic stem cells. The RPE cell sheets to be transplanted into the patient require a specially engineered patch using a new delivery device developed by the London Project Team to be inserted into the posterior of the eye to assess clinical effect.

[2] Regulation (EC) No 1394/2007 of the European Parliament and of the Council of November 13, 2007 on advanced therapy medicinal products and amending Directive 2001/83/EC and Regulation (EC) No 726/2004.

[3] http://www.thelondonproject.org/OurVision/TheProject/?id=1166

TABLE 4.6   ATMP types and key characteristics[a]

Type of ATMP	Key characteristics	Example (combined ATMP)
Gene therapy medicinal product	Comprises an active substance that contains or consists of a recombinant nucleic acid used in or administered to human beings with a view to regulating, repairing, adding, or deleting a genetic sequence	*Mydicar*® (Celladon Corp., San Diego, CA)
	Therapeutic, prophylactic, or diagnostic effect relates directly to the recombinant nucleic acid sequence it contains or to the product of genetic expression of this sequence	Gene therapy for heart failure that uses a benign virus to insert the Serca2a gene into heart cells
Somatic cell therapy medicinal product	Contains or consists of cells or tissues that have been subject to substantial manipulation so that biological characteristics, physiological functions, or structural properties relevant for the intended clinical use have been altered or of cells or tissues that are not intended to be used for the same essential function(s) in the recipient and the donor	*Prochymal*® (Osiris Therapeutics, Columbia, MD)
	Is presented as having properties for or is used in or administered to human beings with a view to treating, preventing, or diagnosing a disease through the pharmacological, immunological, or metabolic action of its cells or tissues	Infusion of human mesenchymal stem cells
		Currently being evaluated in clinical trials for several indications, including acute graft-versus-host disease (GVHD) and also Crohn's disease

(Continued)

TABLE 4.6 ATMP types and key characteristics

Type of ATMP	Key characteristics	Example (combined ATMP)
Tissue-engineered product	Contains or consists of engineered cells or tissues	*MACI®* (Genzyme Biosurgery, Boston, MA)
	Is presented as having properties for or is used in or administered to human beings with a view to regenerating, repairing, or replacing a human tissue	Treatment of articular cartilage defects with matrix-induced autologous chondrocyte implantation
	Cells or tissues shall be considered "engineered" if they fulfill at least one of the following conditions: the cells or tissues have been subject to substantial manipulation, so that biological characteristics, physiological functions, or structural properties relevant for the intended regeneration, repair, or replacement are achieved	

[a]Adapted from Schneider and Celis [4].

Another key distinction of combined ATMPs is a blurring of the line with regards to which component (the active component or device component) contributes to the primary mechanism of action. This is particularly the case in emerging tissue-engineered products. For example, established implantable devices may be combined with cells or tissues to improve patient outcome, making the therapeutic principles much more complex. Patient response to a combination of a medical device with cells or tissues may be different to that seen with either component alone.

The treatment of articular cartilage defects with matrix-induced autologous chondrocyte implantation (MACI®, Genzyme Biosurgery) provides an example of this. The MACI implant uses the patient's own (autologous) cultured cartilage cells (chondrocytes) to repair articular cartilage damages in (knee) joints. Chondrocytes that are taken from a biopsy are cultured to increase the number of cells and then seeded on a biodegradable ACI-Maix membrane. Once the culturing process is complete, the cells seeded on the membrane are returned to the surgeon for implantation. The surgeon will make an incision in the knee and prepare the defect by clearing away any and all damaged tissues. The surgeon will then place and fix the membrane with the cultured cells into the defect. Over several months, these cells create a matrix that covers the articular surface—in effect, replacing the lost cartilage in the knee. Importantly, the MACI matrix is an integral part of this therapeutic intervention; implantation of the chondrocytes alone will not result in a positive clinical outcome.

While combined ATMPs have the potential to radically enhance treatment options, they pose challenges for regulators trying to assess such interventions using conventional principles to assess quality, safety, and efficacy. Regulatory science will need to keep pace with developments in this field to avoid bottlenecks to the development and commercialization of this therapeutic class.

REFERENCES

1. Gopalaswamy S, Gopalaswamy V. *Combination Products: Regulatory Challenges and Successful Product Development.* Boca Raton: CRC Press; 2008.
2. Directive 93/42/EEC.
3. Section 2 of Food and Drug Act Canada.
4. Schneider CK, Celis P. Challenges with advanced therapy medicinal products and how to meet them. *Nat Rev Drug Discov* 2010;9:195–201.

5

LOCAL DELIVERY OF BONE GROWTH FACTORS

Bill McKay, Steve Peckham, and Jared Diegmueller

In recent years, significant resources have been applied to the research and development of growth factors that stimulate bone formation or enhance bone healing. While the body of knowledge on the use of these factors has increased greatly, in many ways, the field is still in the very early stages of investigation. In order to appreciate the accomplishments that have been made and the challenges that await new solutions in bone growth factor delivery, it is important to understand why bone growth factor delivery is needed and how these factors are being used clinically. There are many different scenarios under which bone formation or bone healing is required. The delivery of bone growth factors can be broadly divided into systemic and local applications. While there is a great need for systemic application of bone growth factors to address metabolic bone disorders, this chapter will focus on the local growth factor delivery.

5.1 APPLICATIONS FOR LOCAL DELIVERY OF BONE GROWTH FACTORS

The applications for local delivery of bone growth factors fit into three categories— use as a substitute for autogenous bone grafting (transplanting bone from one part of the body to another), use as an adjunct to standard of care (SOC) to accelerate healing, or use as a prophylactic treatment to prevent future injury. More specifically,

Therapeutic Delivery Solutions, First Edition. Edited by Chung Chow Chan, Kwok Chow, Bill McKay, and Michelle Fung.
© 2014 John Wiley & Sons, Inc. Published 2014 by John Wiley & Sons, Inc.

local bone growth factor delivery has application in repair of traumatic bone injury, stabilization of degenerative joints, replacement of bone removed due to tumor excision, acceleration of fracture repair, and strengthening of bones weakened due to metabolic disorders or drug interactions. Bone growth factors are used in applications across spinal, orthopedic/trauma, and cranial oromaxillofacial surgery.

5.1.1 Autograft Replacement

Each of the three categories for local bone growth factor application will be covered individually to lay the foundation for understanding the needs and challenges that drive the choice of growth factors and delivery system development. Currently, the main use for local bone growth factors is as a replacement for autogenous bone grafting. There are many different scenarios in which bone is taken from one part of the body and placed into an area where new bone formation is required. Taking advantage of the ability of bone to regenerate and remodel, the concept of autogenous bone grafting has been applied for over a century. Examples where bone grafting is needed include spinal trauma leading to a vertebral body fracture, cranial trauma where bones of the skull or mandible require repair, or other skeletal trauma resulting in segmental bone loss. In addition, bones are sometimes fused as a treatment for degenerative conditions. Examples would include fusion of the various segments of the foot and ankle and spinal fusion to achieve spinal stability or deformity correction. In some cases, bone healing is necessary secondary to tumor excision. Among the one million bone grafting procedures performed each year in the United States, the most common use is in spinal fusion procedures. Approximately 300,000 thoracolumbar fusion procedures are performed in the United States each year, each requiring some sort of bone grafting. In trauma applications, there are many fresh fractures and long-bone nonunion surgeries that make use of bone grafting materials. There are also 50,000 hip and knee revisions and over 100,000 cranial oromaxillofacial procedures with bone grafting. With this large number of procedures representing an opportunity for the use of locally delivered growth factors, the resources devoted to the study and development of bone growth factors are easily understood.

Autogenous bone or autograft is the standard against which other bone grafting materials are compared. Over the years, surgeons have identified several sites for autograft harvest including the anterior and posterior iliac crest, the ribs, the proximal and distal tibia, and the mandibular ramus or symphysis [1–5]. The choice of graft site is based on ease of access, the amount of graft material needed, and the procedure being performed. Autograft bone is considered the ideal grafting material since it contains all of the components necessary to support bone formation—molecular signals for bone healing, cells capable of responding to signals and forming bone, and a scaffold on which new bone formation can occur. Autograft is generally considered osteoinductive in that it contains the molecular signals or proteins required to initiate bone formation. However, the amount of osteoinductive proteins is relatively low, and demineralization of the bone is required in order for the proteins to be exposed. Autograft is also osteogenic since it contains bone-forming cells and cells capable of being transformed into bone-forming cells. The viability of bone cells in harvested autograft is dependent on the manner in which the graft is treated and the length of time between harvest and implantation. Finally, autograft is osteoconductive, meaning that the bone acts as a scaffold on which new bone will be formed. Since the patient's own bone is being used, there is no risk of disease transmission or concerns regarding biocompatibility.

That is not to say that autograft is without limitations. First, there is some risk of adverse events associated with the harvest itself. Generally, these risks are thought of as being minor or major and acute or chronic [6, 7]. Major complications with iliac crest harvest would include herniation, vascular injury, deep infection, neurologic injury, hematoma, fracture, and pelvic instability [8–14]. Minor complications include pain at the harvest site, sensory changes, and gait disturbances [15–17]. Major complications with autograft harvest are rare, but the morbidity can be significant. Minor complications are more common with rates reported in the range of 30–40% of patients [6, 17]. This is a somewhat controversial topic with some surgeons reporting significant pain at harvest site for years after the procedure and others claiming the incidence and severity of harvest pain is overstated [18–21].

Independent of concerns regarding residual pain or risk of adverse events, the application of autogenous bone graft is limited in many cases since the volume of available graft material may not be sufficient for the particular application, or there may be issues with autograft bone quality due to metabolic bone disease or patient age. Regardless of where a surgeon stands on the relative morbidity of autograft harvest, it is clear that there is some risk associated with the procedure and the desire for comparable or better results with an economically viable autograft replacement is a real need. For these reasons, a lot of effort has been put into research and development of bone growth factors for use in bone healing or grafting applications.

Growth factors are of interest primarily as a way of eliminating the need to harvest autograft bone. Growth factors may also be used as a means of enhancing autograft or as a graft extender. The need for graft extenders could be due to limitations in the amount of autograft available as a result of previous bone harvest or poor bone quality. A surgeon may also choose to combine autograft with a growth factor if he/she believes that the patient presents a particularly challenging healing environment due to comorbidities or significant soft tissue damage leading to compromised blood supply or in cases where there have already been multiple revision procedures.

5.1.2 Adjunct to Standard of Care (SOC)

While the clinical application of bone growth factors is driven primarily by the desire for a replacement for autograft, this is certainly not the only application that is being explored. Another area of focus for bone growth factors is in use as an adjunct to SOC for bone growth to accelerate bone healing or to increase the likelihood that healing will ultimately occur. For example, most tibia fractures would heal with hardware stabilization and soft tissue management. However, there would be interest from patients and surgeons in a bone growth factor that could be applied to accelerate fracture healing and return the patient to full function sooner. If the growth factor also increased the likelihood of healing, that would be an additional benefit.

5.1.3 Prophylactic Treatment

The previous examples have centered on treating a bone defect, healing a traumatic fracture, or addressing pain or instability through bone fusion. In other words, the treatment is applied after injury has already occurred. The third area of local application for bone growth factors is for prophylactic local treatment of osteoporotic or osteopenic bone as a means to increase bone density and prevent the fracture from happening in the first place. In 2000, there were an estimated nine million osteoporotic fractures worldwide with the most common being hip, forearm, and vertebral fractures [22]. The cost of medical care associated with osteoporosis is $18 billion per year in the United States alone [23]. There are many systemic treatments aimed at reducing the number of osteoporotic fractures; however, adherence to oral antiresorptive therapy is problematic and lack of adherence correlates to increased fracture risk [23, 24]. There may be advantages to treating osteoporosis locally, such as lowering the required dose of medication, limiting systemic side effects, or further reducing fracture risk by using local treatments as an adjunct to systemic therapy. Because the prevalence, morbidity, and mortality of osteoporosis are so significant, there is an opportunity to really improve patients' lives with a successful local treatment.

5.2 TYPES OF BONE GROWTH FACTORS

The most commonly studied growth factors belong to the TGF-beta superfamily of proteins. This family of proteins, with varying degrees of structural homology, includes the bone morphogenetic proteins (BMPs) and growth and differentiation factors (GDFs). The growth factors of today are available thanks to research that began in the middle of the twentieth century in the lab of Marshall Urist, M.D., at UCLA. Dr. Urist published in 1965 on his research demonstrating that demineralized bone would induce new bone formation when placed into a muscle pouch of rats. Dr. Urist postulated that certain factors in the bone could be identified that were responsible for this bone autoinduction [25]. With additional research, it became clear that the growth factors were proteins, and Dr. Urist coined the term "BMPs" to describe them. It was not until the late 1980s that the individual proteins from bone extract were identified and recombinantly produced [26, 27]. This accomplishment ushered in the era of significant nonclinical and clinical research work to understand the mechanism of action, the properties, limitations, and the potential applications for the BMPs and other TGF-beta family growth factors.

Today, the search for bone growth factors has gone far beyond the original targets of the osteoinductive BMPs. Staying within the TGF-beta family, research has been performed on GDFs with most work centered on recombinant human fibroblast growth factor-5 (rhGDF-5). These factors tend to be less potent in terms of osteoin-ductivity in animal models, but they have an ability to induce both bone and soft tissue that make them particularly suited for certain applications, such as periodontal disease where both bone and periodontal ligament must be restored to achieve the desired outcome of regeneration. Another class of recombinant protein that has been studied for bone healing is platelet-derived growth factor (PDGF). PDGF is not osteoinductive or inherently capable of inducing *de novo* bone formation. Instead, PDGF acts as a mitogen and chemotactic factor. Through these processes, PDGF is also indirectly involved in angiogenesis, which can be important to bone as well as soft tissue healing.

Recombinant BMPs such as rhBMP-2 and rhBMP-7 have been shown to be very potent factors inducing bone formation. In nonclinical studies, these proteins form bone as well as or better than autologous bone. As produced for clinical applications, these proteins exist as homodimers consisting of two BMP-2 or BMP-7 chains. There are *in vitro* and *in vivo* data suggesting that heterodimers are more potent bone formers than homodimers. Creating a consistent product with heterodimers adds a new level of complexity to manufacturing; however, the prospect of potentially lowering dose and therefore cost with more potent factors has commercial appeal. Recognizing that these proteins are manufactured versions of naturally occurring proteins that are involved in many aspects of growth and development, the safety implications of implanting more potent growth factors will need to be evaluated. Along these same lines, other research efforts center around combining segments of naturally occurring BMPs to construct new chimeric proteins that may also have increased potency as well as added functions or capabilities that the individual BMPs do not possess. This is an interesting area of research but also an area where the potential risks are not yet well understood.

For any recombinant protein being studied or cell-based system for protein production, it is clear that recombinant protein manufacturing is complex, requiring significant investment to develop and maintain a consistent process. The level of complexity and cost has driven the desire for simpler and less expen-sive bone growth factors. One target has been peptides. In theory, it would be possible to identify peptides that mimic BMP binding to receptors leading to similar receptor-mediated cellular processes. Another target of peptides could be the naturally occurring inhibitors to BMP-induced bone formation. Competitive binding of inhibitors could possibly increase local residence time and enhance the activity of naturally occurring local BMPs or reduce the amount of implanted rhBMP needed for efficacy. Other peptides with functions distinct from the BMP signaling pathway have also been investigated for their impact on bone healing. This would include a peptide of parathyroid hormone (PTH) that is thought to increase osteoblast activity if given at the appropriate dosing regimen and a peptide mimicking the binding site of type I collagen designed to enhance local cell attachment [28, 29]. Neither of these peptides is osteoinductive. They rely on enhancement of local biological activity for their potential effect on facilitating the bone formation process.

Taking the desire for more simple and less costly solutions to another level, there are a number of companies and academic institutions working on research into osteo-inductive or bone growth-promoting small molecules. Much of this research is focused on stimulating the BMP signaling cascades or blocking the action of natural BMP inhibitors using chemical entities to enhance bone healing. These projects are all in the early stages of feasibility and nonclinical development, and efficacy is yet to be determined.

Regardless of the type of bone growth factor being contemplated, they all require a carrier or matrix for localized delivery. The short residence time of the protein at the site of injection makes their independent use unsuitable for treatment. For example, rhBMP-2 injected into nonhuman primates has demonstrated a half-life of 7 minutes [61]. Consequently, most of the therapeutic agent is excreted and passed in urine within 24 h. For efficacious results, bone growth factors must be maintained in the area of healing or fusion because they have no systemic therapeutic effect. Injection of the protein to the treatment site results in a local exposure of hours and is not sufficient to sustain osseous tissue development. Therefore, the protein must be sequestered to obtain longer residence at the site of implantation through the development of a combination device. The systemic elimination of a growth factor after injection in comparison to the local retention of the same growth factor by a carrier, which results in successful osseous tissue formation, is shown in Figure 5.1.

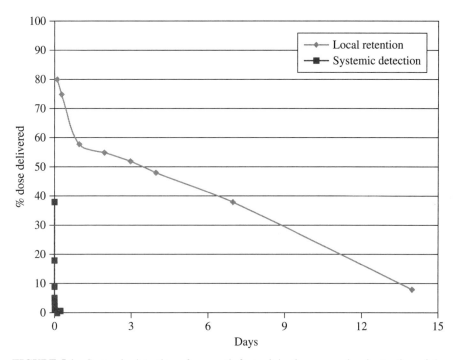

FIGURE 5.1 Systemic detection of a growth factor injection versus local retention of the growth factor by a carrier.

There are many practical considerations that make the difference between success and failure in the clinical application of bone growth factors. A lot depends on the choice of growth factor and its combination with an appropriate carrier matrix. Much of this chapter is devoted to an overview of these considerations and how they have been addressed through product development efforts.

5.3 DELIVERY SYSTEMS

The identification and development of a therapeutic agent for bone grafting alone present significant challenges; however, the rapid clearance of the agent necessitates the use of a carrier [30]. The strengths and weaknesses of particular carriers (or scaffolds) and active agents must be considered, as their interaction in the final combination is critical when developing an efficacious bone grafting treatment. The advantage of a combination product is the ability to use the benefits of individual components to compensate for the shortcomings of the other components. This creates a final product whose composition is optimized for efficacy and whose limitations are minimized.

The environment in which the implant will be placed is also an important consideration. The interactions between the combination product and surrounding tissues and fluids are crucial to the product's performance. Designing a therapeutic protein, carrier combination product that utilizes each component's advantageous interactions with the local environment produces a best-in-class bone grafting device capable of generating bone at the site of implantation. An autograft replacement product that is effective at treating challenging indications such as interbody spinal fusions, posterolateral spinal fusions, joint fusions, healing nonunion fractures, and sinus and alveolar ridge augmentations for dental implant procedures is a valuable surgical tool for clinicians.

Early in the development pathway, there is strong temptation to make the "ideal" therapeutic delivery product containing every desired characteristic and feature. This is appropriate for early brainstorming purposes and developing competing design options; however, efforts to obtain this "ideal" product quickly expose the difficulty in accomplishing this goal. The simplest design solution for therapeutic delivery system for bone grafting is a single material scaffold and the therapeutic protein developed for a particular implant environment. Incorporating additional components or materials usually offers minimal advantages compared to the increased technical and regulatory complexities. Striving for a balance between simplicity of design and incorporation of desired design criteria is necessary. As a means to make the development process less complicated, the design criteria should be separated into a list of "Required" design criteria and a separate list of "Nice to Have" criteria. The "Nice to Have" criteria will have varying probabilities of inclusion due to the prioritization of importance, technology limitations, and negative attributes that conflict with final product design. The final product must be an elegantly balanced design offering optimal safety and efficacy with minimal complexity.

5.3.1 Therapeutic Protein Carrier "Required" Design Criteria for Bone Grafting

The simplest design criteria for a therapeutic protein combination device utilizing a carrier are that it must be both safe and efficacious. These will be a central focus of any regulatory body worldwide. As an example, one of the responsibilities of the U.S. Food and Drug Administration (FDA) is "Protecting the public health by assuring that…medical devices intended for human use are safe and effective" [31]. The design criteria that are central to establishing similar or better safety and efficacy for the device when compared to the SOC must be viewed as a nonnegotiable priority. If the product fails to incorporate any of these "Required" design criteria, ultimate regulatory approval of the product could be in jeopardy.

5.3.1.1 Protein Incorporation and Delivery
As stated previously, the short residence time of growth factors in tissue necessitates the association of the growth factor with a carrier matrix in a combination product. To overcome the short residence time limitation of the protein, a carrier must be capable of sequestering the protein and providing extended delivery of the therapeutic agent for new bone formation to occur. A sufficient amount of protein must bind to the carrier to be capable of influencing a critical mass of bone-forming cells to create the desired tissue volume [30]. Binding of the protein to the carrier ensures proper surgical placement of the therapeutic agent resulting in osseous tissue formation at the desired anatomical location. The carrier must also release the protein over sufficient duration to provide recruitment, proliferation, and differentiation of mesenchymal stem cells to form trabecular bone. One method to accomplish this is via simple diffusion of the protein from the carrier. Simple application of the rhBMP-2 protein to a scaffold at the time of surgery, which follows the Langmuir adsorption isotherm, has demonstrated delivery profiles of the growth factor for a duration of a couple of weeks to a month. These delivery profiles are shown in Figure 5.2 with the application of a growth factor to calcium phosphate (CaP) granules, composite material, and collagen. These durations, with sufficient dosage, have proven effective at obtaining new bone formation, both in preclinical and clinical use [32–35]. However, additional methods of delivery continue to be investigated.

Incorporation of the protein into a polymer or ceramic with protein release upon degradation of the carrier matrix is a popular delivery means that has been explored. This method offers increased duration of delivery yet has shown little increased benefit to final osseous tissue formation. A single, optimal pharmacokinetic delivery profile for the growth factor to obtain osseous tissue formation will be difficult to identify and is unlikely [36]. The delivery profiles of a growth factor in cements are shown in Figure 5.2. The bioactivity of the therapeutic agent must also be maintained during protein incorporation, graft implantation, and extended delivery to obtain sufficient osseous tissue formation and satisfactory efficacy. Coupling a carrier with a bone growth factor provides localized and sustained delivery of the therapeutic protein for bone formation.

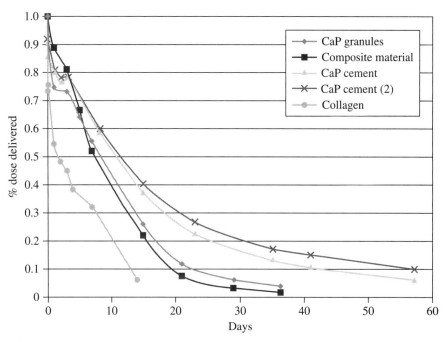

FIGURE 5.2 Local retention of a growth factor by different carriers that utilize both absorption (CaP granules, composite material, and collagen) and incorporation (both CaP cements).

5.3.1.2 *Biocompatibility and Safety*   The materials of a carrier, both independently and collectively, must be biocompatible and safe in order to limit the associated tissue inflammation and preserve cell viability to achieve bone growth, development, and healing. Limited inflammatory response is necessary in the natural healing process and is also necessary in facture repair [37, 38]. ISO 10993 provides guidance for *Biological evaluation of medical devices*. Analysis of a prospective medical device using *ISO 10993 Part 1: Evaluation and testing* will help to determine the recommended biological testing and explain its general governing principles [39]. The remaining collection of guidance documents is a valuable tool in determining the study design of the tests necessary to evaluate the biological response relevant to the safety of the medical device or material. The scope of the evaluations and testing is extensive as seen in Table 5.1. This international standard is updated frequently and is applicable to all medical devices, but should not be seen as all-inclusive. Additional *in vitro* and preclinical work should be considered to investigate further areas of understanding for the therapeutic protein and carrier combination device and its safety, that is, animal models examining the results of misuse of the product.

5.3.1.3 *Biodegradable, Osteoconductive Scaffold*   The carrier also serves as a scaffold utilizing osteoconductive properties to provide a lattice for cell attachment and new bone development. This scaffold should resorb at a similar rate to the formation of new bone to provide ample structure for osseous tissue formation yet

TABLE 5.1   ISO 10993 recommended testing and evaluation

ISO 10993	Biological evaluation of medical devices
Part 1:	Evaluation and testing
Part 2:	Animal welfare requirements
Part 3:	Test for genotoxicity, carcinogenicity, and reproductivity
Part 4:	Selection of test for interactions with blood
Part 5:	Tests for *in vitro* cytotoxicity
Part 6:	Test for local effects after implantation
Part 7:	Ethylene oxide sterilization residuals
Part 8:	Selection and qualification of reference materials for biological tests
Part 9:	Framework for identification and quantification of potential degradation products
Part 10:	Test for irritation and delayed-type hypersensitivity
Part 11:	Test for systemic toxicity
Part 12:	Sample preparation and reference materials
Part 13:	Identification and quantification of degradation products from polymeric medical devices
Part 14:	Identification and quantification of degradation products from ceramics
Part 15:	Identification and quantification of degradation products from metals and alloys
Part 16:	Toxicokinetic study design for degradation products and leachables
Part 17:	Establishment of allowable limits for leachable substances
Part 18:	Chemical characterization of materials

resorb before becoming extensively incorporated in the mineralized tissue. Large amounts of incorporation of the carrier in the osseous formation could lead to a graft/tissue interface area and the creation of structural voids in the new bone matrix resulting in suboptimal mechanical strength of the repaired tissue. Encapsulation of the carrier could also inhibit natural biologic remodeling, which occurs in native bone tissue. Due to the biodegradable nature of these bone graft devices, they have limited load-bearing capability and must be used with mechanical fixation to provide stabilization until sufficient osseous tissue can form. The biodegradation of the carrier scaffold is critical to allowing the body's natural processes to repair and remodel the new bone over time, ultimately resulting in pure host osseous tissue.

5.3.1.4 Porosity   The growth factor carrier needs interconnected porosity for sufficient three-dimensional osseous formations to occur. The porosity increases the surface area for interaction between the protein and the cellular and fluid environment. This enlarged surface area allows for increased protein binding as well as better exposure of the scaffold for degradation. A material mimicking the structure of natural bone tissue with high porosity allows for cell infiltration and the development of vascularization. Pore sizes of 250 μm and larger have been shown to allow for maximum vascularization, and pores in the range of 200–400 μm are ideal for bone ingrowth [40]. The scanning electron microscopy (SEM) image of the scaffold in Figure 5.3 demonstrates the desired interconnected porosity, which biomimics the

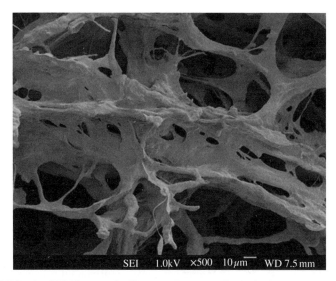

FIGURE 5.3 An SEM image of a fibrous collagen scaffold with interconnected porosity.

natural structure of bone. Porosity alone does not ensure a highly effective bone graft; the scaffold must include interconnectivity to enhance bone ingrowth and osseointegration [40–42]. A porous scaffold provides a three-dimensional template, better protein retention and delivery, better carrier degradation, and better vascularization, all supporting new osseous tissue formation.

Porosity and interconnectivity can either be introduced into the design of the carrier and be present prior to implantation or can occur through a composite carrier with materials of different degradation rates. Engineering porosity into the carrier during manufacturing can be accomplished by varying the void volume of woven or nonwoven sponges via freeze drying, solvent casting and particulate leaching, gas foaming, lyophilization, electrospinning, rapid prototyping, and thermal- or solvent-induced phase separation [30, 43]. Porosity formation with interconnecting pores can also occur after implantation using particulate or fiber leaching, gas foaming, thermal- or solvent-induced phase separation, and hydrolysis or enzymatic degradation of the carrier [30]. A composite carrier composed of materials with fast and slow degradation rates can form a porous carrier after implantation as the fast-degradation material resorbs creating porosity in the remaining slow-degradation material. Many processing techniques have proven successful at creating highly porous, interconnected scaffolds for bone grafting.

5.3.2 Therapeutic Protein Carrier "Nice to Have" Design Criteria for Bone Grafting

After ensuring the therapeutic protein carrier incorporates the required design criteria, there are numerous additional design options that can improve the device, optimizing it for specific indications or differentiating it from competitors. The "Nice to Have" design criteria must be prioritized with a balance of importance to added complexity. Overcomplicating the solution or attempting to add too many design features will result in a final product with a higher risk of safety and/or efficacy concerns and may influence the regulatory approval.

5.3.2.1 Cohesive Scaffold A cohesive carrier offers multiple advantages to granular, loose carriers. The cohesiveness improves the handling characteristics during preparation and implantation at the surgical site. Carriers with a cohesive property also result in less risk of migration of the implant material when compared to loose materials. For example, loose, granular bone grafts including autograft can have a tendency to migrate into unwanted areas. Limiting migration ensures the bone graft develops osseous tissue at the desired location. The use of fibers for cohesiveness also has additional advantages for the scaffold. Fibrous scaffolds better mimic the natural bone tissue structure and provide a better environment for mesenchymal stem cell differentiation [44, 45]. The SEM image in Figure 5.3 demonstrates the desired cohesive, fibrous characteristics of the scaffold. Even the fiber diameter can have an influential role, as diameters of approximately 9 μm and slightly larger have been shown to increase the osteogenic potential of the scaffold [46]. A cohesive carrier offers better handling, resistance to migration, and, if fibrous, better osteogenic potential in support of the osseous tissue formation.

5.3.2.2 Mechanical Integrity The scaffold must endure preparation, handling, implantation, and environmental forces while maintaining physical integrity. The mechanical integrity is important to minimizing micromotion and conserving the ideal osteoconductive and osteogenic environment during early osseous tissue formation. The scaffold accomplishes this by resisting any compressive forces experienced by the environment. For example, bone grafts used in surgical applications adjacent to muscle beds need to be able to limit excessive compression and movement. The scaffold needs to resist these forces on its macrostructure, preserving the implant volume for the desired amount of new osseous tissue formation. The integrity of the carrier is not permanent and is necessary only until degradation of the scaffold and replacement by natural bone tissue. Common processing techniques such as cross-linking are performed during scaffold manufacturing to improve the cohesive integrity of the carrier. The compressive resistance of the carrier is a little more challenging and is most often addressed through the use of a composite where one of the components has sufficient mechanical structure. Maintaining the mechanical integrity for adequate duration is essential to preserving the structure and space for new bone development.

5.3.2.3 Injectable, Flowable, and Settable An injectable, flowable, and settable therapeutic protein carrier is highly desired for bone healing and fusion procedures. The formulation would offer multiple advantages including ease of graft placement, minimally invasive use, and injection delivery to voids or cysts. The settable characteristic also offers all the advantages of a cohesive carrier with sufficient mechanical integrity. The carrier would preserve space, resist migration, and provide an optimal scaffold for bone development. While this formulation holds great promise, it also has numerous technical limitations that must be overcome to offer a viable product.

First identifying an optimal material that is liquid or gelatinous at room temperature yet will set up in the warmer biological environment is challenging, but thermogelling natural polymer composites have been identified [47]. Incorporation of the necessary pore structure is a current serious drawback. Research is ongoing to develop injectable carriers, which will harden and incorporate porosity. One concept is to use fast-degrading materials in a composite to create the necessary porous network in the scaffold. Other researchers are investigating the ability to generate gas bubbles within the scaffold to obtain the desired porosity [48]. Developing an easy-to-use injectable, flowable, and settable scaffold, which incorporates a growth factor, is a highly attractive product solution; however, it may be some time before a design overcomes the numerous technical limitations to offer an ideal treatment solution.

5.3.2.4 Ease of Use   A product combining the growth factor and carrier matrix during manufacturing offers advantages in terms of ease of use. However, incorporation of the growth factor onto a carrier during manufacturing provides possible challenges. In the case of recombinant proteins, sterilization of an unprotected protein incorporated onto a carrier will lead to degradation of the protein and possible antibody effects upon implantation. This is based on the protein fragments and high-molecular-weight aggregates created during the process. If the protein is embedded into a protective material that acts as a carrier, the protein can be shielded from the hazards of the sterilization process. Another choice is the storage of the components separately in lyophilized forms and individual sterile packages. If the components are stored separately, the preparation of the product must support delivery of the protein to the carrier prior to implantation. Ease of use considerations may have little effect on the viability of the product to achieve successful results, but they are still very important. The ease of product preparation and duration of the soak time necessary to ensure the binding of the therapeutic agent to the carrier are areas that should be considered. Modest improvements to the final product can often prove significant in gaining adoption by surgeons.

5.3.3   Common Materials Used in a Therapeutic Protein Carrier for Bone Grafting

To date, several common choices in biomaterials are being utilized that work well as carriers, including natural polymers, ceramics, and synthetic polymers. Each material has advantages and disadvantages as both a carrier and a scaffold for a therapeutic protein combination device. The materials can also be processed into different forms for use as a scaffold, which create different advantages and limitations. A common approach to improve the overall carrier and scaffold characteristics is to use a composite of the materials and forms. The materials used in the carrier must all be biocompatible, offer maximum benefits, and contain few disadvantages.

5.3.3.1 Collagen

5.3.3.1 Collagen   Collagen, a natural component of bone, is a common material for use in scaffolds and has numerous advantageous properties as an element of a bone graft carrier. The organic portion of bone accounts for 25–35% of the tissue by weight, and type I collagen amasses ~90% of the organic component [49]. The fact that type I collagen is a highly conserved peptide across species is also advantageous, allowing for the most common source of raw type I collagen to be bovine while still maintaining biocompatibility. Bovine type I collagen has been used in medicine as a hemostatic agent since the early 1970s [50]. It is also a common material for bone void fillers and therapeutic carriers for osseous tissue formation. Collagen has a high affinity for bone grafting therapeutic proteins such as BMPs [51, 52]. This high affinity makes it an efficient carrier by incorporating the protein for delivery, localizing the therapeutic effects, and supplying sufficient osteoinductive signal for a couple of weeks as the protein diffuses from the surface of the collagen fibers. Collagen is also osteoconductive and serves well as a scaffold for cell attachment and osseous tissue formation. The microstructure of a collagen scaffold is shown in Figure 5.3 and a macroscopic view in Figure 5.4a. As a bioresorbable material, collagen is easily replaced by natural tissue, and all degradation products are bio-compatible. Collagen carriers can be produced in multiple forms including gel,

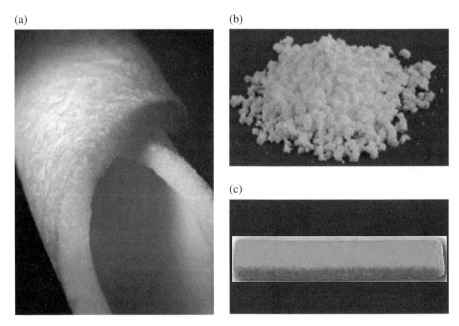

(a)

(b)

(c)

FIGURE 5.4 Collagen (a), a natural polymer; biphasic calcium phosphate (b), a synthetic ceramic; and collagen and ceramic scaffold (c), a composite.

nanofibers, films, and sponges. The collagen used today is highly controlled through sourcing and processing controls. Overall, collagen has numerous advantages as a therapeutic protein carrier for bone grafting, and it is one of the most popular material choices.

5.3.3.2 Other Natural Polymers Additional natural polymer carrier material options exist; however, these are less often utilized. Other naturally occurring polymers that have been studied include hyaluronans, fibrin, chitosan, silk, alginate, and agarose [30, 53]. These materials are biocompatible, are bioresorbable, and can be manufactured to have variable degradation rates. Most are natural components of the extracellular matrix, but none represent as large a percentage of natural bone tissue compared to collagen [30]. Although viable options for bone graft scaffold materials, these materials have not demonstrated advantages in comparison to collagen, and this has led to limited use in commercial products.

5.3.3.3 Ceramic Calcium ceramics are common materials often used in bone void fillers as scaffolds due to their osteoconductive properties. The inorganic phase of bone accounts for 60–70% of the tissue by weight and is comprised mostly of highly carbonate-substituted hydroxyapatite ($Ca_{10}(PO_4)_6OH_2$), a member of the CaP family [49]. Hydroxyapatite (HA), other CaPs, and similar calcium structures such as calcium sulfate are often used as a component in bone void fillers and carriers for therapeutic proteins. BMPs and other growth factors also have a relatively strong affinity for CaPs in an acidic environment [54, 55]. Calcium ceramics serve as a proficient carrier when the mineral contains sufficient porosity and the protein is applied to the surface. The degradation rates of the different calcium ceramics can also be used to tailor delivery of the therapeutic proteins [56]. Tricalcium phosphate (TCP) is a less stable form of CaP and resorbs quickly, while HA is a more stable form of CaP and has a relatively slow resorption rate. Calcium sulfate, with an even more unorganized chemical structure, has a faster degradation rate than TCP. Creating a composite of varying ratios of two, or even all three, of these calcium ceramics will result in carriers of differing degradation rates and delivery of the incorporated therapeutic protein. Biphasic calcium phosphate granules are shown in Figure 5.4b. Ceramics possess impressive compressive strength that can be utilized to resist loading deformation and maintain space for the graft to complete healing or fusion. The limitations include brittleness and poor performance under tensile loads. The calcium ceramics can be manufactured in different forms such as blocks, granules, particulate coatings, and pastes. The extensive benefits of ceramics with relatively few limitations make it a common material choice for therapeutic protein carriers.

5.3.3.4 Synthetic Polymers Biocompatible synthetic polymers have also been used as a delivery vehicle for bone growth factors. The most common are poly-lactide (PLA), polyglycolide (PGA), their copolymer poly-lactide-*co*-glycolide (PLGA), and polyethylene glycol (PEG) and its copolymers [30, 56]. Other polymers studied as carriers include poly(ε-caprolactone), poly(anhydrides), and copolymers comprising poly(ethylene oxide) and poly(butylene terephthalate) [53]. Synthetic polymers offer the advantage that through small modifications in manufacturing the materials, properties can be tailored to change degradation rates and therefore delivery rates of the therapeutic protein. These small manufacturing changes include changing the molecular weight, configuration, conformation of the polymer chains, and the ratio of constituents in a composite. One major disadvantage for the use of synthetic polymers is the degradation pathway often involves hydrolysis and enzymatic cleavage that can cause an inflammatory response and negatively impact osseous tissue formation. Although much promise has been shown by the use of synthetic polymers as carriers for therapeutic proteins in preclinical studies, the limitations of the materials include uncertainty of the *in vivo* reaction, unsubstantiated improved relative effectiveness, and poor resistance to degradation with sterilization. These concerns have impeded their use in commercial products [56].

5.3.3.5 Composites The previously discussed materials can be combined to deliver the best properties for a carrier/scaffold utilized with bone graft proteins. Bioresorbable polymers, either natural or synthetic, are often combined with a ceramic. The ceramic component provides mechanical resistance to compression, while the inclusion of the polymer improves the handling characteristics of the final graft [30, 43]. Often, the most common composite created is bovine type I collagen with HA or TCP incorporated as granules, as shown in Figure 5.4c, or coatings. The collagen, CaP composites are fibrous, are osteoconductive, and contain interconnected porosity. These characteristics best mimic the natural tissue of bone and provide an optimal environment for bone cells. The fact that the collagen, CaP composites are also strong carriers with high affinity for growth factors makes them ideal. The composite also combines collagen, which degrades over weeks, with a CaP, which degrades over months, providing a carrier with favorable degradation properties to support bone formation without substantial incorporation into the new osseous tissue. The ability to provide an osteoconductive scaffold comprising an ideal rate of degradation, with an osteoinductive therapeutic protein signal, to an environment that will provide the necessary osteogenic material is the optimal situation to obtain new bone development.

5.3.4 Protein/Carrier Interaction Considerations

Designing the protein and carrier components of a combination device for bone grafting is not an independent task. The chosen materials and forms of the carrier may affect the efficacious dose and concentration for the growth factor. The degradation rate of the carrier is dependent on the cellular activity generated by the therapeutic protein. A more effective growth factor or increased dose and/or concentration can cause amplified cellular response that will degrade the carrier more rapidly. In addition, the combination of the protein and carrier is also integral to initiating the natural inflammatory response to achieve native integration of the newly generated osseous tissue into the existing bone surrounding the implant. The interaction of the protein and carrier is a critical consideration during product design and should be heavily investigated through preclinical study models.

5.4   CURRENT STATUS OF BONE GROWTH FACTOR-CONTAINING PRODUCTS

Regulatory approval of bone growth factor technology varies depending on geographic location and mode of delivery. In the United States, bone growth factors delivered locally via a surgical procedure were determined to be classified as "devices" back in the mid-1990s because of their local-acting effects and lack of observed systemic effects. Subsequently, they fell under the classification of "combination devices," and therefore, these products began undergoing dual review by the FDA's Center for Devices and Radiological Health (CDRH) and Center for Drug Evaluation and Research (CDER) branches. Bone growth factors delivered percutaneously with a needle into closed fractures, however, have always been regulated as "drugs." As a general rule, Canada, Mexico, South America, Australia, Korea, China, India, South Africa, and a few other middle-Eastern countries have similar regulations to the United States. In Europe, all bone growth factors are regulated as "medicinals" (i.e., drugs).

The number of bone growth factor technologies currently on the U.S. market is limited, but others are in various stages of development. Table 5.2 lists the different technologies and their status.

Among the commercially available bone growth factors, only two possess osteoinductive properties, rhBMP-2 (INFUSE Bone Graft, Medtronic) and rhBMP-7 (OP-1 or Opgenra, Stryker/Olympus). INFUSE Bone Graft was approved via a PMA in 2002 for anterior lumbar interbody spinal fusion (ALIF) in tapered titanium-threaded cages [35, 57]. It was subsequently approved in 2004 for tibia fresh fractures and in 2007 for sinus elevation and extraction socket bone grafting [58–60]. INFUSE Bone Graft was approved at an rhBMP-2 concentration of 1.5 mg/ml delivered on a type I bovine collagen sponge. The rhBMP-2 solution is uniformly applied to the collagen sponge at the time of surgery, to which it inherently binds after waiting a minimum of 15 min prior to implantation. Subsequently, the rhBMP-2 is released from the surgical site over a period of several weeks [61]. Based on years of preclinical research with rhBMP-2, it is believed that local concentration is the predominant factor for ensuring consistent *de novo* bone formation and therefore efficacy. Definitively determining whether it is the local rhBMP-2 concentration, total dose delivered, or sustained growth factor released is very difficult to assess in preclinical studies. The clinical trials of the tibia fresh fracture and sinus elevation indications involved evaluation of two different rhBMP-2 concentrations, 0.75 and 1.50 mg/ml. Both clinical trials demonstrated clinical efficacy of the higher 1.5 mg/ml concentration but not for the lower 0.75 mg/ml concentration. A dose–response trend was observed with the lower concentration but was not statistically different from the control patient populations. The total rhBMP-2 dose delivered in these surgical procedures was a result of the required volumetric amount of bone graft to fill the

TABLE 5.2 Summary of bone growth factor technologies commercially available and in development

Technology	Carrier	Company	Status in the United States	Indication(s)
rhBMP-2	Collagen sponge	Medtronic	PMA approval for three indications (2002, 2004, 2007)	1. Spinal ALIF lumbar fusion 2. Tibia fresh fractures 3. Sinus and extraction socket
rhBMP-7	Collagen powder/CMC	Stryker/Olympus	HDE for two indications (2001, 2004)	1. Tibia nonunions 2. Revision spinal PL lumbar fusion
P-15 peptide	Sintered bovine bone granules	CERAMED	PMA approved	Periodontal bone grafting
rhPDGF-BB	TCP ceramic granules	Cerapedics BioMimetic Therapeutics/ Osteohealth	Pivotal IDE PMA approved	Spinal ACDF cervical fusion Periodontal bone grafting
PTH analog PTH analog	Fibrin Fibrin + ceramic	BioMimetic Therapeutics Kuros Biosurgery/Baxter	PMA submitted Phase II clinical trials	Foot and ankle fusion Tibial plateau and open shaft fractures
B2A peptide	20/80 HA/TCP ceramic granules	BioSET	Phase II clinical trial	Spinal TLIF lumbar fusion
rhFGF-2	Cross-linked gelatin	Kaken Pharmaceutical	Phase II clinical trials	Periodontal bone grafting Tibia closed fracture
rhGDF-5	Collagen sponge with HA particles	DePuy (Johnson & Johnson)	Phase II clinical trial	Spinal 360 lumbar fusion
rhGDF-5	TCP ceramic granules	Scil Technology GmbH	Phase II clinical trial	Periodontal and sinus bone grafting

152

bone defect, and a dose versus efficacy correlation was not observed as seen with concentration versus efficacy.

OP-1 (Stryker/Olympus) has not been approved via a PMA, but instead as a Humanitarian Device Exemption (HDE) in 2001 [62]. A prospective randomized clinical trial had been conducted for tibia nonunions and submitted in a PMA but was ultimately disapproved by the FDA due to failure to meet the study's original primary efficacy endpoint [63]. This same data was used to obtain the HDE approval, which requires demonstration of safety but not efficacy, such that the probable benefit outweighs the risk of use. The HDE limits the product use to 4000 cases per year for tibia nonunions in which autograft had already been tried and failed. A prospective randomized clinical trial was also conducted with OP-1 for spinal posterolateral fusion procedures, which also had its PMA disapproved due to failure to meet the study's original primary efficacy endpoint [64, 65]. Similarly, these data were used to obtain a second HDE in 2004 for previously failed posterolateral spinal fusions in which autograft was unavailable [66]. The OP-1 protein is provided at a concentration of approximately 0.9 mg/ml on a collagen-containing powder carrier (derived from bovine bone) with some carboxymethylcellulose (CMC) to temporarily hold the powder material together during implantation [64]. Vaccaro et al. have described the tendency of the carrier to migrate medially in a posterolateral fusion application due to its lack of compression resistance to the forces experienced by the posterior muscle bed. The lack of cross-linking in development of the carrier, to provide cohesiveness, may have deleterious effects on the protein delivery and overall clinical effectiveness. No dosing studies had been conducted to determine if higher concentrations would be more effective. This technology was subsequently sold to Olympus in 2011, who rebranded it as Opgenra.

Only two other bone growth factor-containing products are commercially available, P-15 (PepGen, CERAMED) and recombinant human platelet-derived growth factor (rhPDGF) (GEM 21, BioMimetic Therapeutics/Osteohealth). Both of these products were approved for treating periodontal bone defects around teeth.

PepGen (CERAMED) contains a peptide called P-15 that was approved in 1999 for treating intrabony periodontal osseous defects due to moderate or severe periodontitis [67]. P-15 is a synthetic 15-amino acid peptide that mimics the cell-binding domain of type I collagen, whose mode of action is described as binding circulating biological agents. Since it is not osteoinductive like the BMPs, it falls into the osteoconductive category of bone grafting materials. The P-15 is bound to a granular HA ceramic carrier derived from sintered bovine bone particles. Its release kinetics is unknown. The periodontal clinical trial demonstrated 0.8 mm of new bone growth over the demineralized bone allograft control patient population at 6 months [68].

This same P-15 technology is being clinically evaluated in a spinal anterior cervical discectomy and fusion (ACDF) procedure (i-Factor, Cerapedics) [69]. Historically, cervical fusion procedures have very high fusion rates using current techniques. Enrollment has been completed and follow-up is ongoing.

GEM 21 (BioMimetic Therapeutics/Osteohealth) contains an rhPDGF that was approved in 2004 for treating intrabony periodontal bone defects, furcation periodontal defects, and gingival recession associated with periodontal defects [70].

rhPDGF is not osteoinductive like the BMPs, but instead is a general mitogen that is believed to cause cell proliferation of any nearby cells. Therefore, its mode of action is not limited to promoting just bone tissue formation, but it could promote other local tissue formation as well that may or may not be beneficial for bone formation. The clinical trial demonstrated 0.9 mm of new bone growth over the TCP ceramic alone control patient population at 6 months [71].

This rhPDGF protein is being evaluated in a clinical trial for hindfoot fusions, and a PMA is under review by the FDA (Augment, BioMimetic Therapeutics) [72]. Results from a multicenter randomized feasibility study of 20 patients suggested that the rate of radiographic union, time to full weight-bearing, and outcomes scores between the Augment and ABG subjects appear comparable [73]. Again, this protein is not an osteoinductive morphogen like the BMPs but a mitogen. The rhPDGF in this product is delivered on a granular TCP ceramic carrier that was found to release approximately 50% during the first 30 min after implantation with approximately 10% remaining by 72 h [74]. The PMA for this indication has been submitted to FDA, and the review is ongoing.

Four other bone grafting products are in various stages of clinical evaluation.

A PTH analog (KUR-111, KUR-113, Kuros Biosurgery/Baxter) is being evaluated in phase II dose-finding clinical trials for tibial plateau fractures and acute open tibial shaft fractures [75, 76]. PTH is a well-known compound involved in the bone remodeling process, but does not possess osteoinductive properties. PTH can have both anabolic and catabolic effects of bone formation depending on dose and rate of delivery. The PTH analog in this product is delivered in Baxter's fibrin-based biomatrix (TISSEEL fibrin sealant) and binded by Kuros' synthetic polymer technology. The release kinetics is unknown, but is expected to be over several days as the clot is enzymatically broken down, resorbed by the body, and remodeled into new tissue.

A novel 23-amino acid peptide B2A peptide (AMPLEX, BioSET) that binds to the rhBMP-2 receptors in an attempt to facilitate new bone formation was developed, but is not osteoinductive itself. This peptide is delivered on a 20/80 HA/TCP granular ceramic and used to augment autograft bone. Therefore, this product is not being developed as an autograft replacement but an autograft supplement. It is being clinically evaluated in a pilot spinal fusion dosing study using a transforaminal lumbar interbody fusion (TLIF) surgical technique [77]. A phase II dosing study with B2A peptide is ongoing. No interim clinical data have been disclosed.

Recombinant human fibroblast growth factor-2 (rhFGF-2) is a nonspecific growth factor that is known to cause cell proliferation, similar to PDGF, but is also not osteoinductive. Clinical trials with FGF-2 (Trafermin, Kaken Pharmaceutical) have been initiated for treating periodontal bone defects around teeth and long-bone closed fractures [78]. The protein is delivered in a cross-linked gelatin that releases the FGF-2 protein over a relatively short period of time with a local half-life of less than 2 days. In a phase II clinical trial, this product resulted in a reduction in median time to healing of 22%, 100 days versus 128 days in placebo control patients [79].

rhGDF-5 belongs to the same TGF-beta family of proteins as BMP-2 and does possess some mild osteoinductivity, meaning that it has the ability to induce new bone formation at a nonbony site. DePuy conducted a phase II dosing study with the protein using a collagen sponge carrier that contained micron-sized HA particles. The study consisted of a combined interbody and posterolateral spinal fusion in which the rhGDF-5 product was placed in both anatomical locations. Phase II clinical trials in sinus elevation procedures and periodontal disease bone defects around teeth demonstrated statistically significant new bone formation over TCP ceramic control groups [80, 81]. Results were not consistent between the spine and sinus elevation studies. As with all combination bone growth products, the anatomical location and carrier attributes can impact clinical outcomes.

It is clear from these products that have been developed and clinically evaluated that the bone growth factor carrier and release kinetics from the carrier have a profound impact on performance. The FDA-approved product, INFUSE Bone Graft(Medtronic) possesses a highly osteoinductive bone growth factor and a cohesive carrier that can act as a scaffold for new bone formation as well as release the growth factor over an extended period of time. These appear to be critical properties to ensure efficacy. In contrast, nonosteoinductive/less osteogenic-specific bone growth factors and carriers that dissipate from the bone grafting site quickly and have short release kinetics have demonstrated minimal efficacy and been unable to achieve FDA approval. These data will help provide direction on the development of future next generation products.

5.5 FUTURE RESEARCH AREAS

Private and public laboratories continue to push the forefront of bone growth factor and delivery technologies. These technologies exist at various stages in the research and development pipeline: discovery, development, clinical testing, and regulatory review. The progression of products through this pipeline oftentimes leads to collaboration between academic research labs and commercial development teams. These continued research efforts attempt to identify and develop new, disruptive bone growth factor and grafting technologies that improve the efficacy of treatment for patients.

5.5.1 Optimized Release Profiles

At least two decades of research on osteoinductive growth factors such as BMP and GDF has revealed that their presence over an extended period of time (months) is not required. A relatively short exposure period of a few days to weeks is all that is needed to "kick-start" the bone formation process. Once the repair process has been initiated, the body's natural repair cycle takes over and ultimately leads to mature remodeled bone. In fact, research has demonstrated that in some cases delayed delivery of rhBMP-2 may not be as effective as earlier application of the protein. Having said this, it may be possible to hone in on an even more precise release period to allow the dose to be lowered to establish a minimally effective dose. Establishing such an optimal release period is not trivial because developing a carrier with slightly different growth factor release rates requires an extensive trial-and-error development process in both animals and humans. In addition, the carrier itself can also have a profound effect on facilitating new bone formation as mentioned previously. The biocompatibility, degradation properties, degradation breakdown products, and osteoconductivity of the carrier will all influence bone formation. So the optimal release profile will also vary depending on the carrier. Even more confounding is that once an optimal release profile is determined for a particular animal species, it will most likely not translate or be the same for higher-order species since the rate of bone formation is so much slower. This is evident by the fact that higher doses of osteoinductive growth factors and slower resorbing carriers are required for higher-order species. Identifying the optimal growth factor release profile would require evaluation of many doses formulated into many different carriers that would be cost prohibitive. Products will slowly evolve as companies identify potentially superior products justifying the time (minimum of 10–15 years) and expense (well over $200M) to bring a combination product to market.

Nonosteoinductive growth factors will have much different optimal release profiles since they are not bone morphogenetic factors and only indirectly affect the repair process. Their release profile will depend on their particular mechanism of action and most likely require even longer sustained release profiles at more precise dosing, thus making it even more challenging to identify an optimal release profile.

5.5.2 Growth Factor Combinations

Bone growth factor research has determined that several growth factors are upregulated during the bone repair process and in a temporal fashion [82, 83]. For example, in a rat femoral fracture model, it was shown that a number of different BMPs are upregulated and then downregulated over several weeks. Theoretically, it should be possible to design a product that contains more than one growth factor to enable lowering the dose of each. Even further yet, the carrier could be designed to temporally release the different growth factors over time to mimic what is observed in the animal models, but again, this profile may be different in humans in which the bone formation process is slower.

Given the difficulty, companies have experienced obtaining FDA approval of bone grafting products with only one growth factor; it would be expected to be exponentially more difficult for a product containing more than one growth factor. To gain FDA approval of a product with a combination of growth factors, the clinical trial would require several arms to evaluate each growth factor separately as well as together to confirm the need for such a combination product formulation. So unless absolutely required, single growth factor formulations would be the simpler way to proceed.

5.5.3 Addition of Synergistic Compounds

Perhaps a more promising area of research is the addition of a compound that enhances the activity of a particular growth factor. For example, a compound that better binds a bone growth protein to its carrier reduces the enzymatic degradation or improves its biological activity by enhancing its anabolic effect or reducing any natural inhibitory cytokines. Examples of such technologies are undergoing research at this time.

5.6 CONCLUSIONS

Bone grafting research has been occurring for decades due to surgeon's desire to avoid having to harvest autograft bone from the patient. First commercial products involved just osteoconductive CaP ceramics to supplement the volume of autografting procedures but evolved into the use of bone growth factors to actually replace autograft. Several bone growth factors have been identified over the last three decades and are in various stages of research and development. Only the BMPs and GDF-5 proteins have osteoinductive properties to drive the bone formation process. Other growth factors facilitate the bone formation via indirect participation in the biological process.

The delivery carrier and release rate of the bone growth factor is very critical in its ultimate clinical efficacy. Various carriers with various growth factor release rates have been utilized. To date, only one technology has been demonstrated in a large prospective randomized clinical trial to be equivalent to autograft bone and received FDA approval, which represented a landmark achievement in orthopedic medicine. Like any new technology, one should not expect it to be perfect at first, but it will continue to evolve over time.

REFERENCES

1. Alt V, Meeder PJ, Seligson D, Schad A, Atienza C Jr. The proximal tibia metaphysis: a reliable donor site for bone grafting? Clin Orthop Relat Res 2003;414:315–321.
2. Booij A, Raghoebar GM, Jansma J, Kalk WW, Vissink A. Morbidity of chin bone transplants used for reconstructing alveolar defects in cleft patients. Cleft Palate Craniofac J 2005;42(5):533–538.
3. Clavero J, Lundgren S. Ramus or chin grafts for maxillary sinus inlay and local onlay augmentation: comparison of donor site morbidity and complications. Clin Implant Dent Relat Res 2003;5(3):154–160.
4. Laurie SW, Kaban LB, Mulliken JB, Murray JE. Donor-site morbidity after harvesting rib and iliac bone. Plast Reconstr Surg 1984;73(6):933–938.
5. Sawin PD, Traynelis VC, Menezes AH. A comparative analysis of fusion rates and donor-site morbidity for autogeneic rib and iliac crest bone grafts in posterior cervical fusions. J Neurosurg 1998;88(2):255–265.
6. Banwart JC, Asher MA, Hassanein RS. Iliac crest bone graft harvest donor site morbidity. A statistical evaluation. Spine 1995;20(9):1055–1060.
7. Younger EM, Chapman MW. Morbidity at bone graft donor sites. J Orthop Trauma 1989; 3(3):192–195.
8. Chan K, Resnick D, Pathria M, Jacobson J. Pelvic instability after bone graft harvesting from posterior iliac crest: report of nine patients. Skeletal Radiol 2001;30(5):278–281.

9. Fowler BL, Dall BE, Rowe DE. Complications associated with harvesting autogenous iliac bone graft. Am J Orthop 1995;24(12):895–903.

10. Kargel J, Dimas V, Tanaka W, Robertson OB, Coy JM, Gotcher J, Chang P. Femoral nerve palsy as a complication of anterior iliac crest bone harvest: Report of two cases and review of the literature. Can J Plast Surg 2006;14(4):239–242.

11. Kuhn DA, Moreland MS. Complications following iliac crest bone grafting. Clin Orthop Relat Res 1986;209:224–226.

12. Kurz LT, Garfin SR, Booth RE Jr. Harvesting autogenous iliac bone grafts. A review of complications and techniques. Spine 1989;14(12):1324–1331.

13. Lim EV, Lavadia WT, Roberts JM. Superior gluteal artery injury during iliac bone grafting for spinal fusion. A case report and literature review. Spine 1996;21(20):2376–2378.

14. Zijderveld SA, ten Bruggenkate CM, van Den Bergh JP, Schulten EA. Fractures of the iliac crest after split-thickness bone grafting for preprosthetic surgery: report of 3 cases and review of the literature. J Oral Maxillofac Surg 2004;62(7):781–786.

15. Baqain ZH, Anabtawi M, Karaky AA, Malkawi Z. Morbidity from anterior iliac crest bone harvesting for secondary alveolar bone grafting: an outcome assessment study. J Oral Maxillofac Surg 2009;67(3):570–575.

16. Heary RF, Schlenk RP, Sacchieri TA, Barone D, Brotea C. Persistent iliac crest donor site pain: independent outcome assessment. Neurosurgery. 2002;50(3):510–516; discussion 516–517.

17. Robertson PA, Wray AC. Natural history of posterior iliac crest bone graft donation for spinal surgery: a prospective analysis of morbidity. Spine 2001;26(13):1473–1476.

18. Delawi D, Dhert WJ, Castelein RM, Verbout AJ, Oner FC. The incidence of donor site pain after bone graft harvesting from the posterior iliac crest may be overestimated: a study on spine fracture patients. Spine 2007;32(17):1865–1868.

19. Howard JM, Glassman SD, Carreon LY. Posterior iliac crest pain after posterolateral fusion with or without iliac crest graft harvest. Spine J 2011;11(6):534–537.

20. Silber JS, Anderson DG, Daffner SD, Brislin BT, Leland JM, Hilibrand AS, Vaccaro AR, Albert TJ. Donor site morbidity after anterior iliac crest bone harvest for single-level anterior cervical discectomy and fusion. Spine 2003;28(2):134–139.

21. Skaggs DL, Samuelson MA, Hale JM, Kay RM, Tolo VT. Complications of posterior iliac crest bone grafting in spine surgery in children. Spine 2000;25(18):2400–2402.

22. Johnell O, Kanis JA. An estimate of the worldwide prevalence and disability associated with osteoporotic fractures. Osteoporos Int 2006;17(12):1726–1733.

23. Black DM, Delmas PD, Eastell R, Reid IR, Boonen S, Cauley JA, Cosman F, Lakatos P, Leung PC, Man Z, Mautalen C, Mesenbrink P, Hu H, Caminis J, Tong K, Rosario-Jansen T, Krasnow J, Hue TF, Sellmeyer D, Eriksen EF, Cummings SR, HORIZON Pivotal Fracture Trial. Once-yearly zoledronic acid for treatment of postmenopausal osteoporosis. N Engl J Med 2007;356(18):1809–1822.

24. Rabenda V, Mertens R, Fabri V, Vanoverloop J, Sumkay F, Vannecke C, Deswaef A, Verpooten GA, Reginster JY. Adherence to bisphosphonates therapy and hip fracture risk in osteoporotic women. Osteoporos Int 2008;19(6):811–818.

25. Urist MR. Bone: formation by autoinduction. Science 1965;150(3698):893–899.

26. Celeste AJ, Iannazzi JA, Taylor RC, Hewick RM, Rosen V, Wang EA, Wozney JM. Identification of transforming growth factor beta family members present in bone-inductive protein purified from bovine bone. Proc Natl Acad Sci U S A 1990;87(24):9843–9847.

27. Wang EA, Rosen V, Cordes P, Hewick RM, Kriz MJ, Luxenberg DP, Sibley BS, Wozney JM. Purification and characterization of other distinct bone-inducing factors. Proc Natl Acad Sci U S A 1988;85(24):9484–9488.

28. Deal C, Gideon J. Recombinant human PTH 1-34 (Forteo): an anabolic drug for osteoporosis. Cleve Clin J Med. 2003;70(7):585–586, 589–590, 592–584 passim.

29. Qian JJ, Bhatnagar RS. Enhanced cell attachment to anorganic bone mineral in the presence of a synthetic peptide related to collagen. J Biomed Mater Res 1996; 31(4):545–554.

30. Seeherman H, Wozney JM. Delivery of bone morphogenetic proteins for orthopedic tissue regeneration. Cytokine Growth Factor Rev 2005;16(3):329–345.

31. FDA. 2012. FDA Basics: What does the FDA do? FDA Fundamentals. Available at http://www.fda.gov/AboutFDA/Transparency/Basics/ucm194877.htm. Accessed February 13, 2014.

32. Boden SD, Moskovitz PA, Morone MA, Toribitake Y. Video-assisted lateral intertransverse process arthrodesis. Validation of a new minimally invasive lumbar spinal fusion technique in the rabbit and nonhuman primate (rhesus) models. Spine 1996;21(22):2689–2697.

33. Sandhu HS, Toth JM, Diwan AD, Seim HB 3rd, Kanim LE, Kabo JM, Turner AS. Histologic evaluation of the efficacy of rhBMP-2 compared with autograft bone in sheep spinal anterior interbody fusion. Spine March 15, 2002;27(6):567–575.

34. Boden SD, Martin GJ Jr., Horton WC, Truss TL, Sandhu HS. Laparoscopic anterior spinal arthrodesis with rhBMP-2 in a titanium interbody threaded cage. J Spinal Disord April, 1998;11(2):95–101.

35. Burkus JK, Gornet MF, Dickman CA, Zdeblick TA. Anterior lumbar interbody fusion using rhBMP-2 with tapered interbody cages. J Spinal Disord Tech October, 2002; 15(5):337–349.

36. Li RH, Wozney JM. Delivering on the promise of bone morphogenetic proteins. Trends Biotechnol July, 2001;19(7):255–265.

37. Simon AM, Manigrasso MB, O'Connor JP. Cyclo-oxygenase 2 function is essential for bone fracture healing. J Bone Miner Res June, 2002;17(6):963–976.

38. Simon AM, O'Connor JP. Dose and time-dependent effects of cyclooxygenase-2 inhibition on fracture-healing. J Bone Joint Surg Am March, 2007;89(3):500–511.

39. Standardization ECF. Biological evaluation of medical devices. Vol ISO 10993. Brussels: European Committee for Standardization; 2003.

40. Druecke D, Langer S, Lamme E, Pieper J, Ugarkovic M, Steinau HU, Homann HH. Neovascularization of poly(ether ester) block-copolymer scaffolds in vivo: long-term investigations using intravital fluorescent microscopy. J Biomed Mater Res A January 1, 2004;68(1):10–18.

41. Karageorgiou V, Kaplan D. Porosity of 3D biomaterial scaffolds and osteogenesis. Biomaterials September, 2005;26(27):5474–5491.

42. Yang S, Leong KF, Du Z, Chua CK. The design of scaffolds for use in tissue engineering. Part I. Traditional factors. Tissue Eng December, 2001;7(6):679–689.

43. Mourino V, Boccaccini AR. Bone tissue engineering therapeutics: controlled drug delivery in three-dimensional scaffolds. J R Soc Interface February 6, 2010;7(43):209–227.

44. Kang YM, Kim KH, Seol YJ, Rhee SH. Evaluations of osteogenic and osteoconductive properties of a non-woven silica gel fabric made by the electrospinning method. Acta Biomater January, 2009;5(1):462–469.

45. Ni P, Fu S, Fan M, Guo G, Shi S, Peng J, Luo F, Qian Z. Preparation of poly (ethylene glycol)/polylactide hybrid fibrous scaffolds for bone tissue engineering. Int J Nanomedicine 2011;6:3065–3075.

46. Takahashi Y, Tabata Y. Effect of the fiber diameter and porosity of non-woven PET fabrics on the osteogenic differentiation of mesenchymal stem cells. J Biomater Sci Polym Ed 2004;15(1):41–57.

47. Wang L, Stegemann JP. Thermogelling chitosan and collagen composite hydrogels initiated with beta-glycerophosphate for bone tissue engineering. Biomaterials May, 2010; 31(14):3976–3985.

48. Khan Y, Yaszemski MJ, Mikos AG, Laurencin CT. Tissue engineering of bone: material and matrix considerations. J Bone Joint Surg Am Febraury, 2008;90(Suppl 1):36–42.

49. Kaplan FS, Hayes WC, Keaveny TM, Boskey A, Einhorn TA, Iannotti JP. Form and function of bone. In: Simon SR, editor. *Orthopaedic Basic Science*. Rosemont, IL: American Academy of Orthopaedic Surgeons; 1994. p 127–184.

50. Schonauer C, Tessitore E, Barbagallo G, Albanese V, Moraci A. The use of local agents: bone wax, gelatin, collagen, oxidized cellulose. Eur Spine J October, 2004;13(Suppl 1): S89–S96.

51. Friess W, Uludag H, Foskett S, Biron R, Sargeant C. Characterization of absorbable collagen sponges as rhBMP-2 carriers. Int J Pharm September 30, 1999;187(1):91–99.

52. Friess W, Uludag H, Foskett S, Biron R, Sargeant C. Characterization of absorbable collagen sponges as recombinant human bone morphogenetic protein-2 carriers. Int J Pharm August 5, 1999;185(1):51–60.

53. Ladewig K. Drug delivery in soft tissue engineering. Expert Opin Drug Deliv September, 2011;8(9):1175–1188.

54. Boix T, Gomez-Morales J, Torrent-Burgues J, Monfort A, Puigdomenech P, Rodriguez-Clemente R. Adsorption of recombinant human bone morphogenetic protein rhBMP-2m onto hydroxyapatite. J Inorg Biochem May, 2005;99(5):1043–1050.

55. Kim HD, Wozney JM, Li RH. Characterization of a calcium phosphate-based matrix for rhBMP-2. Methods Mol Biol 2004;238:49–64.

56. Blokhuis TJ. Formulations and delivery vehicles for bone morphogenetic proteins: latest advances and future directions. Injury December, 2009;40(Suppl 3):S8–11.

57. Boden SD, Zdeblick TA, Sandhu HS, Heim SE. The use of rhBMP-2 in interbody fusion cages. Definitive evidence of osteoinduction in humans: a preliminary report. Spine February 1, 2000;25(3):376–381.

58. Govender S, Csimma C, Genant HK, Valentin-Opran A, Amit Y, Arbel R, Aro H, Atar D, Bishay M, Börner MG, Chiron P, Choong P, Cinats J, Courtenay B, Feibel R, Geulette B, Gravel C, Haas N, Raschke M, Hammacher E, van der Velde D, Hardy P, Holt M, Josten C, Ketterl RL, Lindeque B, Lob G, Mathevon H, McCoy G, Marsh D, Miller R, Munting E, Oevre S, Nordsletten L, Patel A, Pohl A, Rennie W, Reynders P, Rommens PM, Rondia J, Rossouw WC, Daneel PJ, Ruff S, Rüter A, Santavirta S, Schildhauer TA, Gekle C, Schnettler R, Segal D, Seiler H, Snowdowne RB, Stapert J, Taglang G, Verdonk R, Vogels L, Weckbach A, Wentzensen A, Wisniewski T. Recombinant human bone morphogenetic protein-2 for treatment of open tibial fractures: a prospective, controlled, randomized study of four hundred and fifty patients. J Bone Joint Surg Am 2002;84-A(12):2123–2134.

59. Triplett RG, Nevins M, Marx RE, Spagnoli DB, Oates TW, Moy PK, Boyne PJ. Pivotal, randomized, parallel evaluation of recombinant human bone morphogenetic protein-2/

absorbable collagen sponge and autogenous bone graft for maxillary sinus floor augmentation. J Oral Maxillofac Surg 2009;67(9):1947–1960.

60. Fiorellini JP, Howell TH, Cochran D, Malmquist J, Lilly LC, Spagnoli D, Toljanic J, Jones A, Nevins M. Randomized study evaluating recombinant human bone morphogenetic protein-2 for extraction socket augmentation. J Periodontol April 2005;76(4): 605–613.

61. FDA. 2002. INFUSE Bone Graft Summary of Safety and Effectiveness Data. Available at http://www.accessdata.fda.gov/cdrh_docs/pdf/P000058b.pdf. Accessed November 13, 2012.

62. FDA. 2001. OP-1 Implant HDE Approval Letter. Available at http://www.accessdata.fda. gov/cdrh_docs/pdf/h010002a.pdf. Accessed November 13, 2012.

63. Friedlaender GE, Perry CR, Cole JD, Cook SD, Cierny G, Muschler GF, Zych GA, Calhoun JH, Laforte AJ, Yin S. Osteogenic protein-1 (bone morphogenetic protein-7) in the treatment of tibial nonunions. J Bone Joint Surg Am 2001;83-A(Suppl 1(Pt 2)): S151–158.

64. Vaccaro AR, Lawrence JP, Patel T, et al. The safety and efficacy of OP-1 (rhBMP-7) as a replacement for iliac crest autograft in posterolateral lumbar arthrodesis: a long-term (>4 years) pivotal study. Spine December 15, 2008;33(26):2850–2862.

65. Vaccaro AR, Patel T, Fischgrund J, Anderson DG, Truumees E, Herkowitz H, Phillips F, Hilibrand A, Albert TJ. A 2-year follow-up pilot study evaluating the safety and efficacy of op-1 putty (rhbmp-7) as an adjunct to iliac crest autograft in posterolateral lumbar fusions. Eur Spine J September, 2005;14(7):623–629.

66. FDA. 2004. OP-1 Putty HDE Approval Letter. Available at http://www.accessdata.fda. gov/cdrh_docs/pdf2/H020008a.pdf. Accessed November 13, 2012.

67. FDA. 1999. PepGen P-15 PMA Approval Letter. Available at http://www.accessdata.fda. gov/cdrh_docs/pdf/P990033a.pdf. Accessed November 13, 2012.

68. Yukna RA, Callan DP, Krauser JT, Evans GH, Aichelmann-Reidy ME, Moore K, Cruz R, Scott JB. Multi-center clinical evaluation of combination anorganic bovine-derived hydroxyapatite matrix (ABM)/cell binding peptide (P-15) as a bone replacement graft material in human periodontal osseous defects. 6-month results. J Periodontol June 1998;69(6):655–663.

69. Clinicaltrials.gov. 2006. An Assessment of P-15 Bone Putty in Anterior Cervical Fusion with Instrumentation (NCT00310440). Available at http://clinicaltrials.gov/ct2/show/ NCT00310440. Accessed November 13, 2012.

70. FDA. 2004. Gen 21S PMA Approval Letter. Available at http://www.accessdata.fda.gov/ cdrh_docs/pdf4/P040013a.pdf. Accessed November 13, 2012.

71. Jayakumar A, Rajababu P, Rohini S, Butchibabu K, Naveen A, Reddy PK, Vidyasagar S, Satyanarayana D, Pavan Kumar S. Multi-centre, randomized clinical trial on the efficacy and safety of recombinant human platelet-derived growth factor with beta-tricalcium phosphate in human intra-osseous periodontal defects. J Clin Periodontol 2011;38(2):163–172.

72. Clinicaltrials.gov. 2011. Augment™ Injectable Bone Graft Compared to Autologous Bone Graft as a Bone Regeneration Device in Hindfoot Fusions (NCT01305356). Available at http://clinicaltrials.gov/ct2/show/NCT01305356. Accessed November 13, 2012.

73. Digiovanni CW, Baumhauer J, Lin SS, Berberian WS, Flemister AS, Enna MJ, Evangelista P, Newman J. Prospective, randomized, multi-center feasibility trial of rhPDGF-BB versus autologous bone graft in a foot and ankle fusion model. Foot Ankle Int April, 2011;32(4):344–354.

74. Young CS, Ladd PA, Browning CF, Thompson A, Bonomo J, Shockley K, Hart CE. Release, biological potency, and biochemical integrity of recombinant human platelet-derived growth factor-BB (rhPDGF-BB) combined with Augment(TM) Bone Graft or GEM 21S beta-tricalcium phosphate (beta-TCP). J Control Release December 16, 2009;140(3):250–255.

75. Clinicaltrials.gov. 2006. Safety and Efficacy of I 0401 in the Treatment of Tibial Plateau Fractures Requiring Grafting (NCT00409799). Available at http://clinicaltrials.gov/ct2/show/NCT00409799. Accessed November 13, 2012.

76. Clinicaltrials.gov. 2007. Adjunctive Therapy to Treat Tibial Shaft Fractures (TSF) (NCT00533793). Available at http://clinicaltrials.gov/ct2/show/NCT00533793. Accessed November 13, 2012.

77. Clinicaltrials.gov. 2008. Pilot Study to Assess Safety/Preliminary Effectiveness of Prefix in Subjects with Degenerative Disc Disease (DDD) Undergoing Spine Fusion Surgery (NCT00798902). Available at http://clinicaltrials.gov/ct2/show/NCT00798902. Accessed November 13, 2012.

78. Clinicaltrials.gov. 2008. Phase 3 Clinical Trial of Periodontal Tissue Regeneration Using Fibroblast Growth Factor-2 (Trafermin) (NCT00734708). Available at http://clinicaltrials.gov/ct2/show/NCT00734708. Accessed November 13, 2012.

79. Kawaguchi H, Oka H, Jingushi S, Izumi T, Fukunaga M, Sato K, Matsushita T, Nakamura K, TESK Group. A local application of recombinant human fibroblast growth factor 2 for tibial shaft fractures: a randomized, placebo-controlled trial. J Bone Miner Res December, 2010;25(12):2735–2743.

80. Koch FP, Becker J, Terheyden H, Capsius B, Wagner W. A prospective, randomized pilot study on the safety and efficacy of recombinant human growth and differentiation factor-5 coated onto beta-tricalcium phosphate for sinus lift augmentation. Clin Oral Implants Res November, 2010;21(11):1301–1308.

81. Stavropoulos A, Windisch P, Gera I, Capsius B, Sculean A, Wikesjo UM. A phase IIa randomized controlled clinical and histological pilot study evaluating rhGDF-5/beta-TCP for periodontal regeneration. J Clin Periodontol 2011;38(11):1044–1054.

82. Schmitt JM, Hwang K, Winn SR, Hollinger JO. Bone morphogenetic proteins: an update on basic biology and clinical relevance. J Orthop Res March, 1999;17(2):269–278.

83. Morone MA, Boden SD, Hair G, Martin GJ Jr, Racine M, Titus L, Hutton WC. The Marshall R. Urist Young Investigator Award. Gene expression during autograft lumbar spine fusion and the effect of bone morphogenetic protein 2. Clin Orthop Relat Res June, 1998;351:252–265.

6

DELIVERY OF INSULIN: FROM GLASS SYRINGES TO FEEDBACK-CONTROLLED PATCH PUMPS

BILL VAN ANTWERP

6.1 INTRODUCTION

Insulin is the oldest biopharmaceutical drug, and like all protein-based drugs, insulin requires relatively complex delivery compared to a small-molecule oral drug that is taken once per day. This chapter focuses on insulin delivery from the days of Banting and Best to the modern era of small wearable pumps that are connected to continuous glucose sensors.

6.2 THE PHYSIOLOGY OF INSULIN DELIVERY IN THE NORMAL HUMAN

In order to understand the special requirements of insulin delivery, it is important to understand how the pancreas functions in a normal human. Insulin in the pancreas is released both in a basal mode, that is, continuously during the day and night, and in a bolus mode in response to a meal. A typical insulin and glucose profile [1] for normal individuals is shown in Figure 6.1.

As seen in the normal profiles, relatively large changes in prandial insulin, from 10 to 70 µU/ml, lead to quite good blood glucose (BG) control. The goal of exogenous insulin delivery then in patients with diabetes is to mimic the insulin delivery by the pancreas and maintain BG in the 80–120 mg/dl (4.4–6.7 mM).

Therapeutic Delivery Solutions, First Edition. Edited by Chung Chow Chan, Kwok Chow, Bill McKay, and Michelle Fung.
© 2014 John Wiley & Sons, Inc. Published 2014 by John Wiley & Sons, Inc.

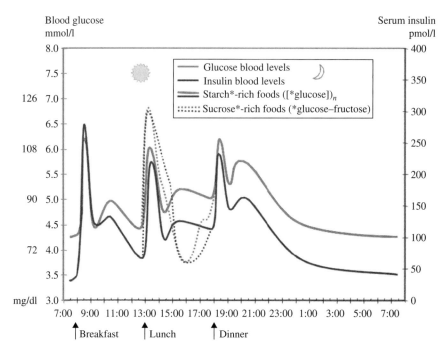

FIGURE 6.1   Healthy volunteer insulin and glucose profiles. Reproduced from Ref. 2, with permission of McStroher, Creative Commons.

Achieving this level of control has proven to be quite a challenge, and the combined efforts of the pharmaceutical industry and the medical device industry have so far not quite achieved normal BG in most patients with diabetes.

6.3   THE EARLY DAYS

Insulin was discovered by Banting [3] and Best in 1921 and commercialized soon after by Eli Lilly. Unfortunately, at that time, the relationship between the pharmacokinetics (PK) of subcutaneous insulin and the relationship to glucose homeostasis was unknown since there were no reasonable assays for either BG or blood levels of insulin. That being said, insulin was still a wonder drug and saved the lives of countless individuals. The delivery of the first commercial insulin was via a glass syringe, sterilized by boiling, and reusable stainless steel needles. The insulin was injected once a day into the subcutaneous tissue. Shortly after the discovery and commercialization of insulin, the most troublesome acute shortcoming of insulin injections was observed, hypoglycemia that can lead to death. It was learned very early in the development of insulin therapy that the dosing of insulin was extremely important; too much and death from hypoglycemia was possible and too little could lead to hyperglycemic coma. The dosing, and the delivery system, needs to be very carefully controlled. Inspection of Figure 6.1 shows clearly that the normal pancreas starts to deliver insulin very rapidly after the BG starts to rise (first-phase insulin response [4] for about the first 10 min) and turns off very quickly as well. The technical difficulty in trying to emulate the normal pancreas is related to the route of administration as well as to the PK of the current insulin formulations, but none of this was known for at least 40 years after the commercialization of insulin. During the early years of insulin delivery, the pharmaceutical companies Lilly, Novo Nordisk, and Hoechst all developed formulations of insulin with varying PK. The need for controlled PK is also evident from Figure 6.1; about half of a nondiabetic's insulin is delivered basally and not in response to a meal. The early insulin formulations were developed accordingly, with a basal insulin (neutral protamine Hagedorn (NPH) or Lente) and rapid-acting insulin (normal) becoming the standards. NPH insulin was invented in 1936 by Hagedorn at Novo Nordisk. At the same time, Eli Lilly developed Lente and Ultralente insulin as basal insulin therapies, and both NPH and Lente/Ultralente were still injected using reusable glass syringes and reusable steel needles. The early syringe market was dominated by patents issued to Luer in France and licensed by Becton Dickinson in the United States. The first insulin syringe was sold by Becton Dickinson in 1924, very shortly after the introduction of insulin itself.

6.4 BETTER SYRINGES AND, FOR THE FIRST TIME, A METER

For the first 30 years after the introduction of insulin, glass syringes were the only method of delivering insulin (Figure 6.2). In the late 1950s after reusable glass syringes led to an outbreak of hepatitis, disposable syringes were introduced to the market-place. The first major use for disposable syringes was for the delivery of Salk's polio

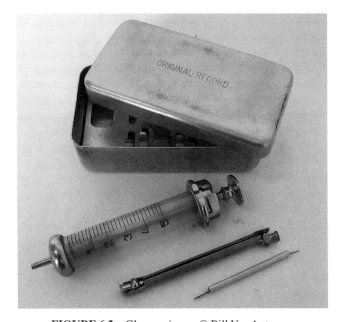

FIGURE 6.2 Glass syringes. © Bill Van Antwerp.

vaccine, although the syringes were still made of glass. By the early 1960s, disposable plastic syringes were available from several manufacturers worldwide, and the life of the type 1 (T1) diabetic patient in particular became a bit simpler. Unfortunately, the improvements in insulin technology were not moving forward as fast as the improvements in syringe technology. During the 1960s, insulin was still extracted from either cow or pig pancreata, and there were significant issues with purity of the insulin used at that time. Over the course of the 1960s and 1970s, purity of the animal-sourced insulins gradually got better, and stable formulations that allowed for reasonable glucose control were available. Additionally, there is a major second component that is critical to the success of insulin delivery: how much insulin does the patient need? In order to answer that question, two pieces of data are required, the amount of carbohydrates in the meal to be eaten and the current BG. Before there were glucose-measuring devices for blood, there were strips that measured BG in the urine. These devices were not very accurate since glucose does not become measurable in the urine until the BG is greater than 180 mg/dl (10 mM). In 1965, Ames Laboratory produced the first BG strips, called Dextrostix, and they revolutionized insulin delivery. To use the Dextrostix, a large drop of blood was obtained from the tip of a finger and placed on the paper surface of the Dextrostix. The paper was coated with an enzyme, glucose oxidase, and a color reagent. Glucose in the blood sample reacted with the enzyme to make hydrogen peroxide, the peroxide oxidized the color reagent, and a blue dye was formed. By comparing the blue color to a preprinted comparison chart, BG was estimated. The device was not very accurate but it was the beginning of the modern era for insulin delivery [5].

In 1970 or so, realizing that reading the Dextrostix was problematical, Ames released a reflectance meter that could automatically read the strips. Precision was improved but strips were still inaccurate and the device was large and expensive. Over the next 10 years, several new companies introduced BG meters including Boehringer, Roche, and LifeScan, all of them using the colorimetric principle. By the end of the 1970s, the situation for insulin delivery was quite simple. There was long-acting insulin that a diabetic injected once or twice per day, and there was shorter-acting insulin that was injected at mealtimes, and there were glucose meters that allowed the patient to monitor his/her BG before and after meals. There was however no great understanding of the benefits and risks of really aggressive BG control for diabetic patients.

Until the late 1970s, the major controversy in diabetes care revolved around the goals of therapy. A small group of physicians believed that good control was essential to eliminating the complications of diabetes, but most physicians still believed that modest BG control was the ideal regimen for patients. In 1976, the American Diabetes Association did a literature review and concluded that good control was beneficial. Opponents of this view said that it was impossible to achieve good BG control without too much hypoglycemia. It was a stalemate since neither side had any data.

Data was soon forthcoming though since during the late 1970s there was a surge of interest in continuous glucose monitoring with the hope that a feedback-controlled insulin system would emerge. This idea was bolstered by the emergence of technology [6] that allowed insulin levels in plasma to be measured, and thus, the normal dynamics of insulin could be understood during both basal and mealtime. Soon, the

PK [7] of normal and basal insulin [8] injections was understood. Once the PK of normal insulin secretion was understood, it was a small theoretical step to trying to emulate normal insulin delivery using mechanical systems. To do that, you needed a way to deliver the insulin with controlled PK and a way to monitor the glucose.

In 1977, Mike Albisser in Toronto filed for a U.S. patent [9] for an artificial beta cell. Mike was clearly 10 years ahead of his time since at the time of filing the patent, neither automatic insulin delivery systems nor continuous BG technology was available. That changed quickly at least for intravenous (IV) delivery of insulin. Almost simultaneously, groups in Osaka [10], Ulm [11], Montpellier [12], Sydney [13], and Toronto [14] published data on glucose-controlled insulin infusion. Most of these systems used homemade continuous glucose sensors and IV insulin, but the Ulm system eventually was licensed to Miles and became a commercial product, the Biostator. The Biostator [15] was a research tool with more than 250 papers published using the technology, not a solution for diabetic patient at home, but has been used extensively to help understand and define the parameters for successful at-home BG-controlled insulin delivery. The Biostator contained pumps for delivery of both insulin and glucose as well as a novel membrane-based glucose monitor. In practice, insulin or glucose was delivered IV during therapy, and the sensor output controlled the patient's BG in a relatively narrow range. The Biostator is still in use today, primarily to control glucose during clamp studies for evaluation of the PK of new insulin formulations.

6.5   THE DCCT STUDY AND WHY CONTROL MATTERS

In the early 1980s, there was still no evidence that good BG control would really make a difference to T1 diabetes patients. T1 diabetes is an autoimmune disease in which total insulin production is destroyed and those patients need insulin to live. In 1983, the U.S. National Institutes of Health/National Institute of Diabetes and Digestive and Kidney Diseases (NIH/NIDDK) funded a 10-year study of more than 1400 patients comparing standard BG control to intensive control. In the standard control group, patients received not more than two injections per day of insulin and used mostly mixed insulin formulations (regular + NPH). In the intensively treated group, patients received up to five injections of insulin per day or could use insulin pumps with frequent BG monitoring up to five times per day. The goal of the intensive group was to push HbA1c (a measure of BG control over the last 2–3 months) as close to normal as possible. Patients were followed for 10 years (and a second 10-year follow-up in the EDIC study), and the clinical endpoints were eye disease, kidney disease, and nerve damage. The findings firmly showed that intensive insulin therapy was beneficial, a 76% reduction in the risk of developing retinopathy, a 50% reduction in the risk of developing significant nephropathy, and a 60% reduction in the risk of severe neuropathy. After the second 10-year evaluation, it was observed that there was a 42% reduction in the risk of any cardiovascular disease event and a 57% reduction in nonfatal or fatal heart attack and stroke. These improvements do not come without a cost though. Intensive therapy leads to more hypoglycemic events

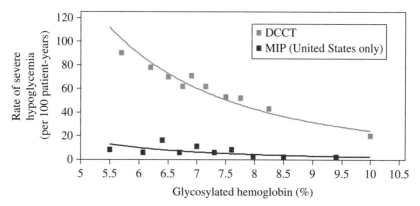

FIGURE 6.3   DCCT data. © Bill Van Antwerp.

in total and a threefold higher incidence of severe hypoglycemia, a hypo event that needed help from a second person.

The data from the DCCT are shown in Figure 6.3.

As seen in the figure at baseline, there was about a 33% chance of a severe hypo event in a single patient per year. Getting to an HbA1c of 6% moved that to about a one event per year per patient. The take-home message is that intensive insulin therapy in T1 patients is extremely beneficial; patients must be diligent about testing BG on a regular basis.

6.6   IMPLANTABLE INSULIN PUMPS: THE BEGINNING OF CONTINUOUS INSULIN DELIVERY

The first published articles on insulin pumps [16] were for fully implantable [17] versions. The implantable devices delivered insulin either directly to the venous system or into the peritoneal cavity. Compared to the subcutaneous injection route, IV access offers a much closer match to the PK of normal insulin delivery compared to the very slow adsorption via the subcutaneous route. The intraperitoneal route offers a large safety advantage over the IV route at the expense of a somewhat slower adsorption but still much faster than subcutaneous infusion [18]. There are several advantages of intraperitoneal insulin over subcutaneous insulin delivery including a better counterregulatory hormone secretion, a much lower incidence of hypogly-cemia and severe hypoglycemia, and a PK profile that is much closer to pancreatic insulin secretion than any other type of device.

Unfortunately, there is a major issue in using implantable pumps. The insulin used for the devices was not physically stable. The early implantable pumps were based on several technologies, peristaltic pumps, pressure-controlled single rate pumps, or fully programmable micro-piston-type pumps. The first implantable [19, 20] insulin delivery pumps were tested in humans in the early 1980s. Shortly thereafter, it became clear that the then available insulin formulations were not sufficiently stable for long-term use and moreover that implantable pumps require insulin of much higher concentration than normal if they were to be clinically acceptable with 90-day refill cycles. By 1984, it was clear that aggregation [21] and fibrillation of insulin in implantable pumps were an important technical issue. Feingold and Kraegen in Australia showed definitively that insulin aggregation was dependent on the materials of construction of the pumps and that controlling the interface between the insulin and the device through intelligent mechanical design and formulation efficacy was key to success. In 1987, Ulrich Grau [22] from Hoechst (now Sanofi) and Chris Saudek at Johns Hopkins presented a new stable insulin formulation of U400 insulin developed by Hoechst specifically for use in implantable pumps, and the use of implantable pumps increased significantly. Implantable pumps eventually were clinically approved for sale in some of the EMEA countries but never were approved by the U.S. FDA for a variety of both commercial and technical reasons. Figure 6.3 shows the most striking clinical outcome of using implantable pumps. In the DCCT, achieving normoglycemia or an HbA1c of about 5.5% leads to a severe hypoglycemia event rate of almost 110 per 100 patient years in the intensively treated group (multiple daily injections or external pumps). In the same patient population, using implantable pumps allowed patients to have normoglycemia with a severe hypo event rate of only 8/100 patient years, which is about a 14-fold improvement. In spite of this clinical success, various technical difficulties such as catheter blockages, battery issues, and the need for an *in vivo* rinse procedure led to the discontinuation of development of the implantable pump.

6.7 EXTERNAL INSULIN PUMPS: THE EARLY DAYS

In comparison to the implantable insulin pumps that were being developed in the early 1970s, external programmable pumps were a bit later to the clinic. The very first insulin pump was developed by Kadish in Beverly Hills but was too big to be thought of as reasonable (Fig. 6.4).

In the late 1970s, John Pickup [23] in England and Bill Tamborlane [24] at Yale were the first to publish human data using continuous subcutaneous infusions (Fig. 6.5 and Fig. 6.6).

Pickup used a relatively large device, the Mill-Hill infuser (Fig. 6.5), for his studies, while Tamborlane used the AutoSyringe (Fig. 6.6) invented by Dean Kamen for his studies. Neither of these devices was purpose built for diabetes but rather based on standard syringe pumps for delivering IV medications. Concomitantly, Franetzki's group [25] in Germany and Irsigler [26] in Vienna published similar studies. All of these very early pumps were basal only to mimic the 50% of pancreatic output between meals and were large and unwieldy ($8 \times 3 \times 2.8$ in., about a pound), and the battery situation was quite dismal at that time. More importantly, the insulin formulations at that time were not of sufficient purity and stability to make CSII very clinically acceptable. In the early 1980s, there were several (more than 20) pump companies that started to supply pumps to the T1 diabetes population. Unfortunately, there was very little understanding of the technical requirements for success, and an initial exuberance about pumping soon led to a significant backlash against pumps. Some of the early pump companies made devices with inadequate safety features, and several cases of pump "runaway" leading to at least

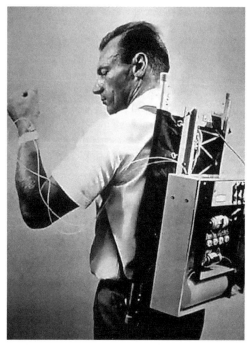

FIGURE 6.4   Backpack pump. © Bill Van Antwerp.

FIGURE 6.5   Early insulin pump. Courtesy of Squidonius, Creative Commons.

one [27] confirmed death occurred. Moreover, there was little awareness of the need for near-sterile technique when inserting catheters, and several infections at infusion sites were reported, some with severe consequences.

Eventually, almost all of the pump companies withdrew from the marketplace including the big insulin companies (Lilly and Novo), and through most of the late 1980s and 1990s, there were only two companies making pumps for the United

FIGURE 6.6 Early insulin pump AutoSyringe. © Bill Van Antwerp.

States and Europe, MiniMed and Disetronic. Eventually, pumps became more reliable and safe, and the rapid acceptance of mobile electronic technology like cell phones, pagers, iPods, etc. has made carrying a mobile electronic medical device much more acceptable. Why did the companies that were initially in the pump business, including some of the largest medical device and pharmaceutical companies, leave the business and what does it take to stay in it? It is clear that the pump business requires something very different than the pharmaceutical business. Of course, the starting point is safe and reliable technology, and despite the belief of many venture capital-based start-up companies, 10 engineers and $10M dollars are not enough to be successful. Once you have the safe and acceptable (to the patient) technology, a company needs to create an ecosystem based on:

1. Training of doctors, nurses, dietitians, pharmacists, and patients in pump therapy
2. Teaching of medical professionals via sponsoring of clinical trials and having a knowledgeable and articulate speakers bureau
3. A group of dedicated and knowledgeable pump trainers for interacting directly with medical practices and mostly with patients
4. Dedicated 24/7/365 help lines with live, highly trained clinical staff including nurse specialists (an online training center is not enough)

5. Direct and responsible interaction with payers to get pumps financed for patients in a reasonable time frame and at an acceptable cost

Modern insulin pumps are all currently based on computer-controlled motors that drive small plastic syringes. The individual components and technologies vary, but in general, pumps have user-changeable batteries, a user-filled plastic reservoir (syringe), and a computer-controlled interface that allows for programming of basal and bolus rates and also typically has memory for events. Some pumps can communicate via wireless technology with continuous glucose sensors. Current generations of pumps have been developed with patient convenience and safety in mind and have proven themselves in the clinic with more than 300,000 patients wearing them 24/7/365 throughout the world.

6.8   WHAT TO DELIVER?

During the 1970s and early 1980s, insulin was procured from beef and pork pancreata and was then purified and formulated into a variety of commercial products. These insulin formulations had varying PK profiles [28]. The profiles ranged from very slow (NPH/Ultralente/Lente, meant to mimic the body's basal profile) to regular (R, meant as a prandial insulin). During the late 1970s, the rapid rise of biotechnology and recombinant DNA technology led to a race to develop a biosynthetic (recombinant) human insulin product. In 1978, Genentech announced the successful production of human insulin in *Escherichia coli*, and in 1982, Lilly introduced Humulin to the marketplace. For a period of time after that, Novo produced a semisynthetic insulin by converting porcine to human insulin via an enzymatic single-amino acid replacement. Ultimately, by 1986, human insulin produced in *E. coli* or yeast has been produced by a number of firms. For pump-based therapies, R or regular human insulin was initially used, but the time to peak blood insulin activity was still about an hour, meaning that the patient had to actually program his/her meal bolus about 30 min before the meal. This could be problematical and everyone in the pump field understood that insulin with faster time courses of action was needed. To understand this, a bit of insulin chemistry is needed. In the pancreas, insulin is created as a hexamer, that is, six insulin molecules held together by zinc. In the animal insulin days, insulin was extracted from the pancreas as the hexamer and subsequently formulated that way. Human insulin is also synthesized as the hexamer and regular insulins are all hexameric. Once injected into the subcutaneous tissue, the hexamer needs to dissociate into monomers before it can be absorbed into the blood. This is a slow process. In order to make faster insulin, the new insulin had to be manufactured to not form stable hexamers. Lilly realized in the early 1990s that human growth hormone resembled insulin in terms of amino acid composition but did not form stable hexamers. Lilly interchanged the B28 and B29 amino acids of insulin to create lispro (Humalog) that was approved in 1996. At about the same time, Novo Nordisk made insulin aspart (NovoLog) by changing B28 from proline to aspartic acid. This was approved in 2000 in the United States. Four years later, Aventis (now Sanofi) had their version of fast-acting insulin (Apidra) approved. All three of these insulin formulations have about the same PK with peak action in about 30–40 min, which is both faster and more predictable than the regular insulins previously used in pumps.

The approach of chemically modifying insulin to create analogs did not end with rapid analogs but continued to create new basal insulin analogs. The most important are insulin glargine (Lantus) and insulin detemir (Novo), which are the first truly basal insulin formulations that can be given successfully only once per day.

6.9 HOW TO DELIVER THE NEW INSULIN: PENS AND PUMPS

Given that insulin formulations were now available that could be given once per day for basal and at mealtime (instead of 30 min prior), new injection technology had to follow. This new injection technology is the prefilled pen. In the early insulin delivery days, insulin was drawn up into a syringe from a glass vial. Today, most insulin is delivered via the pen route using small (30 gauge) needles without the need to carry around a vial. Modern pens typically hold a week's worth of insulin and are disposed of when empty. They are incredibly simple to use; simply turn a dial to get the dose, then insert the needle into the subcutaneous tissue, and press the button. Pens are the choice of most T2 diabetics and many T1 diabetics, but for the ultimate control, insulin pumps are still the best current technology.

As of fall 2012, there are now six insulin pumps available in the United States. These include Animas, Insulet, Medtronic, Roche, Sooil/Dana, and Tandem. All of these pumps except the Omnipod from Insulet use the same basic technology. A syringe (plastic) is filled with insulin and the syringe plunger is pushed via a computer-controlled motor. The insulin is infused into the subcutaneous tissue through a long plastic tube called an infusion set and ultimately through a small plastic cannula that is inserted under the skin. All of the current pumps have multiple possible basal rates. The basal rate was started in the late 1980s when pumps that had 2 rates then moved to 12, then 24, and then 48. Most folks use only one or two, but having multiple rates is still a selling point. All of the pumps have bolus calculators that allow the patient to use the pump to calculate the amount of insulin that is required for a given meal, although the patient still needs to be able to estimate the amount of carbohydrates in the meal. Several pumps also offer an adjustment to the bolus calculator. This is called the insulin (or bolus) on board calculator. An insulin analog once injected into the tissue typically has a duration of action of about 3–4 h [29]. This time duration is variable among patients and can be adjusted for an individual in the pump software. In use, if a patient (or the bolus wizard) determines that a particular meal needs three units of insulin but the insulin on board calculator calculates that there is still one unit "on board," then the bolus will be adjusted to take the insulin action from a previous bolus into account. All of the current pumps also have available software that can take pump and meter data and generate logbooks for the patient and his/her healthcare team to use to help modify future therapy. For the most part, the current generation of insulin pumps is relatively mature as stand-alone devices, but there is a lot of room for newer devices in this marketplace.

Insulet with the Omnipod has created the first of many planned "patch pumps." All of the other current insulin pumps are the size of pagers/smartphones and are attached to the patient with a long infusion set that delivers the insulin through a plastic cannula under the skin. A patch pump is worn directly on the skin and has no infusion set. All of the communication with the patch is done through a hand held controller similar to a smartphone and this controller performs all of the pump programming wirelessly. There are currently at least 10 other patch pump companies that have products either CE marked or in late-stage clinical testing, and it is expected that by the end of 2015, most new insulin pump sales will be for patch-type devices since there will be no need for a large investment in a durable medical device. Current nonpatch pumps cost approximately $5000 in the United States, while patch pumps suitable for a 3-day use might only cost $40. On a daily basis, the pricing is probably the same, but without the need for a significant investment, it is expected that patch pumps will dominate the market.

6.10   THE FUTURE: NEXT 10 YEARS

As mentioned previously, the concept of feedback-controlled insulin delivery where the feedback control was a continuous measure of BG is an old one [30]. The major stumbling block in the early days was the availability of robust continuous glucose [31] sensing technology. The enzyme electrode was invented by Leland Clark [32] in 1962 and was the basis for the common laboratory glucose analyzer. Converting the discovery that glucose oxidase could be used to measure glucose discretely to a continuously working commercial glucose sensor took about 20 years. MiniMed successfully commercialized a continuous BG meter in 1999. In the ensuing years, more than 1800 papers were published describing new ways to measure glucose continuously. Today, there are two commercially available continuous glucose sensors, Medtronic and Dexcom, and both measure glucose in the subcutaneous tissue space using glucose oxidase electrochemical technology.

In order to complete the feedback control system, there are three components necessary. The first is insulin delivery, and as mentioned previously, this is relatively robust technology. The second is continuous glucose sensing, and it is clear that this is almost good enough for feedback control. In Europe, systems are available that will turn off the insulin pump if a hypoglycemic event is predicted, but the U.S. FDA has not yet approved even this level of control suggesting that sensors need to become more reliable. The third and most interesting component is the algorithm required. Looking at Figure 6.1, the time lag between BG rising and insulin delivery from the pancreas is quite short. In the current systems, both glucose measurement and insulin delivery are in the subcutaneous tissue. This leads to two significant problems. One is the delay between blood and tissue glucose and this delay is not fixed but variable. Blood can lead the tissue when BG is going up due to a meal, while tissue (in an insulin-sensitive tissue) can lead the blood when BG is decreasing from insulin action. This lag can be between 5 and 15 min and can depend on a host of variables.

More importantly, in the pancreas, there is a first-phase [4] insulin response that occurs in a very few minutes. In the insulin pump situation, insulin is delivered into the subcutaneous tissue and takes between 30 and 40 min to peak. This is much slower than the *in vivo* situation in a nondiabetic person and leads to significant technical complications in the development of an algorithm to treat diabetes. More importantly, the insulin that is infused into the subcutaneous tissue has a duration of action that is much longer than that delivered by the pancreas directly to the blood.

In summary then, the glucose sensor signal is delayed compared to *in vivo*; the insulin action starts much later and lasts much longer than the insulin action in a nondiabetic patient. This means that there needs to be great caution [33] in developing an algorithm.

The first step in developing a closed-loop system is to develop an open-loop system that uses patient inputs to ameliorate some of the issues. The simplest approach is to have the patient simply signal that a meal is being eaten. The system can then deliver some insulin and do correction bolus later if the meal is too big. This approach and the ability of the pump/sensor to control overnight glycemia (approved in Europe) should improve the control of most T1 diabetics.

6.11 OTHER USES

Currently, insulin pump technology is mainly used by T1 diabetics with estimates of T2 usage ranging from 5% to 12%. Given that most T2 patients are not using prandial insulin, the need for pumps is minimal, but most basal insulin is now delivered via pens. With the rise of the incretins, there are now several agents that can best be delivered via continuous delivery for the noninsulin using T2 patients. In other fields, it has very recently been shown [34] that using insulin pumps to deliver interferon-α to treat hepatitis C is both safe and effective.

6.12 CONCLUSIONS

Concomitant with the discovery of insulin came the realization that delivery of the correct amounts at the correct times was going to be crucial to patient health and safety. From syringes to feedback-controlled insulin pumps in 80 years is an amazing accomplishment for the pharmaceutical and medical device industries. Patients and the medical community eagerly await the improvements to come in the next 80 years. These might include faster and more accurate insulin delivery by changing the adsorption profiles of insulin, oral insulin for basal control, pulmonary insulin with true first-phase insulin response, and ultimately implantable beta cells that can replace pancreatic function while eluding the immune system attack that causes T1 diabetes in the first place. By the end of the century, it is expected that prevention of T1 diabetes will become a clinical reality and that pumps and sensors become historical curiosities. Moreover there was little awareness of the need to fully understand the differences between the pharmacokinetics of pump delivered insulin compared to endogenous insulin. Exogenous insulin, even regular human insulin can not mimic the speed of native insulin.

REFERENCES

1. Owens DR, Zinman B, Bolli GB. Insulins today and beyond. Lancet 2001;358:739.
2. Daly ME, Vale C, Walker M, Littlefield A, Alberti KG, Mathers JC. Acute effects on insulin sensitivity and diurnal metabolic profiles of a high-sucrose compared with a high-starch diet. American Journal of Clinical Nutrition 1998;67(6):1186–1196.
3. Banting FG, Campbell WR, Fletcher AA. Further clinical experience with insulin (pancreatic extracts) in the treatment of diabetes mellitus. Br Med J January 6, 1923;1 (3236):8–12.
4. Gerich JE. Is reduced first-phase insulin release the earliest detectable abnormality in individuals destined to develop type 2 diabetes? Diabetes February 2002;51(Suppl 1): S117–S121.
5. Mazzaferri EL, Skillman TG, Lanese RR, Keller MP. Use of test strips with colour meter to measure blood-glucose. Lancet February 14, 1970;1(7642):331–333.
6. Yalow RS, Berson SA. Immunoassay of endogenous plasma insulin in man. J Clin Invest July 1960;39:1157–1175.
7. Binder C. Absorption of injected insulin: a clinical–pharmacological study. Acta Pharmacol Toxicol 1969;27(Suppl 2):1–84.
8. Berger M, Halban PA, Assal JP, Offord RE, Vranic M, Renold AE. Pharmacokinetics of subcutaneously injected tritiated insulin: effects of exercise. Diabetes January 1979; 28(Suppl 1):53–57.
9. Albisser A, Leibel B. Artificial beta cell. US patent 4,245,634. January 20, 1981.
10. Shichiri M, Kawamori R, Yamasaki Y, Inoue M, Shigeta Y, Abe H. Computer algorithm for the artificial pancreatic beta cell. Artif Organs 1978;2(Suppl):247–250.
11. Pfeiffer EF, Thum C, Clemens AH. The artificial beta cell—a continuous control of blood sugar by external regulation of insulin infusion (glucose controlled insulin infusion system). Horm Metab Res September 1974;6(5):339–342.
12. Mirouze J, Selam JL, Pham TC, Cavadore D. Evaluation of exogenous insulin homoeostasis by the artificial pancreas in insulin-dependent diabetes. Diabetologia May 1977; 13(3):273–278.
13. Kraegen EW, Campbell LV, Chia YO, Meler H, Lazarus L. Control of blood glucose in diabetics using an artificial pancreas. Aust N Z J Med June 1977;7(3):280–286.
14. Albisser AM, Leibel BS, Ewart TG, Davidovac Z, Botz CK, Zingg W. An artificial endocrine pancreas. Diabetes May 1974;23(5):389–396.
15. Fogt EJ, Dodd LM, Jenning EM, Clemens AH. Development and evaluation of a glucose analyzer for a glucose controlled insulin infusion system (Biostator). Clin Chem August 1978;24(8):1366–1372.
16. Blackshear PJ, Dorman FD, Blackshear PL Jr, Varco RL, Buchwald H. The design and initial testing of an implantable infusion pump. Surg Gynecol Obstet January 1972; 134(1):51–56.
17. Thomas LJ, Bessman SP. Prototype for an implantable insulin delivery pump. Proc West Pharmacol Soc 1975;18:393–398.
18. Selam JL, Raymond M, Jacquemin JL, Orsetti A, Richard JL, Mirouze J. Pharmacokinetics of insulin infused intra-peritoneally via portable pumps. Diabet Metab June 1985; 11(3):170–173.

19. Selam JL, Slingeneyer A, Chaptal PA, Franetzki M, Prestele K, Mirouze J. Total implantation of a remotely controlled insulin minipump in a human insulin-dependent diabetic. Artif Organs August 1982;6(3):315–319.

20. Buchwald H, Rupp WM, Rohde TD, Barbosa J, Wigness BD, Dorman FD, McCarthy HB, Goldenberg FJ, Blackshear PJ, Varco RL, Steffes MW, Mauer SM. Implantable insulin pump: insulin infusion in animals and man. Trans Am Soc Artif Intern Organs 1982;28:687–690.

21. Feingold V, Jenkins AB, Kraegen EW. Effect of contact material on vibration-induced insulin aggregation. Diabetologia September 1984;27(3):373–378.

22. Grau U, Saudek CD. Stable insulin preparation for implanted insulin pumps. Laboratory and animal trials. Diabetes December 1987;36(12):1453–1459.

23. Pickup JC, Keen H, Parsons JA, Alberti KG. Continuous subcutaneous insulin infusion: an approach to achieving normoglycaemia. Br Med J January 28, 1978;1(6107): 204–207.

24. Tamborlane WV, Sherwin RS, Genel M, Felig P. Reduction to normal of plasma glucose in juvenile diabetes by subcutaneous administration of insulin with a portable infusion pump. N Engl J Med March 15, 1979;300(11):573–578.

25. Renner R, Hepp KD, Mehnert H, Franetzki M. Continuous intravenous insulin therapy with a miniaturized open-loop system. Horm Metab Res Suppl 1979; (8):186–190.

26. Irsigler K, Kritz H. Long-term continuous intravenous insulin therapy with a portable insulin dosage-regulating apparatus. Diabetes March 1979;28(3):196–203.

27. Centers for Disease Control (CDC). Deaths among patients using continuous subcutaneous insulin infusion pumps–United States. Morb Mortal Wkly Rep 1982;31:80–82, 87.

28. de la Peña A, Riddle M, Morrow LA, Jiang HH, Linnebjerg H, Scott A, Win KM, Hompesch M, Mace KF, Jacobson JG, Jackson JA. Pharmacokinetics and pharmacodynamics of high-dose human regular U-500 insulin versus human regular U-100 insulin in healthy obese subjects. Diabetes Care December 2011;34(12):2496–2501.

29. Bequette BW. Glucose clamp algorithms and insulin time-action profiles. J Diabetes Sci Technol September 1, 2009;3(5):1005–1013.

30. Marliss EB, Murray FT, Stokes EF, Zinman B, Nakhooda AF, Denoga A, Leibel BS, Albisser AM. Normalization of glycemia in diabetics during meals with insulin and glucagon delivery by the artificial pancreas. Diabetes July 1977;26(7):663–672.

31. Gough DA, Andrade JD. Enzyme electrodes. Science April 27, 1973;180 (4084): 380–384.

32. Clark LC Jr. The hydrogen peroxide sensing platinum anode as an analytical enzyme electrode. Methods Enzymol 1979;56:448–479.

33. Percival MW, Wang Y, Grosman B, Dassau E, Zisser H, Jovanovič L, Doyle FJ 3rd. Development of a multi-parametric model predictive control algorithm for insulin delivery in type 1 diabetes mellitus using clinical parameters. J Process Control March 1, 2011;21 (3): 391–404.

34. Roomer R, Bergmann JF, Boonstra A, Hansen BE, Haagmans BL, Kwadijk-de Gijsel S, van Vuuren AJ, de Knegt RJ, Janssen HL. Continuous interferon-α2b infusion in combination with ribavirin for chronic hepatitis C in treatment-experienced patients. Antivir Ther 2012;17(3):509–517.

SECTION 4

ADVANCES AND INNOVATIONS IN CELLULAR AND STEM CELL THERAPEUTIC DELIVERY

7

ENDOCRINE THERAPEUTIC DELIVERY: PANCREATIC CELL TRANSPLANT AND GROWTH

MICHELLE FUNG, DAVID THOMPSON, BREAY W. PATY,
GARTH WARNOCK, ZILIANG AO, MARK MELOCHE,
R. JEAN SHAPIRO, PAUL KEOWN, STEPHEN G.F. HO,
BRUCE VECHERE, JAMES D. JOHNSON, AND GRAYDON MENEILLY

7.1 INTRODUCTION

Islet transplantation for diabetes mellitus is an ideal platform for development and innovation in cell-based therapy in endocrinology. The loss of beta-cell function in diabetes mellitus results in hyperglycemia and diabetes-related complications such as neuropathy, retinopathy, nephropathy, and premature vascular disease. These complications occur despite intensive medical therapy with insulin injections. The Diabetes Control and Complications Trial showed that intensive glycemic control with insulin injections reduced retinopathy by 34–76%, decreased microalbuminuria by 35%, and reduced neuropathy by 65% but that it was associated with a two- to threefold increase in hypoglycemia [1]. The goal for cellular therapy in the form of human islet transplantation in patients with loss of beta-cell function is to restore beta-cell function by normalizing glycemic control, provide endogenous C-peptide effects, and hopefully reduce complications related to hyperglycemia.

Therapeutic Delivery Solutions, First Edition. Edited by Chung Chow Chan,
Kwok Chow, Bill McKay, and Michelle Fung.
© 2014 John Wiley & Sons, Inc. Published 2014 by John Wiley & Sons, Inc.

7.2 HISTORY

Human islet transplantation first began in the early 1970s with autologous islet transplantation [2–4]. Preclinical studies of islet transplantation, initially with pancreatic microfragments followed by highly purified islet clusters, demonstrated reproducible insulin independence in animal models [5]. These outcomes were noted to correlate with a critical mass of islets per unit body weight [6]. Advances in islet cell separation in large animal models guided the development of processes applicable to human pancreatic islet cell isolation. The technology of pancreatic tissue digestion, mechanical dissociation, and islet cell purification is presently used in laboratories throughout the world [7–9].

Simultaneous with improvements in cell separation were innovations in low-temperature preservation of pancreatic islets by cryopreservation [10]. Preclinical research models with highly purified islets in large animal models of diabetes demonstrated that freshly isolated islets or cryopreserved islets transplanted into the spleen could both reliably produce normoglycemia.

In 1989, the international community gathered to discuss and brainstorm solutions to key obstacles in islet transplantation. Following this meeting, the first islet transplant was performed in Canada as a simultaneous islet–kidney (SIK) transplant in a patient with type 1 diabetes mellitus (DM1) and end-stage diabetic nephropathy. The patient received 5000 islets per kilogram of body weight via portal embolization and immunosuppressive therapy with cyclosporine, glucocorticoid steroids, and azathioprine. Sustained insulin secretion was demonstrated but disappointingly did not result in insulin independence [11].

Clinical studies using cryopreserved islets pooled from multiple donors were initiated in order to achieve greater islet mass for transplantation. Insulin independence in a patient receiving a combined kidney–islet transplant using cryopreserved islets was demonstrated [12], and this represented the initial success worldwide of insulin independence following autologous islet transplantation persisting over 1 year [13].

Despite this success, reversal of insulin dependence could not be reliably achieved in islet transplant recipients. Ongoing advances in islet transplantation were necessary to address this shortcoming. Improvements in separation of islets from the human pancreas, the development and availability of more reliable collagenase enzymes, and the identification of favorable organ donor characteristics were significant contributions to better outcomes observed in subsequent clinical studies [14, 15].

Another breakthrough came from the successful application of a glucocorticoid-free immunosuppression protocol in solitary kidney transplant recipients [16]. This finding was followed by the development of the Edmonton protocol, a glucocorticoid-free islet transplantation protocol using the anti-interleukin (IL)-2 receptor blocker daclizumab for induction immunosuppression, followed by combination therapy with sirolimus and tacrolimus. This immunosuppression protocol was used in islet transplant recipients receiving a target 10,000 islets/kg body weight from multiple donors. The results of this landmark trial were published in 2000 and showed successful allogeneic islet transplantation in seven consecutive subjects with labile DM1 who could stop insulin therapy for a median of 7 months' follow-up [17].

These innovations in human allogeneic islet transplantation [17–19] provided a rationale for the development of various transplant centers worldwide that are now studying the reproducibility of the Edmonton protocol, the efficacy of new protocols, the potential risks, and the varying definitions of success of islet cell transplantation as a therapy.

7.3 INDICATIONS

Indications for islet transplantation alone (ITA) include conditions involving loss of beta-cell function without insulin resistance. *Autologous* ITA is performed in recurrent pancreatitis and pancreatic trauma, immediately following the pancreatectomy. It provides the benefit of endogenous beta-cell function post pancreatectomy using one's own islets without immunosuppressive medication and results in varying degrees of glycemic control depending on the numbers of islets transplanted [3, 4, 20–24]. *Allogeneic* ITA is performed predominantly in patients with DM1 and not in type 2 diabetes mellitus where insulin resistance is present. Other indications for islet transplantation include simultaneous islet kidney (SIK) and islet after kidney (IAK) transplantation.

7.4 HUMAN PANCREATIC ISLET PROCESSING AND TRANSPLANTATION

7.4.1 Islet Composition

There are an estimated one million islets in a healthy human pancreas, representing approximately 1–2% of total pancreatic mass. Islets are composed of at least five different cell types (alpha, beta, delta, pancreatic polypeptide, and epsilon cells) and secrete a large number of soluble factors [25]. The beta cells secrete insulin, and optimal function requires complex autocrine and paracrine interaction and innervation between the different islet cell subtypes and preserved islet architecture [25, 26]. With the advent of genomic analysis, it has become clear that islet paracrine signaling is more complex than previously thought and the elucidation of the intercellular signaling networks within islets controlling beta-cell function and survival is an active area of investigation [27].

7.4.2 Pancreas Retrieval

Islet transplantation uses islets from cadaveric donor pancreata obtained through organ donors after cerebral death. In allogeneic islet transplantation, donor-to-recipient blood type and Rh factor matching are required. Living allogeneic islet transplantation is currently not feasible due to surgical risk and risk of diabetes mellitus in the donor and inadequate islet yield from a partial pancreas using currently available isolation techniques.

In allogeneic islet transplantation, human pancreata are obtained with consent from adult heart-beating cadaver organ donors, using a protocol identical to that for whole-organ pancreas transplantation. The protocol includes en bloc dissection with the "no touch" technique and hypothermic *in situ* vascular perfusion with University of Wisconsin (UW) solution, immediate surface cooling of the pancreas to 4°C with ice slush in the lesser omental sac, and sterile transportation at 4°C to the processing lab. The target maximum cold-storage time is less than 12 h [28]. Local organ donor procurement programs [14] significantly reduce pancreatic hypothermic cold-storage time prior to islet isolation and optimize islet isolation from the exocrine and ductal cells of the pancreas.

Statistics from the Collaborative Islet Transplant Registry (CITR) indicate that less than half (37%) of the pancreas procurement procedures took place at the same place as the islet processing and transplant location. The mean time from cross-clamp to pancreas recovery was around 44 min (SD 22), and the mean cold ischemia time was around 7.3 h (range 1–27) among reporting centers. UW and two-layer methods (UW solution/perfluorochemical) were the most common (85%) methods used for pancreas preservation. Other preservation solutions used in conjunction or in absence of UW solution and/or perfluorochemical included histidine–tryptophan–ketoglutarate (HTK), Euro-Collins, Celsior, Institute Georges Lopez (IGL)-1, Solution de Conservation des Organes et des Tissus (SCOT), and extracellular-type trehalose-containing Kyoto (ET-Kyoto) solutions [29].

7.4.3   Islet Isolation

The goal of islet isolation is to collect the highest number of islets from the donor pancreas while maintaining islet morphology and function and in the smallest volume of tissue to be embolized into the recipient's liver. Islets are isolated from pancreatic tissue by enzymatic and mechanical digestion at facilities following good manufacturing practices. Following retrieval and perfusion by controlled ductal perfusion via the pancreatic duct with collagenase (Liberase human islet enzyme; Roche Applied Science, Indianapolis, IN), the pancreas is transferred to a Ricordi dissociation chamber [30]. Islets are purified by continuous density gradient purification using the Ficoll-Hypaque technique [31, 32].

Islet isolation can be further optimized using a repurification protocol to retrieve islets trapped in the exocrine tissue saved after initial purification. Impure tissue fractions are cultured *in vitro* at 22°C in 95% air and 5% carbon dioxide in 15 cm tissue culture plates containing 35 ml of Connaught Medical Research Labs-based media (catalog No. 99-785-cv; Mediatech Inc., Herndon, VA). Cultures are maintained for 12–36 h prior to initiating the repurification protocol. Purified islets are subjected to *in vitro* culture for 12–36 h before transplantation into humans. This repurification protocol results in significantly improved overall islet yields and enables 90% of retrieved pancreata to supply a minimum of 250,000 islet equivalents (IEs) per donor pancreas without adversely affecting islet viability [33].

The CITR reports most (91%) of islet isolation procedures took place at the same institution as the transplanting center. Liberase HI was the most common collagenase type used during most islet processing followed by collagenase NB1. All of the pancreata processed used a density gradient for islet purification. About half (54%) of islets were placed in culture, defined as six or more hours in a specially prepared nutrient medium. When cultured, the median culture time was 27 h (range 6–96) [29].

7.4.4 Islet Characterization

An IE is defined based on insulin content, islet morphology, and islet size. Diphenylthiocarbazone (DTZ) is commonly used to identify insulin granules in beta cells. Since beta cells are only one of several other cell types needed to constitute an islet, a morphological assessment, based upon a mean diameter of 150 µm, is used in addition to staining by DTZ, to define an IE.

Islet samples are collected in duplicate, counted, and sized for standard yields of islets equivalent to 150 µm [34]. Samples are subjected to glucose challenge *in vitro* to detect insulin secretion, and results are expressed as a ratio of stimulated to basal insulin release [34]. Islet quality control assessment includes endotoxin assay by enzyme-linked immunosorbent assay, bacterial cultures, characterization of purity by cell composition [35], detection of apoptosis by assay for caspase 3 activity in fluorometric units (Cpp32 Fluorometric Assay Kit; BioVision Incorporated, Mountain View, California), and terminal deoxynucleotidyl transferase-mediated biotin-deoxyuridine 5-triphosphate nick-end labeling (TUNEL) stain (APO-BRDU Kit; BioSource International, Camarillo, California) [36]. It is likely that future work will employ newer live cell imaging approaches that require fewer islets cells and provide more detailed mechanistic information on both apoptotic and nonapoptotic cell death. The final criteria for islet product release includes an islet mass of 5000 IEs per kilogram (body weight of recipient) or more, islet purity of 30% or more, a membrane-integrity viability of 70% or more, packed-tissue volume of less than 10 ml, negative Gram stain, and endotoxin content of five endotoxin units per kilogram (body weight of recipient) or less [28].

Pancreatic cell isolation, purification, culture, viability assessment, and quality control follow current good manufacturing practice standards and FDA/Health Canada regulations. In Canada, all source establishments and establishments that distribute, or import for further distribution, cells, tissues, and organs must register with Health Canada in accordance with the Safety of Human Cells, Tissues, and Organs for Transplantation Regulations (CTO Regulations). For cell-based therapies, the FDA recommends that lots specify quantitative measurements of purity, including viability and function. Specifications should also include measurements of impurity such as other cell types and nonviable cells. Monitoring impurities is a method of documenting consistency from one lot to the next in the manufacturing process.

7.4.5 Islet Transplantation Procedure

Isolated islets are transplanted into recipients by infusion into the portal vein, and the islets embolize in the hepatic sinusoids. Patients receive local anesthesia with 1% lidocaine, and an interventional radiologist inserts a 4 French catheter (Cook Diagnostic and Interventional Products, Bloomington, Indiana) percutaneously into the main portal vein under fluoroscopic guidance. Islet clusters suspended in bags are infused by gravity into recipients while monitoring portal pressures pre-, mid, and post infusion. Catheters are withdrawn, and the hepatic puncture tracks are sealed with a glue made from Histoacryl® (N-Butyl-2-cyanoacrylate, Aesculap, Inc., 3773 Corporate Parkway, Center Valley, PA, USA, www.aesculapusa.com) mixed with a radiographic contrast medium Lipiodol® (ethiodized oil, Guerbet LLC, Bloomington, IN, United States, www.guerbet.com). Patients are observed overnight and discharged home after a mean±SD length of stay of 1.3±0.6 days (range 1–3 days) [33]. Ultrasound Doppler of the portal vein is performed on day 1 following transplant to rule out portal vein thrombosis. The target total islet dose is greater than 10,000 IE/kg and may require anywhere from one to three infusions per patient to achieve this total dose. Patients receive immunosuppressive medications peri- and post transplant according to various research protocols.

7.5   POTENTIAL COMPLICATIONS OF ISLET TRANSPLANTATION

7.5.1   Procedure-Related and Short-Term Complications

Islet transplantation is a minimally invasive procedure; however, complications can occur. Complications include hepatic bleeding during transhepatic portal vein catheterization (12%) [37], but this has become less common with the use of fibrin sealant, absorbable gelatin sponge, or coils to seal the catheter tract on withdrawal of the catheter [38]. Hepatic bleeding into the peritoneal cavity can occur and usually resolves spontaneously but rarely can require corrective laparotomy. The infusion of foreign material into the portal system increases the risk for portal vein thrombosis at a rate of approximately 4% of islet infusions in experienced centers [37]. Low-dose heparin, given prophylactically during and after transplantation, reduces the risk of portal vein thrombosis but increases the risk of bleeding. Elevated transaminases can occur from temporarily damaged liver parenchyma surrounding the new islets. Monitoring of liver enzyme concentrations after transplantation to confirm resolution is recommended.

7.5.2   Longer-Term Complications

Longer-term complications are primarily related to the side effects of systemic immunosuppressive medications. Systemic immunosuppression increases the risk of infection and malignancy, particularly skin cancers and lymphoproliferative disorders. There are also nephrotoxic effects from calcineurin inhibitors used in immunosuppression.

Islet transplantation can result in allosensitization with anti-HLA antibody formation following exposure to donor tissue. Recipients usually receive several islet infusions matched for ABO blood group, and so this risk is increased with each transplant [39]. Although antibodies to donor-derived HLA antigens are detected in only a minority of islet transplant recipients taking immunosuppressive drugs, patients who are taken off these drugs show an increase in these antibodies [39]. This side effect is important in patients with DM1 because the presence of anti-HLA antibodies may reduce the chance of compatibility if a match is needed for solid organ transplantation in the future, specifically kidney transplantation.

7.6 CURRENT OUTCOMES

7.6.1 Measures of Islet Function

The function of transplanted islets is monitored using a variety of metabolic parameters and includes glycosylated hemoglobin (A1c), fasting glucose, stimulated glucose (by mixed meal, intravenous glucose, intravenous arginine), basal and stimulated C-peptide, daily insulin requirement, continuous glucose monitoring system, and mean amplitude of glucose excursions (MAGE) [40], but there is no consensus as to which approach best quantifies functional islet mass. A number of indices have been used to characterize islet function following transplantation. These include the HYPO score [41], lability index [41], beta score [42], basal C-peptide/glucose ratio [43], homeostasis model assessment—functional beta-cell mass (HOMA-B and HOMA-2B%) [44], and homeostasis model assessment—insulin resistance (HOMA-IR) [45]. Others include transplant-estimated function (TEF) [46] and secretory units of islets in transplantation (SUIT) [47].

7.6.2 Effect on Glycemic Control

In 2000, the Edmonton program reported on seven patients 1 year after islet transplantation. The seven recipients were insulin independent for an average of 11 months, and the results of this small study were enthusiastically received [17]. It also became clear, however, that most patients needed two to three donor islet infusions to achieve insulin independence and that insulin independence was not sustained.

A report on the 5-year outcomes in the Edmonton program for 68 patients in 2005 showed that insulin independence was present in about 69% at 1 year, 37% at 2 years, and 7.5% at 5 years. However, C-peptide was detected in 82% of subjects, indicating persistent but insufficient islet graft function at the end of this study [37]. More recently, about 64% of a cohort of 14 patients was insulin independent, and 83% had detectable C-peptide at 2 years of follow-up [48]. In another study, about half of patients remain insulin independent at 15 months [49].

In 2006, the International Trial of the Edmonton Protocol (induction with anti-IL-2R antibody, maintenance with sirolimus and tacrolimus) for islet transplantation was reported for 36 subjects with DM1 who underwent islet transplantation at nine international sites. The primary endpoint was defined as insulin independence with adequate glycemic control at 1 year after the recipients had received their final islet infusion. Data showed that 44% of the recipients reached their primary endpoint, 28% had partial islet cell function, and 28% had completely lost graft function after 1 year. Around 58% achieved insulin independence, but 76% required insulin again at 2 years, and only 31% who achieved the primary endpoint remained insulin independent at 2 years [28].

In 2007, an interim analysis of 23 completed transplants (2003–2006) in Vancouver, British Columbia, showed that 74% achieved insulin independence with good glycemic control and 40% were insulin-free at 1 year. Glycosylated hemoglobin (HbAIC) was significantly lowered after islet transplantation compared with best medical therapy during 29 months of follow-up. Renal function in both medically treated and islet transplant recipients was similar after 29 months [50].

Data from the CITR from allograft recipients recruited from 1999 to 2008 showed about 27% of recipients were insulin independent at 3 years, C-peptide was detected in about 57%, and 16% of the patient data were missing [29]. A progressive loss of insulin independence with approximately 90% of subjects requiring reintroduction of exogenous insulin has been reported in recent clinical trials based on the Edmonton protocol and some variants of the protocol [17, 28, 37, 48, 51].

Researchers have found that islet transplantation is often not able to achieve long-term insulin independence.

Patients with "partial graft function" have persistent insulin secretion from β cells but require additional oral or subcutaneous antihyperglycemic agents, such as insulin. Indicators of declining islet graft function in patients who have resumed insulin administration include worsening of glycemic control, higher exogenous insulin requirements, and a reduction in C-peptide concentrations. Observations from long-term studies triggered a debate about how to define the "success" of islet transplantation. Historically, the primary goal of islet transplantation has been the ability of donor islets to maintain normal glucose control and removal of the need for exogenous insulin. "Insulin independence" is a comprehensible clinical outcome parameter for success, but success can also be measured in terms of frequency of hypoglycemic episodes and positive effects on diabetes-related complications or quality of life [49].

More recent trials using more potent lymphodepletion (i.e., thymoglobulin, anti-CD3 or anti-CD52 antibodies) and/or biologics (anti-IL-2R, anti-TNF, anti-LFA-1 antibody or cytotoxic T-lymphocyte antigen 4 (CTLA4)-Ig) have shown approximately 50% of insulin independence at 5 years after islet transplantation [48, 52–55].

Long-term partial graft function seems to continue and be expressed clinically by more stable glucose control and lower insulin requirements. Following islet transplantation, there is a reduction in MAGE, reduced insulin requirements, and improvement in A1c [17, 56] in the absence of severe hypoglycemia [57–60]. The prevention of severe hypoglycemia persists long term [37, 41] even if exogenous insulin is required as long as C-peptide is measurable [61, 62]. Improvement in glucagon secretion in response to hypoglycemia is observed [63–68].

7.6.3   Effect on Quality of Life

Improvements in quality of life using standardized psychometric instruments have been demonstrated following islet transplantation [39, 57–60, 69].

7.6.4   Effect on Diabetes-Related Complications

Islet transplantation is reported to improve or stabilize microvascular complications (neuropathy, retinopathy, and nephropathy) and cardiovascular function in prospective observational studies [70–72], but there is no randomized controlled trial to date evaluating the effectiveness of islet transplantation compared to standard medical therapy with respect to outcomes. A retrospective cohort study found that islet transplantation may also prolong the survival of a previous kidney graft [70].

7.7 CHALLENGES AND AREAS OF ONGOING RESEARCH

There are many challenges limiting islet cell transplantation as a therapy. A recent review of islet transplantation identifies several ongoing challenges including a low success rate of islet isolations that yield sufficient islets for transplantation into patients, a low success rate for insulin independence, and toxicity from immunosuppression [73]. There is also emerging controversy regarding indications for islet transplantation.

7.7.1 Rejection and Immunosuppression Toxicity

Acute and chronic transplant rejection and the side effects of immunosuppressive drugs are important causes of islet damage and dysfunction [74, 75]. To address acute allograft rejection, induction immunosuppression is administered at the time of transplant. Currently available induction agents include antilymphocyte (polyclonal and monoclonal) or IL-2 receptor antibodies (basiliximab, daclizumab). To address chronic transplant rejection, maintenance immunosuppression is used following islet transplantation with different classes of drugs to achieve adequate immunosuppression while trying to avoid toxicity. These medications include calcineurin inhibitors (cyclosporine-A and tacrolimus), inhibitors of the molecular target of rapamycin (sirolimus), and purine synthesis inhibitors (mycophenolic acid). Some regimens avoid glucocorticoids, while others use glucocorticoids at induction. Agents that target costimulation pathways in immune cells (CTLA4-Ig) and/or adhesion molecules (LFA-1) are under study [54, 55, 76–78].

7.7.2 Immune Tolerance

Immune suppression to allow immune tolerance and to prevent islet allograft rejection and failure is balanced against the direct toxic effects of immunosuppression on the beta cell. There is research into modifying the host immune response.

Research studies have demonstrated that systemic CTLA4 immunoglobin or local expression of CTLA4 immunoglobin in islet cells can prolong islet allograft survival in rodent and nonhuman primate models [79–81]. CTLA4, also known as cluster of differentiation 152 (CD152), is a protein that plays an important regulatory role in the immune system [82]. CTLA4 is expressed on the surface of helper T cells and transmits an inhibitory signal to T cells [83, 84]. CTLA4 is similar to the T-cell costimulatory protein CD28, and both molecules bind to CD80 and CD86 (also called B7) on antigen-presenting cells. CTLA4 transmits an inhibitory signal to T cells, whereas CD28 transmits a stimulatory signal. Fusion proteins of CTLA4 and antibodies (CTLA4-Ig) have been used in immune tolerance studies. The fusion protein CTLA4-Ig is commercially available as Orencia (abatacept). A second-generation form of CTLA4-Ig known as belatacept is also being tested.

There is research currently studying the roles of activation of B7-H4 pathway to prolong islet graft survival. Researchers from several groups have established that systemic administration of an immunoglobulin fusion protein construct B7-H4 (B7-H4.Ig) inhibits autoreactive and alloreactive T-cell responses in rodents [85–88]. B7-H4 is a member of the B7-CD28 family of negative costimulatory molecules. B7-H4.Ig protein arrests cell cycle progression of activated CD4+ T cells in G0/G1 phase and induces apoptosis of both activated CD4+ and CD8+ T cells. B7-H4.Ig also inhibits the secretion of IFN-c by immune response cells in peripheral blood from diabetic patients when activated by stimulator B-cell-associated antigenic peptides and anti-CD3 antibody. Cell-associated B7-H4.Ig in transfectants of human B cells clearly inhibits the cytotoxicity of B-cell antigen-specific T-cell clones to targeted human B cells. Immobilized human B7-H4.Ig protein inhibits the proliferation of activated T cells from patients with DM1 [89]. B7-H4 is a potential immunosuppressive agent with a greater degree of selectivity to inhibit T-cell-mediated graft rejection. Targeting the B7-H4 pathway may prove to be a better mechanism than targeting CTLA4 to negatively regulate inappropriately activated T-cell responses following islet transplantation.

7.7.3   Medications to Improve Beta-Cell Function

It has been estimated that a large percentage of islets undergo apoptosis during the many steps involved in islet transplantation [90]. Reducing islet cell death during culture, transplantation, and engraftment has the potential to improve clinical outcomes. Focused research is underway to identify key antiapoptotic growth factors that could make isolated islets more resistant to the stresses associated with islet transplantation. Agents to improve islet function and reduce islet apoptosis are under study and include glucagon-like peptide-1 (GLP-1) agonists, dipeptidyl peptidase-4 (DPP-IV) inhibitors, and thiazolidinediones (TZDs). Investigations are currently underway in many academic and industrial laboratories employing high-throughput screening methodologies to identify novel beta-cell protective drugs.

Because GLP-1 has potent stimulatory effects on insulin secretion, some of its antiapoptotic effects may involve autocrine/paracrine insulin signaling [91, 92]. In one study, exendin-4 induced pancreatic and duodenal homeobox gene-1 expression in human fetal islet cell cultures and promoted functional maturation and proliferation of human islet cell cultures transplanted under the rat kidney capsule [93]. Another study found that GLP-1 decreased apoptosis in freshly isolated human islets [94]. Islets that have been engineered to make their own local GLP-1 supply have recently been demonstrated in animal studies [95], and this represents a possible area to exploit therapeutically. Promising *in vitro* and animal studies demonstrate that high concentrations of GLP-1 or GLP-1 agonist decrease islet apoptosis in the face of various insults [96], creating a rationale to incorporate these therapies into clinical islet transplantation study protocols [97, 98]. Endogenous levels of GLP-1 can also be increased by blocking the enzyme DPP-IV, which degrades GLP-1. Several DPP-IV inhibitors have shown promise [99], but it is unclear whether local levels of GLP-1 would be sufficient to prevent apoptosis.

7.7.4   Instant Blood-Mediated Inflammatory Reaction

Instant blood-mediated inflammatory reaction (IBMIR) is characterized by platelet consumption and activation of the coagulation and complement systems. The islets became surrounded by clots and infiltrated with leukocytes causing islet damage. When heparin and a complement inhibitor (SCRI) are added to the system, IBMIR is suppressed and islet damage is reduced [100]. Heparin is added to the transplant medium used during the islet infusion, and low-molecular-weight heparin injections are administered in the posttransplant period to enhance islet engraftment and reduce risk of portal vein thrombosis.

Because of the IMBIR in the portal system, alternative transplant sites are being explored to avoid triggering the IMBIR and the toxic drug levels found in the liver [101–104]. There is ongoing research in immunoisolation techniques to shield islets from immune attack in hopes of achieving sustained islet function following transplantation without requiring immunosuppression [105, 106].

7.7.5   Glucose Toxicity

Neovascularization of new islet clusters and islet engraftment may take up to several weeks following the islet infusion procedure. During this time, it is postulated that hyperglycemia may cause excessive workload and stress for newly transplanted islets and that exogenous insulin may be required and gradually withdrawn according to measured glucose values to maintain normal glucose levels post islet transplantation. There is also evidence of the potent antiapoptotic effects of low doses of insulin in cultured islets [91].

7.7.6   Shortage of Donor Pancreata: Alternative Beta-Cell Sources

There is increasing interest in the use of unlimited alternative sources of transplantable islets, such as xenogeneic or derived from human stem cells [107, 108]. There is also early experimental data suggesting that insulin-producing cells can be obtained from human multipotent stem cells, and great efforts are currently concentrated on developing cellular products with consistent potency and safety for clinical application [102, 103]. Porcine islets offer the potential ability for genetic modification to lack or overexpress specific molecules resulting in reduced immunogenicity for human transplantation; however, there is concern regarding increased zoonotic infection risk.

7.7.7   Difficulty Monitoring Graft Function and Survival

Visualization and monitoring of transplanted islets [109] using noninvasive techniques such as MRI, ultrasound, or PET have not been very successful as a result of insufficient resolution for detecting islet clusters dispersed in the liver. Liver biopsies do not provide adequate graft tissue specimens. Currently available blood tests being studied as ways of identifying rejection and reactivation of autoimmunity [110–113] include biomarkers of immune cell function [114] using flow cytometry [115, 116] and antibody titers, but they are not ideal as they lack specificity. Advances in technology to detect more specific markers of islet dysfunction and immune function may be developed in the future. One promising area involves the use of microRNAs as beta-cell death-associated biomarkers [117].

7.7.8 Indications for Islet Transplantation

There is some controversy regarding which patients with DM1 should be considered and selected for islet transplantation. The ability to reverse brittle diabetes and prevent devastating hypoglycemia has been a solid historical indication for islet transplantation. More recent clinical trials are unique in their proposal to primarily tackle the issue of complications of diabetes [33, 50]. Although promising, it is not known if pancreatic islet transplantation therapies are superior to or more cost effective than current best medical therapy for reducing complications related to diabetes; a randomized trial has not been performed to date to address this question. Future studies will provide important information about effectiveness and costs of current optimal medical care versus islet transplantation and will provide new benchmarks for comparing the efficacy and safety of emerging therapies for diabetes mellitus.

7.8 CONCLUSION

Restoring normal glucose control in patients with diabetes mellitus is the goal of all treatments for diabetes. Islet transplantation has been shown to be beneficial for a specific group of patients with DM1 who have severe glycemic lability, recurrent hypoglycemia, and hypoglycemic unawareness. Results from studies in islet transplantation demonstrate stabilization of glycemic control and endogenous C-peptide effects compared to conventional insulin therapy alone. It restores the function of insulin-secreting beta cells under the influence of neural and paracrine factors that regulate precise insulin secretion.

Islet transplantation has not become a mainstream treatment for DM1 largely because of a shortage of high-quality donor organs for islet isolation, the high costs of laboratory infrastructure required for specialized human islet isolation procedures, and the need for lifelong immunosuppressive agents in transplant recipients. The goal of long-term insulin independence is achieved by a small proportion of patients, an important message to communicate to potential recipients. The lack of randomized controlled trial data comparing islet transplantation to current best medical practice or pancreas transplantation with respect to long-term diabetes-related complications has led to uncertainty about the utility of this procedure [118].

The field overall has not progressed as rapidly as initially expected and has a number of challenges to overcome [119]. Ongoing research addressing the challenges of immune tolerance and beta-cell supply will result in additional advances in islet cell therapy. A number of immunosuppression protocols are under investigation in trials of islet transplantation.

7.9 CURRENT STATUS OF RESEARCH

The current status of islet transplantation research is summarized in Table 7.1

TABLE 7.1 Human islet transplantation studies currently recruiting

Institution	Intervention	Indication	N	Start date	NCT no.
Tokushima University	Islet autotransplantation Single arm, nonrandomized	Benign pancreatic tumor, chronic pancreatitis, traumatic injury of the pancreas	10	2012	JPRN-UMIN000007119
Vancouver Coastal Health	Randomized, strict glucose control at time of ITA	DM1	32	2011	NCT01123122
National Institute of Allergy and Infectious Diseases (NIAID)	Tacrolimus, sirolimus cyclosporine, mycophenolate mofetil, mycophenolic acid Nonrandomized, single group, open label Phase III	DM1	40	2011	NCT01369082
Virginia Commonwealth University	IAK, randomized	DM1 after kidney transplant	10	2011	NCT00784966
University Hospital, Lille	Observational IL-7, fat mass, metabolic profile post islet transplant	DM1, liver, kidney	21	2010	NCT01414660
Dompé s.p.a.	Multicenter, randomized, open label, parallel assignment, pilot study, Phase II Reparixin	DM1	10	2010	NCT01220856
Department of Digestive and Transplantation Surgery, Tokushima University Hospital	Single arm, nonrandomized	DM1	10	2010	JPRN-UMIN000004105

(*Continued*)

193

TABLE 7.1 (Cont'd)

Institution	Intervention	Indication	N	Start date	NCT no.
University Hospital, Grenoble	Trial comparing metabolic efficiency of islet graft to intensive insulin therapy for DM1 treatment TRIMECO Immediate versus 6 months, phase III	DM1, DM1 after kidney transplant	40	2010	NCT01148680
The Japanese Pancreas and Islet Transplantation Association	Single arm, nonrandomized	DM1	20	2010	JPRN-UMIN000003977
University of Chicago	Single group, open label, phase II	DM1, IAK	20	2010	NCT01241864
Fondazione San Raffaele del Monte Tabor	Nonrandomized, single group, open label Alternate site—bone marrow	DM1, post pancreatectomy	8	2009	NCT01345227
University of British Columbia	Sitagliptin, randomized	DM1	12	2009	NCT00853944
UW, Madison Bristol-Myers Squibb	Basiliximab induction with maintenance immunosuppression consisting of belatacept, sirolimus, and mycophenolate, GLP-1 agonist, nonrandomized. SIK transplant	Uremic DM1	8	2009	NCT01033500
Massachusetts General Hospital	IAK Phase I Etanercept Phase II	DM1 after kidney transplant	8	2009	NCT00888628
University of British Columbia	A comparison of islet cell transplantation with medical therapy on the risk of progression of diabetic retinopathy and diabetic macular edema	DM1	40	2009	NCT00853424
University of Alberta	Sitagliptin and pantoprazole, nonrandomized	DM1 post ITA with graft failure	8	2008	NCT00768651
NIAID	Open randomized multicenter study to evaluate safety and efficacy of low-molecular-weight sulfated dextran (LMW-SD) in Islet transplantation	DM1	36	2008	NCT00789308

	Drugs: heparin, mycophenolate mofetil, sirolimus, tacrolimus, cyclosporine, daclizumab, basiliximab				
University of Nebraska NIAID	Steroid-free immunosuppression	DM1	10	2008	NCT00579371
	Belatacept, basiliximab, mycophenolate mofetil	DM1	20	2008	NCT00468403
University of Illinois	University of Chicago protocol Basiliximab 20 mg iv 2 h before transplant and 20 mg iv 2 weeks post transplant; tacrolimus 1 mg p.o. bid adjusted to reach target trough levels of 3–6 ng/ml; sirolimus 0.2 mg/kg loading dose and then 0.1 mg/kg p.o. daily adjusted to reach target trough levels of 10–15 ng/ml during the first 3 months post transplant and 7–10 ng/ml thereafter; etanercept 50 mg iv 1 h before transplant and 25 mg s.c. on days 3, 7, and 10 post transplant; exenatide 5 μg s.c. bid for 1 week and then 10 μg bid for 6 months after each transplant	DM1	50	2007	NCT00679042
University of Minnesota	Raptiva, sirolimus, antithymocyte globulin, nonrandomized	DM1	10	2007	NCT00672204
University of Virginia		DM1 DM1 after kidney transplant	20	2007	NCT00605592
Baylor Research Institute NIAID	Remote site islet processing with culture	DM1 DM1 after kidney transplant	15 48	2007 2007	NCT00530686 NCT00468117
Corline Systems AB	Islets coated with immobilized heparin	DM1	10	2007	NCT00678990

(*Continued*)

TABLE 7.1 (Cont'd)

Institution	Intervention	Indication	N	Start date	NCT no.
University of Miami	Daclizumab, sirolimus, and tacrolimus	DM1	40	2006	NCT00306098
	Infliximab, etanercept				
Fondazione San Raffaele del Monte Tabor	Rapamycin pretreatment (0.1 mg/kg/day) for at least 30 days; induction therapy with ATG (1.5 mg/kg/day for 4 days starting at day 1) and a steroid bolus (methylprednisolone 500 mg, day 1) plus low-dose steroids (prednisone, 10 mg/day) and IL-1 receptor antagonist (100 mg/day) for 2 weeks with ATG and steroid bolus administered only prior to the 1st islet infusion; (iii) maintenance with rapamycin (0.1 mg/kg/day) plus mycophenolate mofetil (2 g/day) Single group, open label Phases I, II	DM1	10	2006	NCT01346085
NIAID	Deoxyspergualin, antithymocyte globulin, daclizumab or basiliximab, sirolimus, tacrolimus, etanercept, nonrandomized	DM1	14	2006	NCT00434850
Emory University	Abatacept, calcineurin-sparing regimen	DM1	20	2005	NCT00276250
City of Hope Medical Center	Sirolimus/tacrolimus/MMF IAK	DM1 after kidney transplant	14	2005	NCT00708604
University of Alberta	Alemtuzumab induction with tacrolimus and MMF maintenance immunosuppression	DM1	12	2005	NCT00175253
University of Illinois	Edmonton protocol	DM1	10	2004	NCT00566813
University of Chicago	Edmonton protocol	DM1	50	2003	NCT00160732
University Hospital, Lille	Phase II nonrandomized IAK	DM1 after kidney transplant	19	2003	NCT01123187

Institution	Protocol/Treatment	Condition	N	Year	NCT Number
University Hospital, Grenoble		Brittle DM1	22	2003	NCT00321256
University of Chicago	Edmonton protocol	DM1	50	2003	NCT00160732
UW, Madison	TZD, Edmonton protocol	DM1	16	2002	NCT00214253
UW, Madison	Edmonton protocol Pioglitazone	DM1	16	2002	NCT00214253
Emory University	Edmonton protocol	DM1	20	2002	NCT00133809
City of Hope Medical Center	Sirolimus/tacrolimus, daclizumab	DM1	20	2002	NCT00706420
National Center for Research Resources (NCRR)	Edmonton protocol	DM1, DM1 after kidney transplant	n/a	2001	NCT00021580
University of Miami	Islet cell transplantation alone and CD34+ enriched donor bone marrow cell infusion in patients with DM1; steroid-free regimen	DM1	6	2000	NCT00315614
National Institutes of Health Clinical Center (CC)	Daclizumab every 2 weeks and FK506 and rapamycin	DM1	6	2000	NCT00006505
NCRR	hOKT3gamma1 (Ala-ala)	DM1	n/a	1998	NCT00008801

REFERENCES

1. The effect of intensive treatment of diabetes on the development and progression of long-term complications in insulin-dependent diabetes mellitus. The Diabetes Control and Complications Trial Research Group. N Engl J Med 1993;329:977–986.
2. Najarian JS, Sutherland DE, Matas AJ, Goetz FC. Human islet autotransplantation following pancreatectomy. Transplant Proc 1979;11(1):336–340.
3. Robertson RP, Lanz KJ, Sutherland DE, Kendall DM. Prevention of diabetes for up to 13 years by autoislet transplantation after pancreatectomy for chronic pancreatitis. Diabetes 2001;50(1):47–50.
4. Teuscher AU, Kendall DM, Smets YF, Leone JP, Sutherland DE, Robertson RP. Successful islet autotransplantation in humans: functional insulin secretory reserve as an estimate of surviving islet cell mass. Diabetes 1998;47(3):324–330.
5. Warnock GL, Cattral MS, Rajotte RV. Normoglycemia after implantation of purified islet cells in dogs. Can J Surg 1988;31(6):421–426.
6. Warnock GL, Dabbs KD, Evans MG, Cattral MS, Kneteman NM, Rajotte RV. Critical mass of islets that function after implantation in a large mammalian. Horm Metab Res Suppl 1990;25:156–161.
7. Warnock GL, Ellis D, Rajotte RV, Dawidson I, Baekkeskov S, Egebjerg J. Studies of the isolation and viability of human islets of Langerhans. Transplantation 1988;45(5): 957–963.
8. Ricordi C, Lacy PE, Scharp DW. Automated islet isolation from human pancreas. Diabetes 1989;38(Suppl 1):140–142.
9. Lake SP, Bassett PD, Larkins A, Revell J, Walczak K, Chamberlain J, Rumford GM, London NJ, Veitch PS, Bell PR. Large-scale purification of human islets utilizing discontinuous albumin gradient on IBM 2991 cell separator. Diabetes 1989;38(Suppl 1): 143–145.
10. Rajotte RV. Islet cryopreservation protocols. Ann N Y Acad Sci 1999;875:200–207.
11. Warnock GL, Kneteman NM, Ryan EA, Evans MG, Seelis RE, Halloran PF, Rabinovitch A, Rajotte RV. Continued function of pancreatic islets after transplantation in type I diabetes. Lancet 1989;2(8662):570–572.
12. Warnock GL, Kneteman NM, Ryan E, Seelis RE, Rabinovitch A, Rajotte RV. Normoglycaemia after transplantation of freshly isolated and cryopreserved pancreatic islets in type 1 (insulin-dependent) diabetes mellitus. Diabetologia 1991;34(1):55–58.
13. Warnock GL, Kneteman NM, Ryan EA, Rabinovitch A, Rajotte RV. Long-term follow-up after transplantation of insulin-producing pancreatic islets into patients with type 1 (insulin-dependent) diabetes mellitus. Diabetologia 1992;35(1):89–95.
14. Lakey JR, Warnock GL, Rajotte RV, Suarez-Alamazor ME, Ao Z, Shapiro AM, Kneteman NM. Variables in organ donors that affect the recovery of human islets of Langerhans. Transplantation 1996;61(7):1047–1053.
15. Lakey JR, Warnock GL, Shapiro AM, Korbutt GS, Ao Z, Kneteman NM, Rajotte RV. Intraductal collagenase delivery into the human pancreas using syringe loading or controlled perfusion. Cell Transplant 1999;8(3):285–292.
16. Cole E, Landsberg D, Russell D, Zaltzman J, Kiberd B, Caravaggio C, Vasquez AR, Halloran P. A pilot study of steroid-free immunosuppression in the prevention of acute rejection in renal allograft recipients. Transplantation 2001;72(5):845–850.

17. Shapiro AM, Lakey JR, Ryan EA, Korbutt GS, Toth E, Warnock GL, Kneteman NM, Rajotte RV. Islet transplantation in seven patients with type 1 diabetes mellitus using a glucocorticoid-free immunosuppressive regimen. N Engl J Med 2000;343(4):230–238.

18. Frank A, Deng S, Huang X, Velidedeoglu E, Bae YS, Liu C, Abt P, Stephenson R, Mohiuddin M, Thambipillai T, Markmann E, Palanjian M, Sellers M, Naji A, Barker CF, Markmann JF. Transplantation for type I diabetes: comparison of vascularized whole-organ pancreas with isolated pancreatic islets. Ann Surg 2004;240(4):631–640.

19. Hering BJ, Kandaswamy R, Harmon JV, Ansite JD, Clemmings SM, Sakai T, Paraskevas S, Eckman PM, Sageshima J, Nakano M, Sawada T, Matsumoto I, Zhang HJ, Sutherland DE, Bluestone JA. Transplantation of cultured islets from two-layer preserved pancreases in type 1 diabetes with anti-CD3 antibody. Am J Transplant 2004;4(3):390–401.

20. Blondet JJ, Carlson AM, Kobayashi T, Jie T, Bellin M, Hering BJ, Freeman ML, Beilman GJ, Sutherland DE. The role of total pancreatectomy and islet autotransplantation for chronic pancreatitis. Surg Clin North Am 2007;87(6):1477–501, x.

21. Bellin MD, Sutherland DE. Pediatric islet autotransplantation: indication, technique, and outcome. Curr Diab Rep 2010;10(5):326–331.

22. Bellin MD, Carlson AM, Kobayashi T, Gruessner AC, Hering BJ, Moran A, Sutherland DE. Outcome after pancreatectomy and islet autotransplantation in a pediatric population. J Pediatr Gastroenterol Nutr 2008;47(1):37–44.

23. Bellin MD, Sutherland DE, Beilman GJ, Hong-McAtee I, Balamurugan AN, Hering BJ, Moran A. Similar islet function in islet allotransplant and autotransplant recipients, despite lower islet mass in autotransplants. Transplantation 2011;91(3):367–372.

24. Robertson RP. Consequences on beta-cell function and reserve after long-term pancreas transplantation. Diabetes 2004;53(3):633–644.

25. Yang YH, Szabat M, Bragagnini C, Kott K, Helgason CD, Hoffman BG, Johnson JD. Paracrine signalling loops in adult human and mouse pancreatic islets: netrins modulate beta cell apoptosis signalling via dependence receptors. Diabetologia 2011;54(4):828–842.

26. Cabrera O, Berman DM, Kenyon NS, Ricordi C, Berggren PO, Caicedo A. The unique cytoarchitecture of human pancreatic islets has implications for islet cell function. Proc Natl Acad Sci USA 2006;103(7):2334–2339.

27. Szabat M, Lynn FC, Hoffman BG, Kieffer TJ, Allan DW, Johnson JD. Maintenance of beta-cell maturity and plasticity in the adult pancreas: developmental biology concepts in adult physiology. Diabetes 2012;61(6):1365–1371.

28. Shapiro AM, Ricordi C, Hering BJ, Auchincloss H, Lindblad R, Robertson RP, Secchi A, Brendel MD, Berney T, Brennan DC, Cagliero E, Alejandro R, Ryan EA, DiMercurio B, Morel P, Polonsky KS, Reems JA, Bretzel RG, Bertuzzi F, Froud T, Kandaswamy R, Sutherland DE, Eisenbarth G, Segal M, Preiksaitis J, Korbutt GS, Barton FB, Viviano L, Seyfert-Margolis V, Bluestone J, Lakey JR. International trial of the Edmonton protocol for islet transplantation. N Engl J Med 2006;355(13):1318–1330.

29. Collaborative Islet Transplant Registry, Sixth Annual Report. Prepared by CITR Coordinating Center, The EMMES Corporation, Rockville, MD, November 1, 2009. Available at https://web.emmes.com/study/isl/reports/CITR%206th%20Annual%20Data%20Report%20120109.pdf. Accessed February 14, 2014.

30. Ricordi C, Lacy PE, Finke EH, Olack BJ, Scharp DW. Automated method for isolation of human pancreatic islets. Diabetes 1988;37(4):413–420.

31. Alejandro R, Strasser S, Zucker PF, Mintz DH. Isolation of pancreatic islets from dogs. Semiautomated purification on albumin gradients. Transplantation 1990;50(2):207–210.

32. Ichii H, Pileggi A, Molano RD, Baidal DA, Khan A, Kuroda Y, Inverardi L, Goss JA, Alejandro R, Ricordi C. Rescue purification maximizes the use of human islet preparations for transplantation. Am J Transplant 2005;5(1):21–30.

33. Warnock GL, Meloche RM, Thompson D, Shapiro RJ, Fung M, Ao Z, Ho S, He Z, Dai LJ, Young L, Blackburn L, Kozak S, Kim PT, Al-Adra D, Johnson JD, Liao YH, Elliott T, Verchere CB. Improved human pancreatic islet isolation for a prospective cohort study of islet transplantation vs best medical therapy in type 1 diabetes mellitus. Arch Surg 2005;140(8):735–744.

34. Ricordi C, Gray DW, Hering BJ, Ricordi C, Gray DW, Hering BJ, Kaufman DB, Warnock GL, Kneteman NM, Lake SP, London NJ, Socci C, Alejandro R. Islet isolation assessment in man and large animals. Acta Diabetol Lat 1990;27(3):185–195.

35. Korbutt GS, Elliott JF, Ao Z, Smith DK, Warnock GL, Rajotte RV. Large scale isolation, growth, and function of porcine neonatal islet cells. J Clin Invest 1996;97(9): 2119–2129.

36. Yang YH, Johnson JD. Multi-parameter, single-cell, kinetic analysis reveals multiple modes of cell death in primary pancreatic beta-cells. J Cell Sci 2013;126(Pt 18): 4286–4295.

37. Ryan EA, Paty BW, Senior PA, Bigam D, Alfadhli E, Kneteman NM, Lakey JR, Shapiro AM. Five-year follow-up after clinical islet transplantation. Diabetes 2005;54(7): 2060–2069.

38. Daly B, O'Kelly K, Klassen D. Interventional procedures in whole organ and islet cell pancreas transplantation. Semin Intervent Radiol 2004;21(4):335–343.

39. Campbell PM, Senior PA, Salam A, Labranche K, Bigam DL, Kneteman NM, Imes S, Halpin A, Ryan EA, Shapiro AM. High risk of sensitization after failed islet transplantation. Am J Transplant 2007;7(10):2311–2317.

40. Service FJ, Molnar GD, Rosevear JW, Ackerman E, Gatewood LC, Taylor WF. Mean amplitude of glycemic excursions, a measure of diabetic instability. Diabetes 1970; 19(9):644–655.

41. Ryan EA, Shandro T, Green K, Paty BW, Senior PA, Bigam D, Shapiro AM, Vantyghem MC. Assessment of the severity of hypoglycemia and glycemic lability in type 1 diabetic subjects undergoing islet transplantation. Diabetes 2004;53(4):955–962.

42. Ryan EA, Paty BW, Senior PA, Lakey JR, Bigam D, Shapiro AM. Beta-score: an assessment of beta-cell function after islet transplantation. Diabetes Care 2005;28(2):343–347.

43. Faradji RN, Monroy K, Messinger S, Pileggi A, Froud T, Baidal DA, Cure PE, Ricordi C, Luzi L, Alejandro R. Simple measures to monitor beta-cell mass and assess islet graft dysfunction. Am J Transplant 2007;7(2):303–308.

44. Wallace TM, Levy JC, Matthews DR. Use and abuse of HOMA modeling. Diabetes Care 2004;27(6):1487–1495.

45. Matthews DR, Hosker JP, Rudenski AS, Naylor BA, Treacher DF, Turner RC. Homeostasis model assessment: insulin resistance and beta-cell function from fasting plasma glucose and insulin concentrations in man. Diabetologia 1985;28(7):412–419.

46. Caumo A, Maffi P, Nano R, Bertuzzi F, Luzi L, Secchi A, Bonifacio E, Piemonti L. Transplant estimated function: a simple index to evaluate beta-cell secretion after islet transplantation. Diabetes Care 2008;31(2):301–305.

47. Noguchi H, Yamada Y, Okitsu T, Iwanaga Y, Nagata H, Kobayashi N, Hayashi S, Matsumoto S. Secretory unit of islet in transplantation (SUIT) and engrafted islet rate

(EIR) indexes are useful for evaluating single islet transplantation. Cell Transplant 2008;17(1–2):121–128.

48. Vantyghem MC, Kerr-Conte J, Arnalsteen L, Sergent G, Defrance F, Gmyr V, Declerck N, Raverdy V, Vandewalle B, Pigny P, Noel C, Pattou F. Primary graft function, metabolic control, and graft survival after islet transplantation. Diabetes Care 2009;32(8): 1473–1478.

49. Robertson RP. Islet transplantation a decade later and strategies for filling a half-full glass. Diabetes 2010;59(6):1285–1291.

50. Fung MA, Warnock GL, Ao Z, Keown P, Meloche M, Shapiro RJ, Ho S, Worsley D, Meneilly GS, Al Ghofaili K, Kozak SE, Tong SO, Trinh M, Blackburn L, Kozak RM, Fensom BA, Thompson DM. The effect of medical therapy and islet cell transplantation on diabetic nephropathy: an interim report. Transplantation 2007;84(1):17–22.

51. Hering BJ, Kandaswamy R, Ansite JD, Eckman PM, Nakano M, Sawada T, Matsumoto I, Ihm SH, Zhang HJ, Parkey J, Hunter DW, Sutherland DE. Single-donor, marginal-dose islet transplantation in patients with type 1 diabetes. JAMA 2005;293(7):830–835.

52. Froud T, Baidal DA, Faradji R, Cure P, Mineo D, Selvaggi G, Kenyon NS, Ricordi C, Alejandro R. Islet transplantation with alemtuzumab induction and calcineurin-free maintenance immunosuppression results in improved short- and long-term outcomes. Transplantation 2008;86(12):1695–1701.

53. Tan J, Yang S, Cai J, Guo J, Huang L, Wu Z, Chen J, Liao L. Simultaneous islet and kidney transplantation in seven patients with type 1 diabetes and end-stage renal disease using a glucocorticoid-free immunosuppressive regimen with alemtuzumab induction. Diabetes 2008;57(10):2666–2671.

54. Posselt AM, Bellin MD, Tavakol M, Szot GL, Frassetto LA, Masharani U, Kerlan RK, Fong L, Vincenti FG, Hering BJ, Bluestone JA, Stock PG. Islet transplantation in type 1 diabetics using an immunosuppressive protocol based on the anti-LFA-1 antibody efalizumab. Am J Transplant 2010;10(8):1870–1880.

55. Posselt AM, Szot GL, Frassetto LA, Frassetto LA, Masharani U, Tavakol M, Amin R, McElroy J, Ramos MD, Kerlan RK, Fong L, Vincenti F, Bluestone JA, Stock PG. Islet transplantation in type 1 diabetic patients using calcineurin inhibitor-free immunosuppressive protocols based on T-cell adhesion or costimulation blockade. Transplantation 2010;90(12):1595–1601.

56. Froud T, Ricordi C, Baidal DA, Hafiz MM, Ponte G, Cure P, Pileggi A, Poggioli R, Ichii H, Khan A, Ferreira JV, Pugliese A, Esquenazi VV, Kenyon NS, Alejandro R. Islet transplantation in type 1 diabetes mellitus using cultured islets and steroid-free immunosuppression: Miami experience. Am J Transplant 2005;5(8):2037–2046.

57. Leitao CB, Tharavanij T, Cure P, Pileggi A, Baidal DA, Ricordi C, Alejandro R. Restoration of hypoglycemia awareness after islet transplantation. Diabetes Care 2008;31(11): 2113–2115.

58. Johnson JA, Kotovych M, Ryan EA, Shapiro AM. Reduced fear of hypoglycemia in successful islet transplantation. Diabetes Care 2004;27(2):624–625.

59. Tharavanij T, Betancourt A, Messinger S, Cure P, Leitao CB, Baidal DA, Froud T, Ricordi C, Alejandro R. Improved long-term health-related quality of life after islet transplantation. Transplantation 2008;86(9):1161–1167.

60. Poggioli R, Faradji RN, Ponte G, Betancourt A, Messinger S, Baidal DA, Froud T, Ricordi C, Alejandro R. Quality of life after islet transplantation. Am J Transplant 2006; 6(2):371–378.

61. Alejandro R, Barton FB, Hering BJ, Wease S. 2008 Update from the Collaborative Islet Transplant Registry. Transplantation 2008;86(12):1783–1788.

62. Pileggi A, Ricordi C, Kenyon NS, Froud T, Baidal DA, Kahn A, Selvaggi G, Alejandro R. Twenty years of clinical islet transplantation at the Diabetes Research Institute—University of Miami. Clin Transplant 2004:177–204.

63. Paty BW, Ryan EA, Shapiro AM, Lakey JR, Robertson RP. Intrahepatic islet transplantation in type 1 diabetic patients does not restore hypoglycemic hormonal counterregulation or symptom recognition after insulin independence. Diabetes 2002;51(12):3428–3434.

64. Rickels MR, Schutta MH, Markmann JF, Barker CF, Naji A, Teff KL. {beta}-Cell function following human islet transplantation for type 1 diabetes. Diabetes 2005;54(1): 100–106.

65. Rickels MR, Schutta MH, Mueller R, Markmann JF, Barker CF, Naji A, Teff KL. Islet cell hormonal responses to hypoglycemia after human islet transplantation for type 1 diabetes. Diabetes 2005;54(11):3205–3211.

66. Campbell PM, Salam A, Ryan EA, Senior P, Paty BW, Bigam D, McCready T, Halpin A, Imes S, Al Saif F, Lakey JR, Shapiro AM. Pretransplant HLA antibodies are associated with reduced graft survival after clinical islet transplantation. Am J Transplant 2007; 7(5):1242–1248.

67. Paty BW, Senior PA, Lakey JR, Shapiro AM, Ryan EA. Assessment of glycemic control after islet transplantation using the continuous glucose monitor in insulin-independent versus insulin-requiring type 1 diabetes subjects. Diabetes Technol Ther 2006; 8(2):165–173.

68. Rickels MR, Naji A, Teff KL. Acute insulin responses to glucose and arginine as predictors of beta-cell secretory capacity in human islet transplantation. Transplantation 2007; 84(10):1357–1360.

69. Toso C, Shapiro AM, Bowker S, Dinyari P, Paty B, Ryan EA, Senior P, Johnson JA. Quality of life after islet transplant: impact of the number of islet infusions and metabolic outcome. Transplantation 2007;84(5):664–666.

70. Fiorina P, Venturini M, Folli F, Losio C, Maffi P, Placidi C, La Rosa S, Orsenigo E, Socci C, Capella C, Del Maschio A, Secchi A. Natural history of kidney graft survival, hypertrophy, and vascular function in end-stage renal disease type 1 diabetic kidney-transplanted patients: beneficial impact of pancreas and successful islet cotransplantation. Diabetes Care 2005;28(6):1303–1310.

71. Fiorina P, Gremizzi C, Maffi P, Caldara R, Tavano D, Monti L, Socci C, Folli F, Fazio F, Astorri E, Del Maschio A, Secchi A. Islet transplantation is associated with an improvement of cardiovascular function in type 1 diabetic kidney transplant patients. Diabetes Care 2005;28(6):1358–1365.

72. Warnock GL, Thompson DM, Meloche RM, Meloche RM, Shapiro RJ, Ao Z, Keown P, Johnson JD, Verchere CB, Partovi N, Begg IS, Fung M, Kozak SE, Tong SO, Alghofaili KM, Harris C. A multi-year analysis of islet transplantation compared with intensive medical therapy on progression of complications in type 1 diabetes. Transplantation 2008; 86(12):1762–1766.

73. Stock PG, Bluestone JA. Beta-cell replacement for type I diabetes. Annu Rev Med 2004;55:133–156.

74. Chatenoud L. Chemical immunosuppression in islet transplantation—friend or foe? N Engl J Med 2008;358(11):1192–1193.

75. Harlan DM, Kenyon NS, Korsgren O, Roep BO. Current advances and travails in islet transplantation. Diabetes 2009;58(10):2175–2184.

76. Turgeon NA, Avila JG, Cano JA, Hutchinson JJ, Badell IR, Page AJ, Adams AB, Sears MH, Bowen PH, Kirk AD, Pearson TC, Larsen CP. Experience with a novel efalizumab-based immunosuppressive regimen to facilitate single donor islet cell transplantation. Am J Transplant 2010;10(9):2082–2091.

77. Fotino C, Pileggi A. Blockade of leukocyte function antigen-1 (LFA-1) in clinical islet transplantation. Curr Diab Rep 2011;11(5):337–344.

78. Badell IR, Russell MC, Thompson PW, Turner AP, Weaver TA, Robertson JM, Avila JG, Cano JA, Johnson BE, Song M, Leopardi FV, Swygert S, Strobert EA, Ford ML, Kirk AD, Larsen CP. LFA-1-specific therapy prolongs allograft survival in rhesus macaques. J Clin Invest 2010;120(12):4520–4531.

79. Gainer AL, Suarez-Pinzon WL, Min WP, Swiston JR, Hancock-Friesen C, Korbutt GS, Rajotte RV, Warnock GL, Elliott JF. Improved survival of biolistically transfected mouse islet allografts expressing CTLA4-Ig or soluble Fas ligand. Transplantation 1998; 66(2):194–199.

80. Gainer AL, Korbutt GS, Rajotte RV, Warnock GL, Elliott JF. Expression of CTLA4-Ig by biolistically transfected mouse islets promotes islet allograft survival. Transplantation 1997;63(7):1017–1021.

81. Levisetti MG, Padrid PA, Szot GL, Mittal N, Meehan SM, Wardrip CL, Gray GS, Bruce DS, Thistlethwaite JR Jr, Bluestone JA. Immunosuppressive effects of human CTLA4Ig in a non-human primate model of allogeneic pancreatic islet transplantation. J Immunol 1997;159(11):5187–5191.

82. Wang S, Chen L. Co-signaling molecules of the B7-CD28 family in positive and negative regulation of T lymphocyte responses. Microbes Infect 2004;6(8):759–766.

83. Rothstein DM, Sayegh MH. T-cell costimulatory pathways in allograft rejection and tolerance. Immunol Rev 2003;196:85–108.

84. Linsley PS, Brady W, Urnes M, Grosmaire LS, Damle NK, Ledbetter JA. CTLA-4 is a second receptor for the B cell activation antigen B7. J Exp Med 1991;174(3):561–569.

85. Choi IH, Zhu G, Sica GL, Strome SE, Cheville JC, Lau JS, Zhu Y, Flies DB, Tamada K, Chen L. Genomic organization and expression analysis of B7-H4, an immune inhibitory molecule of the B7 family. J Immunol 2003;171(9):4650–4654.

86. Prasad DV, Richards S, Mai XM, Dong C. B7S1, a novel B7 family member that negatively regulates T cell activation. Immunity 2003;18(6):863–873.

87. Sica GL, Choi IH, Zhu G, Tamada K, Wang SD, Tamura H, Chapoval AI, Flies DB, Bajorath J, Chen L. B7-H4, a molecule of the B7 family, negatively regulates T cell immunity. Immunity 2003;18(6):849–861.

88. Zang X, Loke P, Kim J, Murphy K, Waitz R, Allison JP. B7x: a widely expressed B7 family member that inhibits T cell activation. Proc Natl Acad Sci USA 2003;100(18): 10388–10392.

89. Ou D, Wang X, Metzger DL, Ao Z, Pozzilli P, James RF, Chen L, Warnock GL. Suppression of human T-cell responses to beta-cells by activation of B7-H4 pathway. Cell Transplant 2006;15(5):399–410.

90. Davalli AM, Scaglia L, Zangen DH, Hollister J, Bonner-Weir S, Weir GC. Vulnerability of islets in the immediate posttransplantation period. Dynamic changes in structure and function. Diabetes 1996;45(9):1161–1167.

91. Johnson JD, Bernal-Mizrachi E, Alejandro EU, Han Z, Kalynyak TB, Li H, Beith JL, Gross J, Warnock GL, Townsend RR, Permutt MA, Polonsky KS. Insulin protects islets from apoptosis via Pdx1 and specific changes in the human islet proteome. Proc Natl Acad Sci USA 2006;103(51):19575–19580.

92. Mehran AE, Templeman NM, Brigidi GS, Lim GE, Chu KY, Hu X, Botezelli JD, Asadi A, Hoffman BG, Kieffer TJ, Bamji SX, Clee SM, Johnson JD. Hyperinsulinemia drives diet-induced obesity independently of brain insulin production. Cell Metab 2012; 16(6):723–737.

93. Movassat J, Beattie GM, Lopez AD, Hayek A. Exendin 4 up-regulates expression of PDX 1 and hastens differentiation and maturation of human fetal pancreatic cells. J Clin Endocrinol Metab 2002;87(10):4775–4781.

94. Farilla L, Bulotta A, Hirshberg B, Li Calzi S, Khoury N, Noushmehr H, Bertolotto C, Di Mario U, Harlan DM, Perfetti R. Glucagon-like peptide 1 inhibits cell apoptosis and improves glucose responsiveness of freshly isolated human islets. Endocrinology 2003;144(12):5149–5158.

95. Wideman RD, Yu IL, Webber TD, Verchere CB, Johnson JD, Cheung AT, Kieffer TJ. Improving function and survival of pancreatic islets by endogenous production of glucagon-like peptide 1 (GLP-1). Proc Natl Acad Sci USA 2006;103(36):13468–13473.

96. Yusta B, Baggio LL, Estall JL, Koehler JA, Holland DP, Li H, Pipeleers D, Ling Z, Drucker DJ. GLP-1 receptor activation improves beta cell function and survival following induction of endoplasmic reticulum stress. Cell Metab 2006;4(5):391–406.

97. Fung M, Thompson D, Shapiro RJ, Warnock GL, Andersen DK, Elahi D, Meneilly GS. Effect of glucagon-like peptide-1 (7–37) on beta-cell function after islet transplantation in type 1 diabetes. Diabetes Res Clin Pract 2006;74(2):189–193.

98. Ghofaili KA, Fung M, Ao Z, Meloche M, Shapiro RJ, Warnock GL, Elahi D, Meneilly GS, Thompson DM. Effect of exenatide on beta cell function after islet transplantation in type 1 diabetes. Transplantation 2007;83(1):24–28.

99. Pospisilik JA, Stafford SG, Demuth HU, Brownsey R, Parkhouse W, Finegood DT, McIntosh CH, Pederson RA. Long-term treatment with the dipeptidyl peptidase IV inhibitor P32/98 causes sustained improvements in glucose tolerance, insulin sensitivity, hyperinsulinemia, and beta-cell glucose responsiveness in VDF (fa/fa) Zucker rats. Diabetes 2002;51(4):943–950.

100. Bennet W, Groth CG, Larsson R, Nilsson B, Korsgren O. Isolated human islets trigger an instant blood mediated inflammatory reaction: implications for intraportal islet transplantation as a treatment for patients with type 1 diabetes. Ups J Med Sci 2000; 105(2):125–133.

101. Merani S, Toso C, Emamaullee J, Shapiro AM. Optimal implantation site for pancreatic islet transplantation. Br J Surg 2008;95(12):1449–1461.

102. Cantarelli E, Piemonti L. Alternative transplantation sites for pancreatic islet grafts. Curr Diab Rep 2011;11(5):364–374.

103. Cantarelli E, Melzi R, Mercalli A, Sordi V, Ferrari G, Lederer CW, Mrak E, Rubinacci A, Ponzoni M, Sitia G, Guidotti LG, Bonifacio E, Piemonti L. Bone marrow as an alternative site for islet transplantation. Blood 2009;114(20):4566–4574.

104. Rafael E, Tibell A, Ryden M, Lundgren T, Sävendahl L, Borgström B, Arnelo U, Isaksson B, Nilsson B, Korsgren O, Permert J. Intramuscular autotransplantation of pancreatic islets in a 7-year-old child: a 2-year follow-up. Am J Transplant 2008;8(2):458–462.

105. Basta G, Calafiore R. Immunoisolation of pancreatic islet grafts with no recipient's immunosuppression: actual and future perspectives. Curr Diab Rep 2011;11(5): 384–391.

106. Basta G, Montanucci P, Luca G, Boselli C, Noya G, Barbaro B, Qi M, Kinzer KP, Oberholzer J, Calafiore R. Long-term metabolic and immunological follow-up of nonimmunosuppressed patients with type 1 diabetes treated with microencapsulated islet allografts: four cases. Diabetes Care 2011;34(11):2406–2409.

107. Baiu D, Merriam F, Odorico J. Potential pathways to restore beta-cell mass: pluripotent stem cells, reprogramming, and endogenous regeneration. Curr Diab Rep 2011; 11(5):392–401.

108. Hansson M, Madsen OD. Pluripotent stem cells, a potential source of beta-cells for diabetes therapy. Curr Opin Investig Drugs 2010;11(4):417–425.

109. Medarova Z, Moore A. Non-invasive detection of transplanted pancreatic islets. Diabetes Obes Metab 2008;10(Suppl 4):88–97.

110. Hilbrands R, Huurman VA, Gillard P, Velthuis JH, De Waele M, Mathieu C, Kaufman L, Pipeleers-Marichal M, Ling Z, Movahedi B, Jacobs-Tulleneers-Thevissen D, Monbaliu D, Ysebaert D, Gorus FK, Roep BO, Pipeleers DG, Keymeulen B. Differences in baseline lymphocyte counts and autoreactivity are associated with differences in outcome of islet cell transplantation in type 1 diabetic patients. Diabetes 2009;58(10):2267–2276.

111. Huurman VA, Velthuis JH, Hilbrands R, Tree TI, Gillard P, van der Meer-Prins PM, Duinkerken G, Pinkse GG, Keymeulen B, Roelen DL, Claas FH, Pipeleers DG, Roep BO. Allograft-specific cytokine profiles associate with clinical outcome after islet cell transplantation. Am J Transplant 2009;9(2):382–388.

112. Roep BO, Stobbe I, Duinkerken G, van Rood JJ, Lernmark A, Keymeulen B, Pipeleers D, Claas FH, de Vries RR. Auto- and alloimmune reactivity to human islet allografts transplanted into type 1 diabetic patients. Diabetes 1999;48(3):484–490.

113. Stobbe I, Duinkerken G, van Rood JJ, Lernmark A, Keymeulen B, Pipeleers D, De Vries RR, Glass FH, Roep BO. Tolerance to kidney allograft transplanted into Type I diabetic patients persists after in vivo challenge with pancreatic islet allografts that express repeated mismatches. Diabetologia 1999;42(11):1379–1380.

114. Lacotte S, Berney T, Shapiro AJ, Toso C. Immune monitoring of pancreatic islet graft: towards a better understanding, detection and treatment of harmful events. Expert Opin Biol Ther 2011;11(1):55–66.

115. Han D, Xu X, Baidal D, Leith J, Ricordi C, Alejandro R, Kenyon NS. Assessment of cytotoxic lymphocyte gene expression in the peripheral blood of human islet allograft recipients: elevation precedes clinical evidence of rejection. Diabetes 2004;53(9): 2281–2290.

116. Han D, Leith J, Alejandro R, Bolton W, Ricordi C, Kenyon NS. Peripheral blood cytotoxic lymphocyte gene transcript levels differ in patients with long-term type 1 diabetes compared to normal controls. Cell Transplant 2005;14(6):403–409.

117. Erener S, Mojibian M, Fox JK, Denroche HC, Kieffer TJ. Circulating miR-375 as a biomarker of beta-cell death and diabetes in mice. Endocrinology 2013; 154(2):603–608.

118. Khan MH, Harlan DM. Counterpoint: clinical islet transplantation: not ready for prime time. Diabetes Care 2009;32(8):1570–1574.

119. Serup P, Madsen OD, Mandrup-Poulsen T. Islet and stem cell transplantation for treating diabetes. BMJ 2001;322(7277):29–32.

8

CELL-BASED BIOLOGIC THERAPY FOR THE TREATMENT OF MEDICAL DISEASES

Man C. Fung and Debra L. Bowen

8.1 INTRODUCTION

Cell-based biologic therapy involves the use of human cells (individual or grouped cells—tissue/organs) to prevent or treat medical problems. In the broadest sense, it is applied to a wide range of medical therapies including stem cell therapy, immunotherapy, gene therapy, regenerative medicine, tissue engineering, and cell-based cancer vaccines [1–5]. From the first human blood transfusion in the nineteenth century to the current bone marrow and tissue/organ transplants, tissue banking, and reproductive *in vitro* fertilization, cell-based therapies are integral to the practice of modern medicine. Unlike early research success, which was primarily attributable to trial and error, modern cell-based therapies are founded upon sophisticated laboratory research and robust science.

A common practice in cell-based therapy involves harvesting the cells from the host, ex vivo processing of these cells by stimulation and expansion with growth factors, and then reinfusing them back into the host for therapeutic uses (autologous host) [6, 7]. Since stem cell therapy is already covered in Chapter 9, this chapter will focus on biological cell therapies other than stem cells. However, xeno cell transplantation is outside the scope of this chapter and will also be excluded.

Therapeutic Delivery Solutions, First Edition. Edited by Chung Chow Chan, Kwok Chow, Bill McKay, and Michelle Fung.

8.2 ENGINEERED OR PROCESSED HUMAN CELLS AS TREATMENT MODALITIES

In addition to stem cell therapy, there are other modalities utilized to effect cell-based biologic therapies. Among these, cell-based treatment employing surgical harvesting and ex vivo cell culture techniques are utilized in orthopedic and burn therapies [8–13]. Used for many years, cartilage and bone replacements from cultured living cell grafts are important for orthopedic applications in acute injury as well as for chronic arthritic conditions [8–10]. In addition, the use of cultured fibroblasts for burn or diabetic ulcers is also a common practice [11–13]. Other novel approaches are being studied, which may allow organ regeneration or replacement.

For example, ISTO Technologies in Missouri is one of the many orthobiologic companies focused on cell-based technology for the repair and regeneration of damaged cartilage in joints and vertebrae [14]. ISTO Technologies has a product that uses a living tissue graft developed from juvenile chondrocytes in the form of minced cartilage tissue for focal articular cartilage repair. It consists of particulate natural articular cartilage with tissue recovered from juvenile human donor joints. The company maintains that juvenile chondrocytes produce cartilage much better than the adult counterpart. The company has a 225-patient randomized phase III trial to compare its product against microfracture therapy, which is the current standard of care for cartilage repair [15]. The key objectives are to observe whether a living cell graft leads to regenerative hyaline cartilage for the restoration of cartilage defects, reestablishment of joint function, and relief of arthralgia.

Besides the use of biological cell therapy in orthopedic conditions, there are several examples of the use of host-derived cells in other fields. For example, in the field of hepatology, Baermed of Switzerland has initiated a small pilot phase I study using a hepatocyte matrix implant as treatment for end-stage liver disease [16]. This is a novel experimental approach combining living cells with a biomatrix technology. Liver tissue from a patient is first harvested during a small liver resection, along with a pancreatic biopsy to remove some pancreatic tissue. The mixed tissues are then sent to a specialized cell culture laboratory where cells are processed in a perfusion procedure and prepared on several plates of matrices (4 mm thick, 20 mm diameter). After culture and expansion with various undisclosed supportive growth factors, the biotissue is then surgically implanted into the patient, but near the mesentery of the small intestine rather than the liver. The cells are then allowed to take on the capillaries of the patient and connect to the vascular system and continue to grow and multiply during the next 2–4 weeks. The carrier matrix, made of formaldehyde-free self-dissolving polymers, is expected to ultimately dissolve, and the implanted hepatocytes are expected to develop into functional liver cell tissue in the new site and to restore normal hepatic function in the patient. This study began in 2011. Results will not be expected for several years.

Another quickly evolving field is in biomedical engineering and regenerative medicine. An improved understanding of cell plasticity now enables the development of unconventional platforms for new cellular and gene therapy strategies such as transdifferentiation. An obvious benefit for such an approach is that autologous transplants do not have to depend upon donor supply, which is in short supply for many organs. Second, it also eliminates the concern of immune rejection and/or the need for the patient to be on chronic immunosuppression if receiving allogeneic tissue. Third, even if the transplant does not result in a full replacement of the original tissue/organ, transdifferentiated tissue can still be used as a temporary measure to maintain a patient until the affected organ regenerates (e.g., acute liver failure) or until a donor organ ultimately becomes available.

Transdifferentiation involves transforming differentiated cells into completely distinct phenotypes, unlike stem cell therapies where the source of cells is omni- or pluripotent [17–19]. One example is the transdifferentiation of isolated hepatocytes into insulin-producing beta cells for the treatment of diabetes (see also Chapter 7). When unique cell types are needed, such technologies have the potential for managing diseases, such as beta pancreatic cells for diabetes, dopamine-generating cells in the substantia nigra for Parkinson's disease, and myocardial cells for heart disease.

Tissue regeneration is feasible. It is well known that skin and gastrointestinal (GI) cells continuously regenerate; the liver has been shown to regenerate even after removal of up to 70% of the organ [20], and axonal outgrowth of a severed nerve can occur. However, mammals do not regenerate body tissues as readily as do lower species. For example, the urodele amphibian has been shown to regenerate many different organs such as the limbs, lens, tail, etc. [21]. Nonetheless, recent research using novel bioengineering techniques and gene therapies may make transdifferentiated cell transplants feasible for humans in the future.

Conversion of hepatic to pancreatic tissue (or the reverse) using a transgenic approach has been reported in the literature [22, 23]. *Pdx1* (pancreatic and duodenal homeobox 1) is the gene used to convert liver to pancreas. This gene is a key transcription factor for initial pancreatic development, β-cell maturation, and the subsequent maintenance of the β-cell phenotype [24, 25]. *Pdx1* is restricted to β cells in the pancreas where it binds to the insulin promoter. Inactivation of the *pdx1* gene in mice results in loss of insulin-producing cells and the development of diabetes in these animals [25]. Using a transgenic approach, adenoviral delivery of *Pdx1* to mice has been shown to induce insulin production and to control hyperglycemia induced by streptozotocin [26]. Using a different approach of exposing permeabilized myofibroblasts to insulinoma cell extracts, transient expression of *Pdx1* and insulin production has been demonstrated for up to 4 weeks [27].

Investigators have used different models to convert tissue into pancreatic exocrine- and endocrine-producing cells. Horb et al. [28], using human hepatoma cells and transgenic tadpole models, demonstrated expression of an activated form of *Pdx1* in liver cells. Sumazaki et al. [29], using hes1 knockout mice, also demonstrated successful conversion of a developing biliary system into a functional pancreatic tissue. In fact, the full array of endocrine cells (glucagon, insulin, somatostatin, and pancreatic polypeptide) was found to be produced in the ectopic tissue. While this research is still in its infancy, the examples illustrate the potential to transplant autologous, transdifferentiated, functional cells to treat targeted diseases in the future.

A hybrid between transdifferentiation and stem cells is the recent success in converting human skin cells directly into other progenitor cells for tissue repairs.

An example is that Dr. Lu et al. of University of Wisconsin recently showed that primate fibroblasts can be reprogrammed under special condition directly into induced neural progenitor (iNP) cells, bypassing the need to first become pluripotent stem cells [30]. Upon implantation into specific neural tissues in the animal model, these iNP cells were shown to be able to further differentiate into region- and function-specific neural cells (i.e., neurons, astrocytes, and oligodendrocytes) with implication for future potential therapy such as spinal cord injury or amyotrophic lateral sclerosis (ALS).

8.3   CELL-BASED IMMUNOTHERAPY

This field evolved based on observations from original research in both animal models and human trials, which demonstrated that immune cell infiltration of primary tumor tissue was associated with better prognoses and survivals in cancer [31, 32]. Several approaches have been developed, and cancer immunotherapy has steadily progressed to clinical therapies. These approaches include adoptive transfer of tumor-specific cytotoxic T lymphocytes (CTLs), natural killer cells (NK cells), and dendritic cell (DC)-based vaccines [33–35]. While much of this research has been applied to cancer treatments, cell-based immunotherapies are also applicable to infectious and immunologic diseases.

8.3.1   Adoptive Immunity: LAK, CTL, and TIL

Concurrent with the observation that patients with immune cell infiltration in their tumors are associated with better survival, there was interest in employing adoptive immunotherapy to treat cancer [31, 32]. Adoptive cell transfer utilizes T-cell-based cytotoxic responses to eliminate tumor cells. In general, immune T cells that have a natural or genetically engineered reactivity to a patient's cancer are reactivated ex vivo and reinfused back into the patient [33–35]. These activated immune cells are either injected directly into the tumor or administered intravenously to the patient. Intratumor injection should theoretically permit a higher number of effector cells at the tumor site, but results with intravenous administration appear to be comparable to date [36].

Lymphokine-activated killer (LAK) cells were the first cells utilized [37, 38]. If lymphocytes are cultured in the presence of the cytokine interleukin 2 (IL-2), the resultant effector cells are cytotoxic to tumors. LAK cells are obtained by cultivating peripheral lymphocytes in the presence of IL-2, yielding a mixed population with different subsets of T cells and NK cells [39]. A limitation of these therapies has been that the cytolytic properties of these polyclonal cells are not specific, and the results in oncologic conditions have been variable [37–39]. Even so, some studies suggest that the addition of LAK cells may help improve the efficacy of tumor-specific monoclonal antibodies [40–42]. LAK cells express human Fc receptors and may help mediate antibody-dependent cell-mediated cytotoxicity (ADCC). Early results demonstrated that addition of LAK cells and IL-2 to an anti-CD20 antibody in lymphoma resulted in enhanced ADCC activity, inducing greater efficacy than with the antibody alone [40]. Similar observations were also seen for other antibody–LAK combinations for other tumors [41, 42].

Adoptive immunity using CTLs has also been studied [43–45]. Ex vivo antigenic stimulation of peripheral blood mononuclear cells (PBMC) is a common approach used to generate CTL. Autologous tumor cells (ATC), either intact or fragment derivatives, are often used as the source of antigen stimulation, which can induce polyclonal expansion of CD8 and CD4 cells. Again, IL-2 supports the growth and expansion of these cells before reinfusing back into the patient [46, 47]. Another approach is the use of CTL obtained from tumor-infiltrating lymphocytes (TIL) [32]. A similar approach is used to collect lymphocytes from the lymph nodes or PBMC after peripheral injection of irradiated ATC using the growth factor GM-CSF for support and stimulation [48]. Whatever the source, TILs are usually expanded ex vivo using IL-2, anti-CD3, and alloreactive feeder cells. These T cells are then reinfused back into the patient, along with continuous exogenous administration of IL-2, to further boost their anticancer activity.

Adoptive immunity for the treatment of melanoma appears to have been more successful than its use for other cancers [49], perhaps due to the more immunogenic nature of the cancer itself and/or to the availability of several well-characterized tumor antigens (e.g., gp100, MART-1, NY-ESO-1). As tumor antigens are identified and better characterized for other tumors, this approach should be amenable for other tumor types. Lastly, besides metastasis, adoptive immunity seems to work better in adjuvant use after surgery, perhaps because the tumor burden is lower in those settings [50].

However, due to the lack of standardization of techniques utilized, the types and quantity of cells infused, and the preconditioning regimens employed, it is difficult to compare outcomes using different approaches. Standardized *in vitro* tests (e.g., killing assays directed against tumor cells or phenotyping of the effector cells used) can potentially assist in gauging the treatment efficiency of various regimens. Lastly, since most studies are conducted as small pilot studies without randomization against a standard of care, more clinical work is needed to elucidate optimal treatment conditions.

There are several challenges inherent in developing therapies for adoptive immunity. For example, most approaches require IL-2. Unfortunately, IL-2 use (especially in higher doses) can lead to significant toxicities, particularly to the kidney and lung [51, 52]. With increased experience in using adoptive immunity, optimal dosing of IL-2, either its administration (e.g., subcutaneously) or dosing (timing, quantity, and frequency), can modulate this toxicity.

Another challenge is the durability of response. Initial adoptive cell transfer studies revealed that persistence of the transferred cells *in vivo* was brief as evidenced by the analysis of tumor samples taken during relapse [53, 54]. It was later discovered that before reinfusion of the stimulated immune cells to a patient, lymphocyte depletion [55–58] in the recipient helps eliminate regulatory T cells (Tregs) as well as other endogenous lymphocytes (which compete with the transferred cells for homeostatic cytokines, especially IL-15) [55, 56]. Lymphodepletion is generally accomplished through a nonmyeloablative chemotherapeutic regimen (e.g., cyclophosphamide, fludarabine) and/or with total body irradiation prior to transfer of the stimulated immune cells [57, 58]. Lymphodepletion is important to the persistence of infused cells and positively correlated with more favorable treatment outcomes [59–62]. However, lymphodepleting regimens are themselves associated with potentially serious toxicity [58].

Further to the issue of durability, replicative capacity of infused T cells and their ability to persist as memory cells can have a major impact on treatment outcomes. One lymphokine, IL-21, was found to have the ability to elicit increased antigen-specific CTL when T cells are exposed to it during initial stimulation [63–66]. IL-21 can enable the production of a population of CTL with greater avidity and expansion potential. IL-21 exposure also leads to generation of helper-independent CTL effectors with a distinctive memory phenotype capable of autocrine IL-2 production [64]. These effects on CTL appear to originate from the ability of IL-21 to suppress or eliminate Tregs with a resulting higher quantity and frequency of tumor-specific T cells [66]. Thus, use of IL-21 has the potential to minimize the need to grow CTL to very large numbers (1–10 billions), which should reduce the time and cost for CTL preparation.

Other options to enhance durability include vaccine and the concomitant use of immunomodulators. Animal models have suggested that the concomitant use of adoptive T-cell transfer along with vaccination may improve efficacy. As a potential substitute for IL-2, IL-15 has been studied along with tumor vaccine to boost memory T cells [67].

Meanwhile, recent FDA approval of the novel immunomodulator, ipilimumab (anti-CTLA4 antibody), may permit another option. Combined use of CTL and ipilimumab may enhance the function and survival of adoptive T cells, lowering the threshold for the induction of endogenous T-cell responses to tumor antigens.

There are other measures that have been attempted to enhance responses. For example, use of superagonist altered peptide ligands (APLs) as a complementary means of enhancing T-cell responses to tumor-associated self-antigens has been studied extensively [68]. It was demonstrated that the superagonist APLs often elicit robust antitumor CTL responses, while the native tumor-associated epitope does not. Unfortunately, the ability of a given analogue to act as a superagonist varies from patient to patient. This implies that a comprehensive panel of potential superagonist APLs may be needed to individualize efficient tumor responses, which may limit the utility of this approach.

Lastly, use of adoptive immunity for certain tumors may pose other challenges. For example, in high-grade glioma, a potential limitation arose with corticosteroids given to patients to reduce edema post surgery. Steroids induce lymphopenia along with an abnormal rise of circulating CD14+/HLA-DR low/neg monocytes, thereby preventing the generation of a large number of effector cells using adoptive approaches [69]. This reduces treatment efficacy. Also, CD14+/HLA-DR low/neg monocytes cannot fully differentiate into mature DCs, which could be problematic if concomitant use of a DC-based vaccination were desired.

In summary, while adoptive immunity has been extensively studied, achieving an effective clinical response has not been consistently demonstrated. Further

TABLE 8.1   Cytotoxic T-cell (CTL) studies conducted by a commercial sponsor

Sponsor	Product	Indication	N	Start time	Trial no.
BMS	TIL + ipilimumab	Melanoma	10	2012	NCT-01701674
Chiron	CTL + DC vaccine	Melanoma	98	2006	NCT-00338377
CellMedica	CMV-sp T cells	CMV infection	36	2010	NCT-01220895
CellMedica	CMV-sp T cells	CMV infection	90	2008	NCT-01077908

optimization is necessary in order to gain confidence with these therapies. It can be anticipated that combined modalities incorporating vaccination, optimized pre- and postinfusion conditions, and manipulation of the tumor immune microenvironment are necessary for effective antitumor responses and long-term immunologic memory.

Finally, while research in adoptive immunotherapy has been commonly conducted in academic settings, industrialization of these technologies will bring them to an expanded audience and permit more patients to benefit from these innovative approaches. To that end, the news that a biotech, Genesis Biopharma, is working in collaboration with the National Cancer Institute is an auspicious beginning [70]. Following this theme, the focus of this chapter concentrates mostly on development activities from the biotech/pharmaceutical industry for these technologies, wherever applicable. Ongoing studies by industry sponsors from clinicaltrials.gov [71] are listed in Table 8.1. While historically development of CTL is mostly in melanoma, there are two ongoing studies in cytomegalovirus (CMV) infection.

8.3.2   T-Cell Receptor (TCR) Gene Therapy

One recent major advance in adoptive cell therapy is represented by gene manipulation of autologous T cells in order to acquire strict specificity toward the target, which results in potent lytic activity, significant proliferation, increased survival, and a persistent antitumor memory state in the host. It can be achieved by either high-avidity genetically engineered T-cell receptors (TCRs) or through chimeric antigen receptor (CAR) gene therapy technology [72]. Although most of the research focus has been in oncology, this technique may also apply to infectious diseases or immunologic conditions.

For TCR therapies, genetically engineered T cells are created by transfecting a patient's cells with a retrovirus vector containing a copy of a genetically engineered high-avidity TCR gene specialized to recognize targeted antigens [73–75]. Because it lacks key genes required for replication, the retrovirus vector is subsequently unable to reproduce within the cell after it is integrated into the human genome, and thus, the new TCR gene remains stable within the T cell. Genetically transduced T cells will then express the new TCR on its surface and use the new TCR to engage the peptide antigen target on cancer or infected cells and destroy them.

A key challenge for TCR technology, as it is for many other adoptive therapies, is that most naturally occurring TCRs are of low affinity [76, 77]. This is because human T cells need to balance recognizing foreign antigens while not attacking self-derived peptides (both are presented by the same MHC molecules). Moreover, most peptides on tumor surfaces are derived from self-antigens so the tumor cells are already selected to present low antigen levels. TCRs are expressed as membrane-anchored proteins that go through a selection process in the body and only T cells expressing TCRs of low affinity survive. On the other hand, viral peptide antigens can undergo mutation to avoid recognition after the initial infection (e.g., HIV), reducing the ability of effector T cells to neutralize the infection. As a result, many T cells do not recognize the foreign antigen presented.

Consequently, one recent strategy is to generate T cells with higher capability to recognize foreign antigens by artificially constructing TCRs with greater antigen-binding affinities, such as mutations of a residue in the constant regions of the α and β chain of TCR and/or isolation of high-affinity variants of natural TCRs [78, 79]. A transgenic approach is then utilized to insert a gene encoding the new TCR with greater antigen specificity and affinity into the T cells, followed by expansion of these manipulated T cells, and then reinfused back to the patient to treat the disease.

The TCR approach has two obvious advantages over conventional adoptive cell therapy. First, it allows the use of a set of TCR genes with known effectiveness in a large patient population. Second, it may overcome the intensive process of *in vitro* generation of large numbers of specific T cells. On the other hand, limitations of TCR include the fact that target recognition by TCR is MHC restricted allowing its use in patients with only certain haplotypes and a potential for mispairing introduced chains with endogenous TCR subunits, which may lead to reduced TCR surface expression and potentially lower activity [80, 81]. These (and other) limitations can be overcome by the use of CAR (see Section 8.3.3). In 2003, Morgan et al. [82] demonstrated the first successful adoptive cell transfer of lymphocytes transduced with retrovirus encoding TCRs in patients with metastatic melanoma. Since then, many TCR studies have been conducted [83, 84]. Table 8.2 lists several ongoing studies by industry sponsors in this area [71].

TABLE 8.2   TCR gene therapy studies conducted by a commercial sponsor

Sponsor	Product	Indication	N	Start time	NCT no.
Adaptimmune	TCR (*NY-ESO-1*)	Synovial sarcoma	10	2011	NCT-01343043
Adaptimmune	TCR (*MAGE-3* or *NY-ESO-1*)	Melanoma	12	2011	NCT-01350401
Adaptimmune	TCR (*MAGE-3/6* or *NY-ESO-1*)	Melanoma	12	2011	NCT-01352286
Adaptimmune	TCR (*WT-gag/α6-gag*)	HIV	48	2009	NCT-00991224
Altor	ALT801 (IL2 bound to p53 TCR)	Urothelial cancer	76	2011	NCT-01326871
Altor	ALT801 (IL2 bound to p53 TCR)	Melanoma	25	2010	NCT-01029873
VIRxSYS	VRX496 (HIV env antisense) TCR	HIV	40	2007	NCT-00622232

8.3.3 Chimeric Antigen Receptor (CAR) Gene Therapy

One of the barriers to the widespread use of cellular therapy has been the usual MHC restriction of antigen recognition. CAR can be engineered to bypass this requirement. CARs are engineered receptors, which graft an artificial specificity (earlier studies used the Fab antigen-binding site of an antibody) against a known target onto a T cell frequently using a retroviral vector [85–87]. Early CARs were also known as T-bodies. Using recent iterations of this novel approach, specific targeted T cells can be generated against antigen-expressing cancers for use in adoptive cell therapy. Recent clinical studies have shown encouraging results in various malignancies including several patients with relapsed/refractory acute lymphoblastic leukemia who had rapid remission after treatment with CAR-modified T cells with CD19 [88–92].

Although little is known about the initial activation of CARs, they orient the activity of T cells toward specific targets expressed on the cancer cell surface. The technology is achieved by gene insertion of a CAR, artificial molecules containing antibody-derived fragments against a specific target joined to a potent signaling TCR-derived domain that activates the manipulated T cells [93, 94]. The most popular configuration is the fusion of a single-chain variable fragment (scFv) from a monoclonal antibody with the CD3-zeta transmembrane and endodomain. (Other possible antigen-binding moieties include signaling portions of hormone or cytokine molecules and extracellular domains of membrane receptors.) When the scFv recognizes its target, a zeta signal is transmitted, and the CAR-transformed T cells recognize and kill tumors (or other cells) that express its target.

The ectodomain is the extracellular part of this artificial TCR and is composed of a signal peptide, an antigen recognition region, and a spacer sequence. The scFv is generated by synthetic DNA technology and usually produced by fusing the variable portions of an immunoglobulin light and heavy chain using a flexible linker. A signal peptide, which precedes the scFv, serves to direct the nascent protein into the endoplasmic reticulum as well as the subsequent surface expression. The flexible spacer allows the scFv to orient in various directions and facilitates antigen binding. Optimal T-cell activation depends on the relative length of the hinge region spacer and the distance of the epitope from the target cell membrane [95]. Derived mainly from the original molecule of the signaling endodomain, the transmembrane domain protrudes into the T cell and transmits the intended signal. Despite removal of its constant regions and the introduction of linker peptides, the chimeric domain can retain the specificity of the original immunoglobulin. Gene transfer is typically accomplished via retrovirus infection, which allows RNA encoding into DNA via reverse transcriptase after infection. (These retroviruses lack key genes so they cannot replicate after they infect the cell.) Figure 8.1 illustrates how a typical CAR is constructed.

There are several advantages to CARs [85–87]. These engineered T cells can exhibit specific lysis toward tumor cells with cytokine secretion, leading to sustained tumor cell lysis beyond that of monoclonal antibodies. The perforin/granzyme killing mechanism of T cells may be effective against cells that are resistant to antibody or

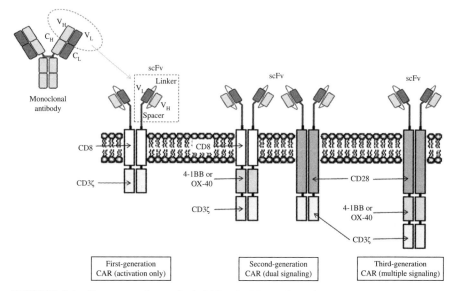

FIGURE 8.1   Structure of a typical CAR. 4-1BB or OX-40, costimulatory molecules belonging to the TNF/nerve growth factor super family of receptors; CD28, cluster of differentiation 28, provides costimulatory signal; CD3ζ, cluster of differentiation 3ζ, generates an activation signal in T lymphocytes; CD8, cluster of differentiation 8, a transmembrane glycoprotein that serves as a coreceptor for the TCR; C_H, constant region of heavy chain; C_L, constant region of light chain; scFv, single-chain variable fragment; V_H, variable region of heavy chain; V_L, variable region of light chain.

complement, while cytokine secretion can help to recruit other components of the immune system, amplifying and enhancing antitumor efficacy. Furthermore, CAR-engineered T cells have been shown to display superior tumor penetration and homing abilities. The CAR approach also overcomes a critical limitation of conventional TCR. Target recognition by CAR is non-MHC restricted and independent of antigen processing, permitting use in patients with different haplotypes and bypassing tumor escape from MHC molecule downregulation mechanisms. In addition, CARs could be targeted toward molecules like glycolipids or carbohydrates, not just peptides. Because random or potentially harmful specificities are highly unlikely (which may occur with transduced TCRs forming hybrids with an endogenous TCR), this approach has major advantages over TCR.

CARs can be generated against a wide range of surface molecules expressed by cancers. Hematologic malignancies are especially suitable for CAR due to the strong expression of specific antigens on the tumor cell surface. For example, CD19 antigen is a common target, and many studies are being conducted in various B-cell malignancies. Upon transduction with anti-CD19, CARs, and antigen recognition, cytokine-induced killer (CIK) effector cells demonstrate strong antitumor activity. In addition to the zeta chain, second- or third-generation CARs containing costimulatory molecule(s) have demonstrated superior *in vivo* effects [96, 97]. For example, CAR-expressing T cells containing the costimulatory CD28 endodomain (+/− 4-1BB or OX-40) show more growth and persistence compared with CAR-expressing T cells encoding the zeta endodomain alone [87, 97] (see also Fig. 8.1).

Nevertheless, the downsides to using CARs [85–87] include that only surface antigens can be recognized (unlike TCR) and the presence of soluble antigen shed by tumors can compete with binding to and killing of malignant cells. Finally, use of CAR containing costimulatory or growth-promoting molecules could theoretically lead to CAR-expressing T cells persisting beyond the desired time frame.

In case of unexpected reactivity of transduced cells, the introduction of suicide genes has been proposed. Various suicide gene strategies have been explored, most involving adding genes such as *HSV-TK*, *mTMPK*, or *CD20* [98, 99]. Uncontrolled activity could then be terminated by the use of ganciclovir, zidovudine, and ritux-imab, respectively, as agents, which can kill the transduced cells expressing one of these suicide genes. Another option is the use of the inducible Casp9-CID suicide system [100], based on a fusion protein consisting of caspase 9 and FK506 binding protein. This would permit conditional dimerization by an external chemical inducer when termination of activity is necessary. Finally, the use of nonviral methods of transducing T cells with the CAR transposon/transposase system has been intro-duced [101]. This approach is based on the nucleofection/electroporation of plasmid DNA, which enable the nuclei of cells to be stably transduced without use of retro-viral vectors. Each approach has drawbacks and an optimal strategy remains to be elucidated.

Overall, CARs represent a novel approach for the treatment of cancers with encouraging results in phase I/II studies in various malignancies, especially B-cell leukemia. Even so, treatment-related toxicities, including systemic inflammatory reactions, cytokine storm, and the possibility of tumor lysis syndrome and occasional mortality, including several children with severe combined immunodeficiency whose hematopoietic stem cells were treated with common gamma chain receptor genes and who experienced insertional mutagenesis with resultant leukemia [88–92], have led to cautious patient selection, careful consideration of cell quantities and schedules, and thoughtful optimization of lymphodepleting regimens. Further development of the suicide gene checkoff mechanism may also be helpful in avert-ing off-target adverse events. CAR regimens, which are still in the early stages of development, are being conducted primarily within academic centers with little industry involvement. In addition to the study by the biotech Immunocore, Table 8.3 also includes several academic conducted CAR studies ($N \geq 30$) for illustration purposes.

8.3.4 Natural Killer Cell (NK Cell) Therapy

NK cells are neither T nor B lymphocytes, but a distinctive lymphocyte that bridges innate and adaptive immune systems [102]. The name is due to the observation made during early research on cell-mediated cytotoxicity in the 1970s when researchers discovered a small population of large granular lymphocytes that were able to lyse cancer cells seemingly without prior sensitization. Thus, this type of lymphocyte was capable of spontaneous ("natural") cytotoxicity. NK cells play a

TABLE 8.3 Car gene therapy studies conducted by an academic sponsor

Sponsor	Product	Indication	N	Start time	Trial no.
Baylor	CAR-Her2	Sarcoma	36	2009	NCT-00902044
City of Hope National Medical Center	CAR-CD19	NHL	57	2011	NCT-0318317
City of Hope National Medical Center	CAR-CD19	NHL	30	2013	NCT-0815749
Fred Hutchinson	CAR-CD19	B-cell malignancies	30	2011	NCT-01475058
King's College (London)	CAR-TIE28Z	Head and neck cancer	30	2013	NCT-0818323
NCI	CAR-CD19	B-cell malignancies	40	2009	NCT-00924326
NCI	CAR-CD19	B-cell malignancies	36	2010	NCT-01087294
NCI	CAR-VEGFR2	Metastatic CA	118	2010	NCT-01218867
NCI	CAR-EGFRvIII	GBM	160	2011	NCT-01454596
NCI	CAR-mesothelin	Metastatic cancers	136	2012	NCT-01583686
NCI	CAR-CD19	B-cell malignancies	48	2012	NCT-01593696
University College (London)	CAR-CD19	ALL	30	2012	NCT-01195480
Immunocore	CAR (IMCgp100)	Melanoma	50	2010	NCT-01211262

key role in early host defense against infections and cancer [103–106]. They kill cells by releasing enzymes (perforin and granzyme) that cause a target cell to undergo apoptosis. Although morphologically similar to other lymphocytes, they do not express TCR, CD3, or surface immunoglobulin receptors. NK cells recognize glycolipid antigen and exert their effector function by direct killing besides producing cytokines and chemokines, leading to a cascade of additional adaptive immune responses. Their function is tightly regulated by activating and inhibitory receptors expressed on their surfaces, by cytokines and chemokines, and also through cross talk with other immune cells.

Since NK cells have been shown to lyse cancer cells, they were studied extensively in cancer immunotherapy in the 1980s but with limited clinical success. In recent years, renewed interest of studying NK cells emerged as further elucidation of their biology and function became available [107–110]. One major challenge in NK cell research is the low quantity of NK cells that can be generated for infusion. However, a breakthrough occurred recently when IL-15 coadministered with a glucocorticoid was shown to greatly expand and enhance the function of NK cells [111–113]. Activation with IL-15 and cortisol will not only enhance proliferation of NK cells, but it can also allow these cells to retain their functional potential and be protected from apoptosis. This approach of expansion allows the NK cells to preserve viability, retain high expression of their activating receptors (NKG2D and NKp46), and be functionally intact to attack targeted cells. They also have enhanced migratory ability to move toward tumor or tumor-draining lymph nodes, as evidenced by the increased expression of CXCR3, CXCR4, and CD62L. The overall result is the potential for large-scale production of highly active NK cells [114, 115].

Preconditioning to eliminate Tregs by Ontak™ (denileukin diftitox) or anti-CD25 mAb has demonstrated better outcomes for the adoptive therapy in general. To that end, elimination of Tregs before NK cell infusion might favor the generation of potent antitumor T-cell responses in a cascade of events in the integrated adaptive immune system [116]. The costimulation with IL-15 and cortisol of NK cells leads to expanded quantity of cells with good results and certainly seems to be a viable approach moving forward. This signals that NK cell transfer may be an alternative or complement to current approaches in cancer immunotherapy. Table 8.4 identifies ongoing NK cell therapy studies from industry sponsors.

TABLE 8.4 NK cell studies conducted by a commercial sponsor

Sponsor	Product	Indication	N	Start time	Trial no.
Altor	Donor NK cells (plus ALT-801)	AML	68	2011	NCT-01478074
Binex	TK cells (NK)	Gastric	94	2009	NCT-00854854
Miltenyi	NK cell infusion	Pediatric cancers	12	2008	NCT-00582816
Miltenyi	Donor NK cells	AML	51	2013	NCT-01639456
NKBio	Biocell NK mixture	DLBCL	276	2007	NCT-00846157

8.3.5   T-Regulator Cell Therapy

Tregs (formerly known as suppressor T cells) are critical to the maintenance of immunologic tolerance [117–119]. The major functions of Treg are to suppress autoreactive T cells (which escape negative selection in the thymus) and to shut down T-cell-mediated immunity at the end of an immune reaction. There are two main classes of CD4+ Tregs: thymically derived natural Treg (nTreg) and peripherally generated adaptive Treg (aTreg) [120]. Naturally occurring Tregs arise in the thymus and are linked to interactions between developing T cells with DCs activated by thymic stromal lymphopoietin. They can be distinguished from other T cells by the intracellular molecule, FoxP3. In fact, *FoxP3* gene mutation can prevent Treg cell development. aTreg cells originate in the circulation during a normal immune response.

A key function of the immune system is to distinguish self versus nonself. When this discrimination does not occur properly, the immune system destroys cells and tissues, leading to autoimmune diseases [121]. Tregs appear to actively suppress abnormal activation of the immune system, preventing pathologic autoimmunity. In fact, severe autoimmune syndromes emerge in patients with genetic defects in Tregs. Normal Treg contributes to the elimination of pathogens from the body. However, it is known that during some infections such as tuberculosis, malaria, and leishmaniasis, pathogens may manipulate Tregs to immunosuppress the host, enhancing the pathogen's own survival [122, 123].

Our understanding of the immunopathogenesis of HIV infection [124] has been enhanced by recent discoveries concerning the cellular reservoir of HIV-infected cells [CD4 stem-cell memory T-cells (Tscm)] [125] and identification of distinct pathways for CD4 T cell pyroptosis [126]. These discoveries have opened up new potential classes of therapeutics including those that target the host rather than the virus.

Increased Tregs at tumor sites have been demonstrated in several animal models and in humans [127, 128]. Increased Tregs at the tumor site, in tumor-draining lymph nodes, or in the circulation have been observed in patients with many different human malignancies. Some have hypothesized that tumors can actively recruit Tregs or can convert non-Tregs into Tregs. Since inhibition or depletion of Tregs can lead to enhanced antitumor activities, chemotherapies (e.g., cytotoxic) or novel Treg antibodies can be used to suppress Tregs and to enhance anticancer effects [129, 130].

Animal models illustrate that adoptive Treg therapies can be used as a therapeutic option to prevent organ transplant rejection and/or autoimmune disease. Nonetheless, human studies of direct autologous Treg infusion for therapeutic use have been relatively few. Instead, indirect means are used to induce Treg, such as the use of pharmacologic agents, antibodies, synthetic cytokines, and DCs [131]. There are only three academic studies ongoing. There is one phase I study in China to explore whether Treg infusions could induce tolerance in acute rejection of liver transplant in patients [132]. There is another phase I study by UCSF/Juvenile Diabetes Research Foundation to evaluate the feasibility and dose selection of Tregs in slowing diabetes progression and/or reverse new-onset diabetes [133].

There is only one randomized phase II study by the Russian State Medical University, which was initiated in late 2011 [134]. The study is to evaluate the use of autologous CD4+CD25+CD127lowFoxP3+ Treg expanded ex vivo to prevent organ rejection for renal transplants in children. Thirty pediatric patients with end-stage renal failure were randomized to receive either standard immunosuppressive therapy alone or in combination with autologous Treg infusion at days 30 and 180 post transplant. Before the transplant, $70\,ml/1.73\,m^2$ body surface area of blood will be collected twice from the patient, separated by a week. Tregs will be harvested, cleaned, expanded ex vivo, and frozen in liquid nitrogen. On day 30, 2×10^8 autologous Tregs, expanded from a previously frozen sample, will be reinfused to the patient. Levels of Treg in the patient's blood will be assessed by flow cytometry 1 week post infusion. This step will be repeated once on day 180. This is a 3-year study and the endpoint is a comparison of the rate of organ rejections between the two groups upon study completion.

8.3.6 Macrophage Therapy

There are at least three types of antigen-presenting cells, one of which is macrophages (MAC). The three cell types and their distinguishing characteristics are summarized in Table 8.5.

MAC are produced as monocytes undergo differentiation in tissues. MAC function in both innate (nonspecific defense) and adaptive immunity [135–137]. They can be mobile or stationary cells and have numerous functions. While initially thought to be phagocytic cells that clean up pathogens, cancer cells, or cellular debris, they are now known to play a critical role in initiating an immune response as well as in tissue repair. They work with DCs to present antigen and stimulate lymphocytes and other immune cells to respond to pathogens. As secretory cells, MAC are critical to regulating immune responses and inflammation, producing a wide array of enzymes, complement proteins, and regulatory factors. Once a T cell becomes an activated effector cell (upon antigen recognition on an aberrant cell), it will release a mediator lymphokine that stimulates MAC into a more active state.

TABLE 8.5 Antigen-presenting cell types and characteristics

	MAC	DC	B cell
MHC-II expression	Low levels Induced by bacteria and/or cytokines	Always expressed	Always expressed Inducible upon activation
Antigen type: MHC presentation	Extracellular antigens presentation via MHC-II	Intracellular and extracellular antigens presentation via MHC-I/II	Extracellular antigens (bind to specific Ig receptors) presentation via MHC-II
Costimulation (B7 expression)	Low level induced by bacteria and/or cytokines	High level always expressed	Low level induced upon activation
Location	Lymphoid tissue Connective tissue Body cavities	Lymphoid tissue Connective tissue Epithelium	Lymphoid tissue Blood

MAC can be classified into two groups histologically: M1 and M2 [138, 139]. M1 MAC are immune effector cells that engulf and digest infected cells as well as produce many lymphokines. M2 MAC are cells with nonphagocytic activity (e.g., cells involved in wound healing, tissue repair, and shutting down immune activation by producing anti-inflammatory cytokines (IL-10)) [140].

An illustrative example is typified by the observations after prolonged muscle use (e.g., a marathon), which can lead to muscle membrane lysis and inflammation [141, 142]. Phagocytic M1s first appear following the onset of myocyte injury, peak in 24 h, and disappear within 48 h. The nonphagocytic M2 MAC will then appear and peak between days 2 and 4, remaining elevated during muscle rebuilding. They distribute near regenerative fibers. It is thought that M2 MAC release certain chemofactors that affect the growth, repair, and regeneration of muscles. These M2 MAC promoting tissue repair are not muscle specific as they are also observed throughout repair and healing processes of other tissue types.

An interesting ex vivo adoptive MAC transfer study was conducted in 2007 in mice [143]. Researchers isolated MAC from the spleens of BALB/c mice and used lipopolysaccharide to induce M1 and IL-4/IL-13 to induce M2 MAC. These ex vivo prepared MAC were then infused into severe combined immunodeficient (SCID) mice with induced nephropathy by Adriamycin. Both M1 and M2 MAC were found to localize in the area of injury and maintained their phenotypes after even a month. However, mice infused with M1 MAC were found to have more severe functional and histological injury compared to reduced injury in the M2 MAC-transfused mice. The protective effect observed for the adoptively transferred M2 MAC was associated with reduced accumulation and also downregulated chemokine and cytokine expression by the host MAC.

Although animal models illustrate that generating ex vivo cytotoxic MAC against tumors should be plausible, attempts in the 1990s and early 2000s in several human trials were not as promising [144–146]. Despite infusing as high as 3×10^9 autologous MO-derived MAC activated by interferon (IFN)-γ or lipopolysaccharide, many studies across a spectrum of diverse tumors demonstrated only modest results [144]. Consequently, research has since shifted to DC-based adoptive therapy. Given its role in tissue healing, autologous M2 MAC adoptive therapy could reemerge. In 2003, Proneuron Biotechnologies, in collaboration of the Marcus Foundation and the Israel's Binational Industrial Research and Development, conducted a phase II study in 61 patients to examine the use of autologous incubated MAC in patients with spinal cord injuries [147]. Unfortunately, the study was later suspended and results are not available.

8.3.7 Dendritic Cells (DC) and DC-Based Vaccines

One major challenge for adoptive immunotherapy is maintaining transferred T-cell activity for a sufficient period of time [148]. One way to accomplish this would be to use a tumor vaccine to induce tumor-specific effector and memory T cells. However, early vaccines have not been very successful because cancers circumvent the immune system via recruitment of Treg, inflammatory type 2 T cells, and myeloid-derived suppressor cells [149]. Recent advances in understanding the DC system may help to overcome this limitation [150–152].

Effective vaccines have been developed against a variety of infectious agents, but others still lack effective vaccines, for example, HIV, malaria, hepatitis C, and tuberculosis. Breakthroughs in basic immunology have enhanced our understanding of DCs and their role in initiating an immune response to foreign antigens and our recognition that vaccine adjuvants act primarily because they activate DCs. This knowledge has led to a resurgence of interest in vaccines as a tool to prevent, control, or eliminate existing infections and to treat cancer by inducing strong cellular immune responses. However, it has been shown that the most efficient vaccines (e.g., smallpox) and one of the most potent vaccines ever generated against yellow fever (YF-17D) activate multiple DC subsets [153] and lead to integrated immune responses that include both humoral and cellular immunity [154]. Thus, it is generally believed that an integrated immune response may be the most effective approach.

DCs are antigen-presenting cells with high endocytic and phagocytic capacity [155]. They can initiate and maintain immune responses when in contact with an antigen. DCs collect antigen and carry them to lymphoid organs to activate T cells. There are two main subsets in humans: myeloid and plasmacytoid DCs. Plasmacytoid DCs recognize viral nucleic acids through TLR-7 and TLR-9 receptors and secrete IFN. DCs in the blood express three surface molecules: BDCA-1 (CD1c), BDCA-2 (CD303), and BDCA-3 (CD141) [156]. Myeloid DC subsets primarily localize to the skin. In the skin, there are at least two different mDC subsets: Langerhans DCs (epidermis), which preferentially regulate cellular immunity, and CD14+DCs (dermis), which preferentially regulate humoral immunity. Langerhans DC can produce IL-15, which is a growth and maintenance factor for NK and CD8+ T cells. CD14+ DCs produce various interleukins, such as IL-1β, IL-6, IL-8, IL-10, and IL-12; GM-CSF; membrane cofactor protein-1; and tumor growth factor-βs. A long list of chemokines (GM-CSF, IL-1β, IL-4, IL-6, IFN-α, IFN-γ, TNF-α, PGE2, poly I:C) have been found to influence DC maturation and the subsequent capability to elicit T-cell responses [157].

Recent research has shown that DCs can be stimulated ex vivo to activate a cytotoxic response toward an antigen upon reinfusion to the host. In this approach, DCs are harvested from a patient and then either pulsed with an antigen in the form of peptides, proteins, tumor lysates, or mRNAs [158, 159]. Alternatively, DCs can also be allowed to fuse with tumor cells or can be genetically modified to express tumor-associated antigens (TAAs) and/or immune-stimulatory genes with a viral vector. Upon reinfusion back into the patient, these activated cells present tumor antigen to effector lymphocytes, which in turn initiate a cytotoxic response against antigen-expressed tumors. In 2010, Provenge™ (from Dendreon) was the first example of this approach approved by the FDA for prostate cancer [160, 161]. DCs were stimulated with an engineered fusion protein composed of prostatic acid phosphatase (PAP) and GM-CSF (facilitates uptake of the engineered protein by DCs and promotes DC stimulation). Possible future DC-targeted vaccines for both infectious diseases and cancer look quite promising, especially following the observation that active immunization with an HPV recombinant viral capsid protein prevents HPV-positive cervical cancer [162]. There were multiple DC-based clinical studies initiated [150, 163, 164].

There are however some limitations with current DC-based therapies [165, 166]. Firstly, some tumors can evade the immune system via downregulation of surface or intracellular antigens. Secondly, other tumors secrete immunosuppressive cytokines, which convert immature DCs into tolerogenic DCs and recruit Treg. In addition, the *in vivo* vaccine interaction may not be totally efficient. Antigen peptides pulsed onto DCs may only bind transiently to MHC molecules due to variations in peptide-binding affinities, MHC–peptide complex dissociation, or even MHC turnover. Further, identifying TAA peptide epitopes corresponding to the MHC haplotype of the patient can be challenging. Finally, selection of appropriate clinical endpoints to measure effectiveness when a longer horizon immune response must first be elicited is essential. Conventional measures to demonstrate early shrinkage of tumor, a typical response evaluation criterion for solid tumors (RECIST), have not been positive even though survival endpoints have been demonstrated in randomized phase III trials with two DC-based agents. For example, trials of anti-CTLA4 (ipilimumab) demonstrated a twofold improvement in overall survival without indication of early tumor shrinkage [167]. Similarly, in a phase III Provenge™ (sipuleucel-T) prostate cancer trial, a prolonged median survival of 4 months was demonstrated [168].

To address these concerns, insertion of specific TAA genes into DCs has been proposed [169–171]. Transduction of DCs with TAA genes should allow expression of the full-length protein, enabling prolonged antigen presentation. Presentation of multiple or unidentified antigen epitopes relevant to MHC class I (or class II) molecules may also be possible. However, gene introduction efficiency and expression efficiency in DCs are generally quite low when using conventional gene transduction methods, such as adenovirus vector infections, lipofection, or electroporation. Because the MHC class I- and II-restricted peptide presentation to naive T cells occurs in lymphoid tissue, strategies that enhance migration of the infused DCs to lymph nodes may enhance their effectiveness. One concern is that ex vivo DC manipulation may affect lymphoid trafficking. To overcome this, combining DCs with chemokine receptors that facilitate lymph node migration might be a possible approach. Also, combining DCs with cytokines such as IFNγ may enhance DC function.

DC-based vaccines that initiate adaptive immunity have also been developed for infectious diseases [172, 173]. Unlike conventional infectious disease vaccines that (though they may be associated with adjuvants and APC activities) trigger mainly protective antibodies, infections like HIV, malaria, and tuberculosis may require more durable and protective T-cell immunity. For HIV, the gag-p24 protein antigen is introduced into an antibody that targets the DEC-205 antigen uptake receptor on DCs [174–177]. When administered together with an adjuvant (e.g., Poly(I:C) or its analogue), early data showed that HIV gag-p24 within anti-DEC-205 antibody was very immunogenic in animal models as well as healthy human volunteers. Although immunogenicity may not be the key factor for development of an effective vaccine in this disease because it is known that the major HIV viral proteins, which include p24 (core antigen) and gp41 (envelope antigen), are highly immunogenic (providing the basis for most HIV testing), further research is necessary. In any case, DC-targeted vaccines may represent a pathway forward to induce integrated immune responses against antigens of some infectious pathogens. Table 8.6 lists ongoing industry studies using DC-based vaccines.

TABLE 8.6 DC-based vaccine trials conducted by a commercial sponsor

Sponsor	Product	Indication	N	Start time	Trial no.
Argos	AGS-003 (RCC)	RCC	7	2011	NCT-01482949
Chiron	DC vaccine (plus CTL)	Melanoma	98	2006	NCT-00338377
CureTech	DC AML vac (+CT-011 mAb)	AML	35	2010	NCT-01096602
DCPrime BV	DC-One	AML	12	2011	NCT-01373515
Dendreon	Sipuleucel-T	Prostate	90	2011	NCT-01338012
Dendreon	Sipuleucel-T ± DNA vaccine	Prostate	30	2012	NCT-01706458
Dendreon	DN24-02	HER2+ urothelial cancer	180	2011	NCT-01353222
Geron	GRNVAC-1	AML	25	2007	NCT-00510133
Immunicum AB	Combig-DC	RCC	12	2012	NCT-01525017
ImmunoCell	ICT-107	GBM	200	2011	NCT-01280552
Northwest Biotherapeutics	DCVac-L	GBM	240	2006	NCT-00045968
Oncovir	Peptide-pulsed DC	Pancreas	4	2011	NCT-01410968
Prima BioMed	Cvac	Ovarian	1000	2012	NCT-01521143
Trimed Bio	Trivax	GBM	56	2010	NCT-01213407
Argos	AGS-004 (HIV)	HIV	42	2010	NCT-01069809

8.4 CONCLUSION

Recent advances in elucidating the molecular and cellular bases of many diseases, along with a refined understanding of basic immunologic processes and the availability of novel tools for ex vivo alteration of human cells, offer new opportunities for disease management [178, 179]. In addition to its established orthopedic and dermatologic use, cell-based biologic therapies have demonstrated potential utility in treating many diseases including cancer, infection, and autoimmune disorders. Many of these experimental therapies are undergoing human clinical trials to elucidate their use. As further delivery tools and more efficient vectors are developed, successful translation of these experimental therapies into actual standard medical practice will follow.

Our expanding knowledge of basic immunologic mechanisms in the last few decades has ushered in an era of cell-based strategies with potential therapeutic value for the treatment of many challenging diseases. Some of these cell-based therapies have made the transition from the preclinical animal research stage into human clinical trials, and a few therapies have been approved. The feasibility and safety of many of these therapies are still being established. More work must be done to define appropriate patient populations, timing of administration, cell quantities, and cellular compositions to achieve optimal therapeutic interventions. To this end, standardization of research protocols for the isolation, stimulation, expansion, reinfusion, or storage of cells for use in cell-based therapies remains essential [180–185]. In fact, better quality controls to minimize variation (both intra- and interpatient) to produce uniform, standardized products are critical for the regulatory approval and ultimate commercialization of these autologous products. Controlled, randomized trials that evaluate the long-term safety and therapeutic efficacy of these approaches will enhance confidence in novel therapies. Given further improvements in delivery and vector technologies, safer and more efficacious therapies will continue to evolve. There is, indeed, a promising future outlook for cell-based biologic therapies.

REFERENCES

1. Ullrich E, Bosch J, Aigner M, Voelkl S, Kroeger I, Hoffmann P, Kreutz M, Dudziak D, Gerbitz A. Advances in cellular therapy: 6th international symposium on the clinical use of cellular products. *Cancer Immunol Immunother* 2012;61(3):433–443.

2. Sanges D, Lluis F, Cosma MP. Cell-fusion-mediated reprogramming: pluripotency or transdifferentiation? Implications for regenerative medicine. *Adv Exp Med Biol* 2011; 713:137–159.

3. Beeson W, Woods E, Agha R. Tissue engineering, regenerative medicine, and rejuvenation in 2010: the role of adipose-derived stem cells. *Facial Plast Surg* 2011;27(4):378–387.

4. Kohara H, Tabata Y. Tissue engineering technology to enhance cell recruitment for regeneration therapy. *J Med Biol Eng* 2010;30(5):267–276.

5. Anonymous. 6th international symposium on biological therapy of cancer: from basic research to clinical application. *Eur J Cancer* 2001;37(Suppl 3):S1–S116.

6. Keilholz U, Klein HG, Körbling M, Brado B, Carter CS, Cullis H, Galm F, Hunstein W. Peripheral blood mononuclear cell collection from patients undergoing adoptive immunotherapy or peripheral blood-derived stem cell transplantation and from healthy donors. *J Clin Apher* 1991;6(3):131–136.

7. Klein HG. Future technologies in hemapheresis. *J Clin Apher* 1991;6(2):103–105.

8. Sun H, Liu W, Zhou G, Zhang W, Cui L, Cao Y. Tissue engineering of cartilage tendon and bone. *Front Med* 2011;5(1):61–69.

9. Ivkovic A, Marijanovic I, Hudetz D, Porter RM, Pecina M, Evans CH. Regenerative medicine and tissue engineering in orthopaedic surgery. *Front Biosci* 2011;3:923–944.

10. Sakai D. Future perspectives of cell-based therapy for intervertebral disc disease. *Eur Spine J* 2008;17(Suppl 4):452–458.

11. Leclerc T, Thepenier C, Jault P, Bey E, Peltzer J, Trouillas M, Duhamel P, Bargues L, Prat M, Bonderriter M, Lataillade JJ. Cell therapy of burns. *Cell Prolif* 2011; 44(Suppl 1):48–54.

12. Wong T, McGrath JA, Navsaria H. The role of fibroblasts in tissue engineering and regeneration. *Br J Dermatol* 2007;156(6):1149–1155.

13. Lee KH. Tissue-engineered human living skin substitutes: development and clinical application. *Yonsei Med J* 2000;41(6):774–779.

14. ISTO Technologies, Inc. Available at http://www.istotech.com/. Accessed May 31, 2013.

15. Neocartilage Implant Phase III Trial. Available at http://clinicaltrials.gov/ct2/show/ NCT01400607. Accessed May 31, 2013.

16. Hepatocyte Matrix Implant Study Indonesia (HMIIndo). Available at http://clinicaltrials. gov/ct2/show/NCT01335568?term=baermed&rank=2. Accessed May 31, 2013.

17. Efe JA, Yuan X, Jiang K, Ding S. Development unchained: how cellular reprogramming is redefining our view of cell fate and identity. *Sci Prog* 2011;94(Pt 3):298–322.

18. Masip M, Veiga A, Izpisúa Belmonte JC, Simón C. Reprogramming with defined factors: from induced pluripotency to induced transdifferentiation. *Mol Hum Reprod* 2010; 16(11):856–868.

19. Burke ZD, Tosh D. Therapeutic potential of transdifferentiated cells. *Clin Sci* 2005; 108(4):309–321.

20. Fausto N, Campbell JS, Riehle KJ. Liver regeneration. *Hepatology* 2006;43(2 Suppl 1): S45–S53.

21. Tweedell KS. The urodele limb regeneration blastema: the cell potential. *ScientificWorld Journal* 2010;10:954–971.

22. Meivar-Levy I, Ferber S. Adult cell fate reprogramming: converting liver to pancreas. *Methods Mol Biol* 2010;636:251–283.

23. Thowfeequ S, Li WC, Slack JM, Tosh D. Reprogramming of liver to pancreas. *Methods Mol Biol* 2009;482:407–418.

24. Fujimoto K, Polonsky KS. Pdx1 and other factors that regulate pancreatic beta-cell survival. *Diabetes Obes Metab* 2009;11(Suppl 4):30–37.

25. Babu DA, Deering TG, Mirmira RG. A feat of metabolic proportions: pdx1 orchestrates islet development and function in the maintenance of glucose homeostasis. *Mol Genet Metab* 2007;92(1–2):43–55.

26. Ber I, Shternhall K, Perl S, Ohanuna Z, Goldberg I, Barshack I, Benvenisti-Zarum L, Meivar-Levy I, Ferber S. Functional, persistent, and extended liver to pancreas transdifferentiation. *J Biol Chem* 2003;278(34):31950–31957.

27. Håkelien AM, Gaustad KG, Collas P. Transient alteration of cell fate using a nuclear and cytoplasmic extract of an insulinoma cell line. *Biochem Biophys Res Commun* 2004; 316(3):834–841.

28. Horb ME, Shen CN, Tosh D, Slack JM. Experimental conversion of liver to pancreas. *Curr Biol* 2003;13(2):105–115.

29. Sumazaki R, Shiojiri N, Isoyama S, Masu M, Keino-Masu K, Osawa M, Nakauchi H, Kageyama R, Matsui A. Conversion of biliary system to pancreatic tissue in Hes1-deficient mice. *Nat Genet* 2004;36(1):83–87.

30. Lu J, Liu H, Huang CT, Chen H, Du Z, Liu Y, Sherafat MA, Zhang SC. Generation of integration-free and region-specific neural progenitors from primate fibroblasts. *Cell Rep* 2013;3:1580–1591.

31. Gooden MJ, de Bock GH, Leffers N, Daemen T, Nijman HW. The prognostic influence of tumour-infiltrating lymphocytes in cancer: a systematic review with meta-analysis. *Br J Cancer* 2011;105(1):93–103.

32. Oble DA, Loewe R, Yu P, Mihm MC Jr. Focus on TILs: Prognostic Significance of Tumor Infiltrating Lymphocytes in Human Melanoma. *Cancer Immun* 2009;9:3.

33. Grupp SA, June CH. Adoptive cellular therapy. *Curr Top Microbiol Immunol* 2011;344:149–172.

34. Melief CJ, O'Shea JJ, Stroncek DF. Summit on Cell Therapy for Cancer: the Importance of the Interaction of Multiple Disciplines to Advance Clinical Therapy. *J Transl Med* 2011;9:107.

35. Fromm PD, Gottlieb D, Bradstock KF, Hart DN. Cellular therapy to treat haematological and other malignancies: progress and pitfalls. *Pathology* 2011;43(6):605–615.

36. Vauleon E, Avril T, Collet B, Mosser J, Quillien V. Overview of cellular immunotherapy for patients with glioblastoma. *Clin Dev Immunol* 2010;2010:689171.

37. Kuebler JP, Whitehead RP, Ward DL, Hemstreet GP 3rd, Bradley EC. Treatment of metastatic renal cell carcinoma with recombinant interleukin-2 in combination with vinblastine or lymphokine-activated killer cells. *J Urol* 1993;150(3):814–820.

38. Hiserodt JC. Lymphokine-activated killer cells: biology and relevance to disease. *Cancer Invest* 1993;11(4):420–439.

39. Tilden AB, Itoh K, Balch CM. Human lymphokine-activated killer (LAK) cells: identification of two types of effector cells. *J Immunol* 1987;138(4):1068–1073.

40. Berdeja JG, Hess A, Lucas DM, O'Donnell P, Ambinder RF, Diehl LF, Carter-Brookins D, Newton S, Flinn IW. Systemic interleukin-2 and adoptive transfer of lymphokine-activated killer cells improves antibody-dependent cellular cytotoxicity in patients with relapsed B-cell lymphoma treated with rituximab. *Clin Cancer Res* 2007;13(8):2392–2399.

41. Yamaguchi Y, Hironaka K, Okawaki M, Okita R, Matsuura K, Ohshita A, Toge T. HER2-specific cytotoxic activity of lymphokine-activated killer cells in the presence of trastuzumab. *Anticancer Res* 2005;25(2A):827–832.

42. Tyler DS, Stanley SD, Bartlett JA, Bolognesi DP, Weinhold KJ. Lymphokine-activated killer (LAK) cell anti-HIV-1 ADCC reactivity: a potential strategy for reduction of virus-infected cellular reservoirs. *J Surg Res* 1998;79(2):115–120.

43. Andersen MH, Schrama D, Thor Straten P, Becker JC. Cytotoxic T cells. *J Invest Dermatol* 2006;126(1):32–41.

44. Bleackley RC. A molecular view of cytotoxic T lymphocyte induced killing. *Biochem Cell Biol* 2005;83(6):747–751.

45. Wang E, Selleri S, Marincola FM. The requirements for CTL-mediated rejection of cancer in humans: NKG2D and its role in the immune responsiveness of melanoma. *Clin Cancer Res* 2007;13(24):7228–7231.

46. Wang J, Wicker LS, Santamaria P. IL-2 and its high-affinity receptor: genetic control of immunoregulation and autoimmunity. *Semin Immunol* 2009;21(6):363–371.

47. Kurnick JT, Kradin RL. Adoptive Immunotherapy with recombinant interleukin 2, LAK and TIL. *Allergol Immunopathol* 1991;19(5):209–214.

48. Tran KQ, Zhou J, Durflinger KH, Langhan MM, Shelton TE, Wunderlich JR, Robbins PF, Rosenberg SA, Dudley ME. Minimally cultured tumor-infiltrating lymphocytes display optimal characteristics for adoptive cell therapy. *J Immunother* 2008;31(8):742–751.

49. Hershkovitz L, Schachter J, Treves AJ, Besser MJ. Focus on adoptive T cell transfer trials in melanoma. *Clin Dev Immunol* 2010;2010:260–267.

50. Khammari A, Nguyen JM, Pandolfino MC, Quereux G, Brocard A, Bercegeay S, Cassidanius A, Lemarre P, Volteau C, Labarrière N, Jotereau F, Dréno B. Long-term follow-up of patients treated by adoptive transfer of melanoma tumor-infiltrating lymphocytes as adjuvant therapy for stage III melanoma. *Cancer Immunol Immunother* 2007;56 (11):1853–1860.

51. Dutcher J, Atkins MB, Margolin K, Weiss G, Clark J, Sosman J, Logan T, Aronson F, Mier J. Kidney cancer: the cytokine working group experience (1986–2001): Part II. Management of IL-2 toxicity and studies with other cytokines. *Med Oncol* 2001;18(3):209–219.

52. Sundin DJ, Wolin MJ. Toxicity management in patients receiving low-dose aldesleukin therapy. *Ann Pharmacother* 1998;32(12):1344–1352.

53. June CH, Adoptive T. Cell therapy for cancer in the clinic. *J Clin Invest* 2007; 117(6):1466–1476.

54. Dudley ME, Rosenberg SA. Adoptive cell transfer therapy. *Semin Oncol* 2007; 34(6):524–531.

55. Kim HR, Hwang KA, Park SH, Kang I. IL-7 and IL-15: biology and roles in T-cell immunity in health and disease. *Crit Rev Immunol* 2008;28(4):325–339.

56. Bodnár A, Nizsalóczki E, Mocsár G, Szalóki N, Waldmann TA, Damjanovich S, Vámosi G. A biophysical approach to IL-2 and IL-15 receptor function: localization conformation and interactions. *Immunol Lett* 2008;116(2):117–125.

57. Wrzesinski C, Paulos CM, Kaiser A, Muranski P, Palmer DC, Gattinoni L, Yu Z, Rosenberg SA, Restifo NP. Increased intensity lymphodepletion enhances tumor treatment efficacy of adoptively transferred tumor-specific T cells. *J Immunother* 2010;33(1):1–7.

58. Laurent J, Speiser DE, Appay V, Touvrey C, Vicari M, Papaioannou A, Canellini G, Rimoldi D, Rufer N, Romero P, Leyvraz S, Voelter V. Impact of 3 different short-term chemotherapy regimens on lymphocyte-depletion and reconstitution in melanoma patients. *J Immunother* 2010;33(7):723–734.

59. Sener A, Tang AL, Farber DL. Memory T-cell predominance following T-cell depletional therapy derives from homeostatic expansion of naive T cells. *Am J Transplant* 2009; 9(11):2615–2623.

60. Wang LX, Shu S, Plautz GE. Host lymphodepletion augments T cell adoptive immunotherapy through enhanced intratumoral proliferation of effector cells. *Cancer Res* 2005; 65(20):9547–9554.

61. Gattinoni L, Finkelstein SE, Klebanoff CA, Antony PA, Palmer DC, Spiess PJ, Hwang LN, Yu Z, Wrzesinski C, Heimann DM, Surh CD, Rosenberg SA, Restifo NP. Removal of

homeostatic cytokine sinks by lymphodepletion enhances the efficacy of adoptively transferred tumor-specific CD8+ T cells. *J Exp Med* 2005;202(7):907–912.

62. Klebanoff CA, Khong HT, Antony PA, Palmer DC, Restifo NP. Sinks, suppressors and antigen presenters: how lymphodepletion enhances T cell-mediated tumor immunotherapy. *Trends Immunol* 2005;26(2):111–117.

63. Søndergaard H, Skak K. IL-21: roles in immunopathology and cancer therapy. *Tissue Antigens* 2009;74(6):467–479.

64. Li Y, Bleakley M, Yee C. IL-21 influences the frequency, phenotype, and affinity of the antigen-specific CD8 T cell response. *J Immunol* 2005;175(4):2261–2269.

65. Takaki R, Hayakawa Y, Nelson A, Sivakumar PV, Hughes S, Smyth MJ, Lanier LL. IL-21 enhances tumor rejection through a NKG2D-dependent mechanism. *J Immunol* 2005;175 (4):2167–2173.

66. Bucher C, Koch L, Vogtenhuber C, Goren E, Munger M, Panoskaltsis-Mortari A, Sivakumar P, Blazar BR. IL-21 blockade reduces graft-versus-host disease mortality by supporting inducible T regulatory cell generation. *Blood* 2009;114(26):5375–5384.

67. Berzofsky JA. A push-pull vaccine strategy using toll-like receptor ligands, IL-15, and blockade of negative regulation to improve the quality and quantity of T cell immune responses. *Vaccine* 2012;30(29):4323–4327.

68. Abdul-Alim CS, Li Y, Yee C. Conditional Superagonist CTL Ligands for the promotion of tumor-specific CTL responses. *J Immunol* 2010;184(11):6514–6521.

69. Gustafson MP, Lin Y, New KC, Bulur PA, O'Neill BP, Gastineau DA, Dietz AB. Systemic immune suppression in glioblastoma: the interplay between CD14+HLA-Drlo/Neg monocytes, tumor factors, and dexamethasone. *Neuro Oncol* 2010;12(7):631–644.

70. NCI, Genesis Biopharma Sign CRADA Agreement to Develop Cancer Immunotherapies. Published on August 10, 2011. Available at http://www.news-medical.net/news/20110810/ NCI-Genesis-Biopharma-sign-CRADA-agreement-to-develop-cancer-immunotherapies. aspx. Accessed May 31, 2013.

71. ClinicalTrials.gov. Available at http://clinicaltrials.gov/. Accessed May 31, 2013.

72. Park TS, Rosenberg SA, Morgan RA. Treating cancer with genetically engineered T cells. *Trends Biotechnol* 2011;29(11):550–557.

73. Udyavar A, Geiger TL. Rebalancing immune specificity and function in cancer by T-cell receptor gene therapy. *Arch Immunol Ther Exp* 2010;58(5):335–346.

74. Kieback E, Uckert W. Enhanced T cell receptor gene therapy for cancer. *Expert Opin Biol Ther* 2010;10(5):749–762.

75. Schmitt TM, Ragnarsson GB, Greenberg PD. T cell receptor gene therapy for cancer. *Hum Gene Ther* 2009;20(11):1240–1248.

76. Edwards LJ, Evavold BD. T cell recognition of weak ligands: roles of signaling, receptor number, and affinity. *Immunol Res* 2011;50(1):39–48.

77. Schmid DA, Irving MB, Posevitz V, Hebeisen M, Posevitz-Fejfar A, Sarria JC, Gomez-Eerland R, Thome M, Schumacher TN, Romero P, Speiser DE, Zoete V, Michielin O, Rufer N. Evidence for a TCR affinity threshold delimiting maximal CD8 T cell function. *J Immunol* 2010;184(9):4936–4946.

78. Haidar JN, Pierce B, Yu Y, Tong W, Li M, Weng Z. Structure-based design of a T-cell receptor leads to nearly 100-fold improvement in binding affinity for pepMHC. *Proteins* 2009;74(4):948–960.

79. Subbramanian RA, Moriya C, Martin KL, Peyerl FW, Hasegawa A, Naoi A, Chhay H, Autissier P, Gorgone DA, Lifton MA, Kuus-Reichel K, Schmitz JE, Letvin NL, Kuroda MJ. Engineered T-cell receptor tetramers bind MHC-peptide complexes with high affinity. *Nat Biotechnol* 2004;22(11):1429–1434.

80. Jorritsma A, Schotte R, Coccoris M, de Witte MA, Schumacher TN. Prospects and limitations of T cell receptor gene therapy. *Curr Gene Ther* 2011;11(4):276–287.

81. Correia-Neves M, Waltzinger C, Wurtz JM, Benoist C, Mathis D. Amino acids specifying MHC class preference in TCR V alpha 2 regions. *J Immunol* 1999;163(10):5471–5477.

82. Morgan RA, Dudley ME, Yu YY, Zheng Z, Robbins PF, Theoret MR, Wunderlich JR, Hughes MS, Restifo NP, Rosenberg SA. High Efficiency TCR gene transfer into primary human lymphocytes affords avid recognition of melanoma tumor antigen glycoprotein 100 and does not alter the recognition of autologous melanoma antigens. *J Immunol* 2003;171(6):3287–3295.

83. Brenner M. T cell receptors and cancer: gain gives pain. *Nature Med* 2010;16:520–521.

84. Cohen CJ, Li YF, El-Gamil M, Robbins PF, Rosenberg SA, Morgan RA. Enhanced anti-tumor activity of T cells engineered to express T-cell receptors with a second disulfide bond. *Cancer Res* 2007;67(8):3898–3903.

85. Curran KJ, Pegram HJ, Brentjens RJ. Chimeric antigen receptors for T cell immunotherapy: current understanding and future directions. *J Gene Med* 2012;14(6):405–415.

86. Lipowska-Bhalla G, Gilham DE, Hawkins RE, Rothwell DG. Targeted immunotherapy of cancer with CAR T cells: achievements and challenges. *Cancer Immunol Immunother* 2012;61(7):953–962.

87. Shirasu N, Kuroki M. Functional design of chimeric T-cell antigen receptors for adoptive immunotherapy of cancer: architecture and outcomes. *Anticancer Res* 2012;32(6): 2377–2383.

88. Grupp SA, Kalos M, Barrett D, Aplenc R, Porter DL, Rheingold SR, Teachey DT, Chew A, Hauck B, Wright JF, Milone MC, Levine BL, June CH. Chimeric antigen receptor-modified T cells for acute lymphoid leukemia. *N Engl J Med* 2013;368 (16): 1509–1518.

89. Brentjens RJ, Davila ML, Riviere I, Park J, Wang X, Cowell LG, Bartido S, Stefanski J, Taylor C, Olszewska M, Borquez-Ojeda O, Qu J, Wasielewska T, He Q, Bernal Y, Rijo IV, Hedvat C, Kobos R, Curran K, Steinherz P, Jurcic J, Rosenblat T, Maslak P, Frattini M, Sadelain M. CD19-targeted T cells rapidly induce molecular remissions in adults with chemotherapy-refractory acute lymphoblastic leukemia. *Sci Transl Med* 2013;5 (177): 177ra38.

90. Kochenderfer JN, Rosenberg SA. Treating B-cell cancer with T cells expressing anti-CD19 chimeric antigen receptors. *Nat Rev Clin Oncol* 2013;10(5):267–276.

91. Kandalaft LE, Powell DJ Jr, Coukos G. A phase I clinical trial of adoptive transfer of folate receptor-alpha redirected autologous T cells for recurrent ovarian cancer. *J Transl Med* 2012;10:157.

92. Kebriaei P, Huls H, Jena B, Munsell M, Jackson R, Lee DA, Hackett PB, Rondon G, Shpall E, Champlin RE, Cooper LJ. Infusing CD19-directed T cells to augment disease control in patients undergoing autologous hematopoietic stem-cell transplantation for advanced B-lymphoid malignancies. *Hum Gene Ther* 2012;23(5):444–450.

93. Bridgeman JS, Hawkins RE, Hombach AA, Abken H, Gilham DE. Building better chimeric antigen receptors for adoptive T cell therapy. *Curr Gene Ther* 2010;10(2):77–90.

94. Hollyman D, Stefanski J, Przybylowski M, Bartido S, Borquez-Ojeda O, Taylor C, Yeh R, Capacio V, Olszewska M, Hosey J, Sadelain M, Brentjens RJ, Rivière I. Manufacturing Validation of Biologically Functional T Cells Targeted to CD19 Antigen for Autologous Adoptive Cell Therapy. *J Immunother* 2009;32(2):169–180.

95. Almåsbak H, Lundby M, Rasmussen AM. Non-MHC-dependent redirected T cells against tumor cells. *Methods Mol Biol* 2010;629:453–493.

96. Savoldo B, Ramos CA, Liu E, Mims MP, Keating MJ, Carrum G, Kamble RT, Bollard CM, Gee AP, Mei Z, Liu H, Grilley B, Rooney CM, Heslop HE, Brenner MK, Dotti G. CD28 Costimulation improves expansion and persistence of chimeric antigen receptor-modified T cells in lymphoma patients. *J Clin Invest* 2011;121(5):1822–1826.

97. Hombach AA, Abken H. Costimulation by chimeric antigen receptors revisited the T cell antitumor response benefits from combined CD28-OX40 signaling. *Int J Cancer* 2011;129(12):2935–2944.

98. Casucci M, Bondanza A. Suicide gene therapy to increase the safety of chimeric antigen receptor-redirected T lymphocytes. *J Cancer* 2011;2:378–382.

99. Hoyos V, Savoldo B, Quintarelli C, Mahendravada A, Zhang M, Vera J, Heslop HE, Rooney CM, Brenner MK, Dotti G. Engineering CD19-specific T lymphocytes with interleukin-15 and a suicide gene to enhance their anti-lymphoma/leukemia effects and safety. *Leukemia* 2010;24(6):1160–1170.

100. Ramos CA, Asgari Z, Liu E, Yvon E, Heslop HE, Rooney CM, Brenner MK, Dotti G. An inducible caspase 9 suicide gene to improve the safety of mesenchymal stromal cell therapies. *Stem Cells* 2010;28(6):1107–1115.

101. Manuri PV, Wilson MH, Maiti SN, Mi T, Singh H, Olivares S, Dawson MJ, Huls H, Lee DA, Rao PH, Kaminski JM, Nakazawa Y, Gottschalk S, Kebriaei P, Shpall EJ, Champlin RE, Cooper LJ. Piggybac transposon/transposase system to generate CD19-specific T cells for the treatment of B-lineage malignancies. *Hum Gene Ther* 2010; 21(4):427–437.

102. Van Kaer L, Parekh VV, Wu L. Invariant natural killer T cells: bridging innate and adaptive immunity. *Cell Tissue Res* 2011;343(1):43–55.

103. Levy EM, Roberti MP, Mordoh J. Natural Killer Cells in Human Cancer: from Biological Functions to Clinical Applications. *J Biomed Biotechnol* 2011;2011:676198.

104. Lee SK, Gasser S. The role of natural killer cells in cancer therapy. *Front Biosci* 2010;2:380–391.

105. Carlsten M, Malmberg KJ, Ljunggren HG. Natural killer cell-mediated lysis of freshly isolated human tumor cells. *Int J Cancer* 2009;124(4):757–762.

106. Terme M, Ullrich E, Delahaye NF, Chaput N, Zitvogel L. Natural killer cell-directed therapies: moving from unexpected results to successful strategies. *Nat Immunol* 2008; 9(5):486–494.

107. Guo H, Qian X. Clinical applications of adoptive natural killer cell immunotherapy for cancer: current status and future prospects. *Onkologie* 2010;33(7):389–395.

108. Burke S, Lakshmikanth T, Colucci F, Carbone E. New views on natural killer cell-based immunotherapy for melanoma treatment. *Trends Immunol* 2010;31(9):339–345.

109. Sutlu T, Alici E. Natural killer cell-based immunotherapy in cancer: current insights and future prospects. *J Intern Med* 2009;266(2):154–181.

110. Srivastava S, Lundqvist A, Childs RW. Natural killer cell immunotherapy for cancer: a new hope. *Cytotherapy* 2008;10(8):775–783.

111. Moustaki A, Argyropoulos KV, Baxevanis CN, Papamichail M, Perez SA. Effect of the simultaneous administration of glucocorticoids and IL-15 on human NK cell phenotype, proliferation and function. *Cancer Immunol Immunother* 2011;60(12):1683–1695.

112. Suck G, Oei VY, Linn YC, Ho SH, Chu S, Choong A, Niam M, Koh MB. Interleukin-15 supports generation of highly potent clinical-grade natural killer cells in long-term cultures for targeting hematological malignancies. *Exp Hematol* 2011;39(9):904–914.

113. Perez SA, Mahaira LG, Demirtzoglou FJ, Sotiropoulou PA, Ioannidis P, Iliopoulou EG, Gritzapis AD, Sotiriadou NN, Baxevanis CN, Papamichail M. A potential role for hydrocortisone in the positive regulation of IL-15-activated NK-cell proliferation and survival. *Blood* 2005;106(1):158–166.

114. Suck G, Koh MB. Emerging natural killer cell immunotherapies: large-scale ex vivo production of highly potent anticancer effectors. *Hematol Oncol Stem Cell Ther* 2010;3(3):135–142.

115. Cho D, Campana D. Expansion and activation of natural killer cells for cancer immunotherapy. *Korean J Lab Med* 2009;29(2):89–96.

116. Salagianni M, Lekka E, Moustaki A, Iliopoulou EG, Baxevanis CN, Papamichail M, Perez SA. NK Cell adoptive transfer combined with Ontak-mediated regulatory T cell elimination induces effective adaptive antitumor immune responses. *J Immunol* 2011;186 (6): 3327–3335.

117. Sakaguchi S. Regulatory T cells: history and perspective. *Methods Mol Biol* 2011;707:3–17.

118. Byrne WL, Mills KH, Lederer JA, O'Sullivan GC. Targeting regulatory T cells in cancer. *Cancer Res* 2011;71(22):6915–6920.

119. Leguern C. Regulatory T cells for tolerance therapy: revisiting the concept. *Crit Rev Immunol* 2011;31(3):189–207.

120. Campbell DJ, Koch MA. Phenotypical and functional specialization of FOXP3+ regulatory T cells. *Nat Rev Immunol* 2011;11(2):119–130.

121. Wright GP, Ehrenstein MR, Stauss HJ. Regulatory T-cell adoptive immunotherapy: potential for treatment of autoimmunity. *Expert Rev Clin Immunol* 2011;7(2):213–225.

122. Sued O, Quiroga MF, Socías ME, Turk G, Salomón H, Cahn P. Acute HIV seroconversion presenting with active tuberculosis and associated with high levels of T-regulatory cells. *Viral Immunol* 2011;24(4):347–349.

123. Scholzen A, Minigo G, Plebanski M. Heroes or villains? T regulatory cells in malaria infection. *Trends Parasitol* 2010;26(1):16–25.

124. Bowen DL, Lane HC, Fauci AS. Immunopathogenesis of the acquired immunodeficiency syndrome. *Ann Int Med* 1985;103(5):704–709.

125. Buzon MJ, Sun H, Li C, Shaw A, Seiss K, Ouyang Z, Martin-Gayo E, Leng J, Henrich TJ, Li JZ, Pereyra F, Zurakowski R, Walker BD, Rosenberg ES, Yu XG, Lichterfeld M. HIV-1 persistence in CD4+ T cells with stem cell-like properties. *Nat Med* 2014; 20:139–142.

126. Doitsh G, Galloway NLK, Geng X, Yang Z, Monroe KM, Zepeda O, Hunt PW, Hatano H, Sowinski S, Munoz-Arias I, Greene WC. Cell death by pyroptosis drives CD4 T-cell depletion in HIV-1 infection. *Nature* 2014;505:509–514.

127. Liu F, Lang R, Zhao J, Zhang X, Pringle GA, Fan Y, Yin D, Gu F, Yao Z, Fu L. CD8+ Cytotoxic T cell and FOXP3+ regulatory T cell infiltration in relation to breast cancer survival and molecular subtypes. *Breast Cancer Res Treat* 2011;130(2):645–655.

128. Grotenhuis JA, de Vries IJ, Wesseling P, Adema GJ. Prognostic significance and mechanism of Treg infiltration in human brain tumors. *J Neuroimmunol* 2010;225(1–2):195–199.

129. Zhang C, Shan J, Feng L, Lu J, Xiao Z, Luo L, Li C, Guo Y, Li Y. The effects of immunosuppressive drugs on CD4(+) CD25(+) regulatory T cells: a systematic review of clinical and basic research. *J Evid Based Med* 2010;3(2):117–129.

130. Ohkura N, Hamaguchi M, Sakaguchi S. FOXP3+ regulatory T cells: control of FOXP3 expression by pharmacological agents. *Trends Pharmacol Sci* 2011;32(3):158–166.

131. Brusko T, Putnam A, Bluestone J. Human Regulatory T Cells: Role in Autoimmune Disease and Therapeutic Opportunities. *Immunol Rev* 2008;223:371–390.

132. Safety Study of Using Regulatory T Cells Induce Liver Transplantation Tolerance (Treg). Available at http://clinicaltrials.gov/ct2/show/NCT01624077?term=Tregs&rank=6. Accessed May 31, 2013.

133. T1DM Immunotherapy Using CD4+CD127lo/-CD25+ Polyclonal Tregs. Available at http://clinicaltrials.gov/ct2/show/NCT01210664?term=Tregs&rank=5. Accessed May 31, 2013.

134. Treatment of Children with Kidney Transplants by Injection of CD4+CD25+FoxP3+ T Cells to Prevent Organ Rejection. Available at http://clinicaltrials.gov/ct2/show/NCT01446484. Accessed May 31, 2013.

135. Galli SJ, Borregaard N, Wynn TA. Phenotypic and functional plasticity of cells of innate immunity: macrophages, mast cells and neutrophils. *Nat Immunol* 2011;12(11):1035–1044.

136. Mosser DM, Edwards JP. Exploring the full spectrum of macrophage activation. *Nat Rev Immunol* 2008;8(12):958–969.

137. Taylor PR, Martinez-Pomares L, Stacey M, Lin HH, Brown GD, Gordon S. Macrophage Receptors and Immune Recognition. *Annu Rev Immunol* 2005;23:901–944.

138. Murray PJ, Wynn TA. Protective and pathogenic functions of macrophage subsets. *Nat Rev Immunol* 2011;11(11):723–737.

139. Devaraj S, Jialal I. C-reactive protein polarizes human macrophages to an M1 phenotype and inhibits transformation to the M2 phenotype. *Arterioscler Thromb Vasc Biol* 2011; 31(6):1397–1402.

140. Hazeki K, Kametani Y, Murakami H, Uehara M, Ishikawa Y, Nigorikawa K, Takasuga S, Sasaki T, Seya T, Matsumoto M, Hazeki O. Phosphoinositide 3-kinaseγ controls the intracellular localization of CpG to limit DNA-PKcs-dependent IL-10 production in macrophages. *PLoS One* 2011;6(10):e26836.

141. Lu H, Huang D, Saederup N, Charo IF, Ransohoff RM, Zhou L. Macrophages Recruited via CCR2 Produce insulin-like growth factor-1 to repair acute skeletal muscle injury. *FASEB J* 2011;25(1):358–369.

142. Koh TJ, DiPietro LA. Inflammation and Wound Healing: the Role of the Macrophage. *Expert Rev Mol Med* 2011;13:e23.

143. Wang Y, Wang YP, Zheng G, Lee VW, Ouyang L, Chang DH, Mahajan D, Coombs J, Wang YM, Alexander SI, Harris DC. Ex Vivo Programmed macrophages ameliorate experimental chronic inflammatory renal disease. *Kidney Int* 2007;72(3):290–299.

144. Andreesen R, Hennemann B, Krause SW. Adoptive immunotherapy of cancer using monocyte-derived macrophages: rationale, current status, and perspectives. *J Leukoc Biol* 1998;64(4):419–426.

145. Thiounn N, Pages F, Mejean A, Descotes JL, Fridman WH, Romet-Lemonne JL. Adoptive immunotherapy for superficial bladder cancer with autologous macrophage activated killer cells. *J Urol* 2002;168(6):2373–2376.

146. Faradji A, Bohbot A, Schmitt-Goguel M, Roeslin N, Dumont S, Wiesel ML, Lallot C, Eber M, Bartholeyns J, Poindron P. Phase I trial of intravenous infusion of ex-vivo-activated autologous blood-derived macrophages in patients with non-small-cell lung cancer: toxicity and immunomodulatory effects. *Cancer Immunol Immunother* 1991;33(5):319–326.

147. Autologous Incubated Macrophages for Patients with Complete Spinal Cord Injuries. Available at http://clinicaltrials.gov/ct2/show/NCT00073853?term=proneuron&rank=1. Accessed May 31, 2013.

148. Hanson HL, Donermeyer DL, Ikeda H, White JM, Shankaran V, Old LJ, Shiku H, Schreiber RD, Allen PM. Eradication of established tumors by CD8+ T cell adoptive immunotherapy. *Immunity* 2000;13(2):265–276.

149. Liu MA. Cancer vaccines. *Philos Trans R Soc Lond B: Biol Sci* 2011;366(1579): 2823–2826.

150. Bhargava A, Mishra D, Banerjee S, Mishra PK. Dendritic cell engineering for tumor immunotherapy: from biology to clinical translation. *Immunotherapy* 2012;4(7): 703–718.

151. Tuettenberg A, Becker C, Correll A, Steinbrink K, Jonuleit H. Immune regulation by dendritic cells and T cells—basic science, diagnostic, and clinical application. *Clin Lab* 2011;57(1–2):1–12.

152. Mascanfroni ID, Cerliani JP, Dergan-Dylon S, Croci DO, Ilarregui JM, Rabinovich GA. Endogenous lectins shape the function of dendritic cells and tailor adaptive immunity: mechanisms and biomedical applications. *Int Immunopharmacol* 2011;11(7):833–841.

153. Querec T, Bennouna S, Alkan S, Laouar Y, Gorden K, Flavell R, Akira S, Ahmed R, Pulendran B. Yellow fever vaccine YF-17D activates multiple dendritic cell subsets via TLR2, 7, 8, and 9 to stimulate polyvalent immunity. *J Exp Med* 2006; 203(2):413–424.

154. Gaucher D, Therrien R, Kettaf N, Angermann BR, Boucher G, Filali-Mouhim A, Moser JM, Mehta RS, Drake DR 3rd, Castro E, Akondy R, Rinfret A, Yassine-Diab B, Said EA, Chouikh Y, Cameron MJ, Clum R, Kelvin D, Somogyi R, Greller LD, Balderas RS, Wilkinson P, Pantaleo G, Tartaglia J, Haddad EK, Sékaly RP. Yellow fever vaccine induces integrated multilineage and polyfunctional immune responses. *J Exp Med* 2008;205(13):3119–3131.

155. Banchereau J, Briere F, Caux C, Davoust J, Lebecque S, Liu YJ, Pulendran B, Palucka K. Immunobiology of dendritic cells. *Annu Rev Immunol* 2000;18:767–811.

156. Dzionek A, Fuchs A, Schmidt P, Cremer S, Zysk M, Miltenyi S, Buck DW, Schmitz J. BDCA-2, BDCA-3, and BDCA-4: three markers for distinct subsets of dendritic cells in human peripheral blood. *J Immunol* 2000;165(11):6037–6046.

157. Castiello L, Sabatino M, Jin P, Clayberger C, Marincola FM, Krensky AM, Stroncek DF. Monocyte-derived DC maturation strategies and related pathways: a transcriptional view. *Cancer Immunol Immunother* 2011;60(4):457–466.

158. Trumpfheller C, Longhi MP, Caskey M, Idoyaga J, Bozzacco L, Keler T, Schlesinger SJ, Steinman RM. Dendritic cell-targeted protein vaccines: a novel approach to induce T cell immunity. *J Intern Med* 2012;271(2):183–192.

159. Le DT, Pardoll DM, Jaffee EM. Cellular vaccine approaches. *Cancer J* 2010; 16(4):304–310.

160. Thara E, Dorff TB, Pinski JK, Quinn DI. Vaccine therapy with Sipuleucel-T (Provenge) for prostate cancer. *Maturitas* 2011;69(4):296–303.

161. Arlen PM. Prostate cancer immunotherapy: the role for Sipuleucel-T and other immunologic approaches. *Oncology* 2011;25(3):261–262.

162. Herman L, Hubert P, Herfs M, Kustermans G, Henrotin Y, Bousarghin L, Boniver J, Delvenne P. The L1 major capsid protein of HPV16 differentially modulates APC trafficking according to the vaccination or natural infection context. *Eur J Immunol* 2010; 40(11):3075–3084.

163. Ruzevick J, Jackson C, Phallen J, Lim M. Clinical trials with immunotherapy for high-grade glioma. *Neurosurg Clin N Am* 2012;23(3):459–470.

164. Wang J, Liao L, Tan J. Dendritic cell-based vaccination for renal cell carcinoma: challenges in clinical trials. *Immunotherapy* 2012;4(10):1031–1042.

165. Hill M, Segovia M, Cuturi MC. What is the role of antigen-processing mechanisms in autologous tolerogenic dendritic cell therapy in organ transplantation? *Immunotherapy* 2011;3(4 Suppl):12–14.

166. Palucka K, Ueno H, Banchereau J. Recent developments in cancer vaccines. *J Immunol* 2011;186(3):1325–1331.

167. Hodi FS, O'Day SJ, McDermott DF, Weber RW, Sosman JA, Haanen JB, Gonzalez R, Robert C, Schadendorf D, Hassel JC, Akerley W, van den Eertwegh AJ, Lutzky J, Lorigan P, Vaubel JM, Linette GP, Hogg D, Ottensmeier CH, Lebbé C, Peschel C, Quirt I, Clark JI, Wolchok JD, Weber JS, Tian J, Yellin MJ, Nichol GM, Hoos A, Urba WJ. Improved survival with ipilimumab in patients with metastatic melanoma. *N Engl J Med* 2010;363(8):711–723.

168. Higano CS, Schellhammer PF, Small EJ, Burch PA, Nemunaitis J, Yuh L, Provost N, Frohlich MW. Integrated data from 2 randomized, double-blind, placebo-controlled, phase 3 trials of active cellular immunotherapy with sipuleucel-T in advanced prostate cancer. *Cancer* 2009;115(16):3670–3679.

169. Torabi-Rahvar M, Bozorgmehr M, Jeddi-Tehrani M, Zarnani AH. Potentiation strategies of dendritic cell-based antitumor vaccines: combinational therapy takes the front seat. *Drug Discov Today* 2011;16(15–16):733–740.

170. Boudreau JE, Bonehill A, Thielemans K, Wan Y. Engineering dendritic cells to enhance cancer immunotherapy. *Mol Ther* 2011;19(5):841–853.

171. Lee WT. Dendritic cell-tumor cell fusion vaccines. *Adv Exp Med Biol* 2011;713:177–186.

172. Ludewig B. Dendritic cell vaccination and viral infection—animal models. *Curr Top Microbiol Immunol* 2003;276:199–214.

173. Brandonisio O, Spinelli R, Pepe M. Dendritic cells in leishmania infection. *Microbes Infect* 2004;6(15):1402–1409.

174. Niu L, Termini JM, Kanagavelu SK, Gupta S, Rolland MM, Kulkarni V, Pavlakis GN, Felber BK, Mullins JI, Fischl MA, Stone GW. Preclinical evaluation of HIV-1 therapeutic ex vivo dendritic cell vaccines expressing consensus gag antigens and conserved gag epitopes. *Vaccine* 2011;29(11):2110–2119.

175. García F, Routy JP. Challenges in dendritic cells-based therapeutic vaccination in HIV-1 infection workshop in dendritic cell-based vaccine clinical trials in HIV-1. *Vaccine* 2011;29(38):6454–6463.

176. Smed-Sörensen A, Loré K. Targeting dendritic cells for improved HIV-1 vaccines. *Adv Exp Med Biol* 2013;762:263–288.

177. Cruz LJ, Rueda F, Tacken P, Albericio F, Torensma R, Figdor CG. Enhancing immunogenicity and cross-reactivity of HIV-1 antigens by in vivo targeting to dendritic cells. *Nanomedicine* 2012;7(10):1591–1610.

178. Chen C, Loe F, Blocki A, Peng Y, Raghunath M. Applying macromolecular crowding to enhance extracellular matrix deposition and its remodeling in vitro for tissue engineering and cell-based therapies. *Adv Drug Deliv Rev* 2011;63(4–5):277–290.

179. Kasko AM, Wong DY. Two-photon lithography in the future of cell-based therapeutics and regenerative medicine: a review of techniques for hydrogel patterning and controlled release. *Future Med Chem* 2010;2(11):1669–1680.

180. Malyguine A, Strobl S, Zaritskaya L, Baseler M, Shafer-Weaver K. New approaches for monitoring CTL activity in clinical trials. *Adv Exp Med Biol* 2007;601:273–284.

181. Elkord E, Burt DJ. Novel IFNγ ELISPOT Assay for detection of functional carcinoembryonic antigen-specific chimeric antigen receptor-redirected T cells. *Scand J Immunol* 2011;74(4):419–422.

182. Schroten C, Kraaij R, Veldhoven JL, Berrevoets CA, den Bakker MA, Ma Q, Sadelain M, Bangma CH, Willemsen RA, Debets R. T cell activation upon exposure to patient-derived tumor tissue: a functional assay to select patients for adoptive T cell therapy. *J Immunol Methods* 2010;359(1–2):11–20.

183. Liu G, Swierczewska M, Niu G, Zhang X, Chen X. Molecular imaging of cell-based cancer immunotherapy. *Mol Biosyst* 2011;7(4):993–1003.

184. Jha P, Golovko D, Bains S, Hostetter D, Meier R, Wendland MF, Daldrup-Link HE. Monitoring of natural killer cell immunotherapy using noninvasive imaging modalities. *Cancer Res* 2010;70(15):6109–6113.

185. Turtle CJ, Hudecek M, Jensen MC, Riddell SR. Engineered T cells for anti-cancer therapy. *Curr Opin Immunol* 2012;24(5):633–639.

9

DEVELOPMENT OF STEM CELL THERAPY FOR MEDICAL USES

KLAUDYNE HONG AND MAN C. FUNG

9.1 INTRODUCTION

Stem cell therapy is generally defined as the use of stem or progenitor cells to treat a medical condition. Often classified within the field of regenerative medicine, stem cell therapy is tied to the research that gained prominence when McCulloch and Till of the University of Toronto first identified stem cells in mouse in the 1960s [1]. This eventually led to the first successful application of stem cell therapy for a patient with severe combined immunodeficiency, whereby hematopoietic stem cells were harvested from the bone marrow of the patient's sibling and transplanted into the patient [2]. Since then, hematopoietic stem cell transplant (HSCT) has been widely adopted by hematologists and oncologists for many hematologic diseases and malignancies including various anemia, leukemia, lymphoma, and immunologic disorders [3,4]. While most HSCTs are performed for lymphoid and hematologic cancers, there are successes also for other conditions such as amyloidosis, some autoimmune disorders, as well as inborn errors of metabolism [4].

In the 1990s, embryonic stem cells (ESCs) were touted for their omnipotency and unlimited expansion potential [5–7]. ESCs have the potential of differentiating into any and all cell types that can be used to repair or replace disordered or injured body tissues or organs. However, the source of ESCs is rooted in ethical dilemma, and alternate technologies and sources of donor cells were and continue to be developed [7,8]. For example, great progresses for alternative sources such as neonatal

Therapeutic Delivery Solutions, First Edition. Edited by Chung Chow Chan, Kwok Chow, Bill McKay, and Michelle Fung.
© 2014 John Wiley & Sons, Inc. Published 2014 by John Wiley & Sons, Inc.

(nonembryonic) stem cells, adult stem cells, and most recently induced pluripotent stem cells (iPSCs) have opened new doors for therapeutic use [9–14]. This chapter will focus on the recent development of nonembryonic cell research and their potential use in medical treatment.

9.2 HISTORY OF STEM CELL DEVELOPMENT

The history of stem cell development began in the 1960s when McCulloch and Till identified the presence of self-renewing cells in the bone marrow of mice [1]. Since then, there are major research breakthroughs in stem cell research. Due to the ethical controversy of using ESCs, there are alternate developments of non-ESCs from other sources such as adipose, dental, and other body tissues [9–13]. There were also attempts to induce mature skin cells to become omnipotent stem cells by inserting various gene signals [13–15]. Table 9.1 illustrates the key milestones of stem cell developments over the past few decades [1–3,5,9,13,15–22].

TABLE 9.1 Key chronological events of stem cell development

The 1960s–1980s

1963: E McCulloch/J Till (University of Toronto, Toronto, Canada) identified self-renewing cells in bone marrow of mice

1968: HSCT between two siblings was successful in treating a patient with severe combined immunodeficiency

1978: Hematopoietic stem cells were discovered in human cord blood

1981: M Evans/M Kaufman (Cambridge University, Cambridge, UK) and G Martin (University of California San Francisco (UCSF), San Francisco, CA) produced mouse ESCs from embryoblast

1988: The first successful cord blood transplant in a patient with Fanconi's anemia in Paris

The 1990s

1993: First unrelated donor cord blood transplant performed at Duke University, Durham, NC

1997: First cord blood transplant using ex vivo expansion for a patient with chronic myelogenous leukemia

1998: J Thomson et al. (University of Wisconsin, Madison, WI) derived the first human ESC line

1998: J Gearhart (Johns Hopkins University, Baltimore, MD) derived pluripotent stem cell lines from fetal gonadal tissue destined to form germ cells

2001–2005

2001: P Zuk et al. (University of California Los Angeles (UCLA), Los Angeles, CA) identified a stem cell population from adipose stromal tissue

2001: J Cibelli et al. (Advanced Cell Tech, CA) cloned the first early human embryos for the generation of ESC lines

2003: S Shi (US National Institute of Health (US NIH), Bethesda, MD) discovered new source of adult stem cells in children's primary teeth

2003: A Prusa et al. (University of Vienna, Vienna, Austria) reported discovery of pluripotent stem cells in amniotic fluids

(Continued)

TABLE 9.1 (Cont'd)

2005: C McGuckin et al. (Kingston University, Surrey, UK) identified cord blood-derived embryonic-like stem cells

2005: H Keirstead et al. (University of California Irvine (UC Irvine), Irvine, CA) restored partial ability for spinal cord-injured rats to walk by infusing human neural stem cells

2006–present

2006: N Forraz/C McGuckin (Newcastle University, Newcastle, UK) created the first artificial hepatocytes using cord blood stem cells

2007: A Atala et al. (Wake Forest University, Winston-Salem, NC) used AFSC and transformed them into multiple cell types

2007: S Mitalipov (Oregon Health and Sciences University, Portland, OR) reported creation of a primate stem cell line through somatic cell nuclear transfer

2007: K Takahashi/S Yamanaka (Kyoto University, Kyoto, Japan) and J Yu et al. (University of Wisconsin, Madison, WI) independently reported creation of human *iPSCs*

2008: S Yamanaka et al. (Kyoto University, Kyoto, Japan) generated nontumorigenic pluripotent stem cells from adult mouse liver and stomach

2008: C Centeno et al. (Regenerative Sciences Inc., Broomfield, CO) performed successful cartilage repair using autologous adult MSCs

2008: K Eggan (Harvard University, Boston, MA)/C Henderson (Columbia University, New York) created neuron cells from iPSCs derived from fibroblasts of two ALS patients

2008: S Conrad et al. (Tübingen University, Germany) generated pluripotent stem cells from adult human testis

2009: A Nagy et al. (Samuel Lunenfeld Institute, Canada) used nonviral approach (electroporation) for gene insertion to create iPSCs

2009: D Kim et al. (Harvard University, Boston, MA) reported creation of iPSCs from fibroblasts by direct delivery of reprogramming proteins fused with a cell-penetrating peptide, eliminating the needs of virus or DNA transfection as gene transfer vector

2010: P Madeddu et al. (Bristol University, Bristol, UK) extracted stem cells from blood vessels removed from bypass surgeries and used them to stimulate growth of new arteries in patients

2011: Weinberg (Harvard University, Boston, MA) reported plasticity in mammary epithelial cells that spontaneously converted to stemlike cells

2011: K Eto (Kyoto University, Kyoto, Japan) reported the use of iPSCs to create immortal megakaryocytic cell lines for platelet generation

2012: J Hickman et al. (University of Central Florida, Orlando, FL) created brain cells from cord blood stem cells

2012: Osiris's Prochymal® (adult MSCs) was approved in Canada and New Zealand for GVHD

2013: V Giampapa (CellHealth Institute, Montclair, NJ) found adult stem cells can be reprogrammed to act as younger versions of themselves

9.3 SOURCES OF NONEMBRYONIC STEM CELLS

There are three major sources of nonembryonic, nonfetal stem cells: neonatal, adult, and reprogrammed stem cells. For neonatal-related stem cells, the focus will be amniotic stem cells or umbilical cord-/placenta-sourced stem cells. While it is possible to also harvest stem cells from tissues or organs from an aborted or miscarried fetus, such practice is not included in this review due to a similar controversy like ESCs. The following description provides more detailed information about these two sources:

9.3.1 Neonatal-Related Stem Cells

9.3.1.1 Amniotic Fluid Stem Cells (AFSCs) Stem cells derived from amniotic fluid are a new source of cells that have therapeutic potential in diseases both pre- and postnatally [23–35]. Amniotic fluid cells consist of a heterogeneous cell population of exfoliated fetal and amniotic cells [31]. In 2003, cells that displayed high plasticity with pluripotent characteristics (expressing Oct-4 marker) were identified in the amniotic fluid [32]. These cells were found to be cryopreservable for cell banking, not tumorigenic, and can be expanded in over 250 population doublings. The potency of these cells seems to be between ESCs and adult stem cells, and they are capable to differentiate into multiple tissue types encompassing each of the three embryonic germ layers [33,34]. For example, human lines from amniotic fluid stem cells (AFSCs) have been induced to differentiate into a broad range of cell types including adipogenic, osteogenic, myogenic, endothelial, neuronal, and even hepatic lineages. Early animal data showed utility of these cells in cardiac and renal diseases as well as for hematopoietic uses [25,35]. Availability of AFSCs opens new opportunities in regenerative medicine because AFSCs are not subjected to teratoma formation or ethical controversy. Thus, they are an appealing alternative to ESCs. In addition, their immunomodulating properties and low immunogenicity make them promising for regeneration of tissues as well as useful in graft-versus-host disease (GVHD). Besides adult diseases, these cells also have the potential for use in prenatal and postnatal therapies [23,27,29,30]. Monoclonal AFSC lines via amniocenteses have been shown to be genomically stable with high proliferative potential. In addition, during the time of newborn delivery, these cells can be collected and expanded after labor (e.g., if an index fetus with congenital malformation needs them in a subsequent postnatal reconstructive surgery). As for inborn errors of metabolism, these cells can function as transgene carriers by carrying the corrected genes transduced by a viral vector. A key advantage for AFSCs is their easy availability. During the prenatal period, amniotic fluid can be accessed via amniocentesis. At birth, the neonatal membranes are usually discarded, making them readily available in large quantity. Isolation of cells from the discarded amniotic fluids is straightforward with low cost. All these make AFSCs an exciting source of stem cells in current research.

9.3.1.2 Umbilical Cord (± Placental) Stem Cells Hematopoietic stem cells from "allogeneic" umbilical cord blood (UCB) donations from public cord blood banks have been used in the treatment of many diseases for a few decades [36–45]. These cells have the advantage of being easily cryopreserved and thawed later without substantial loss of viability. UCB also has the benefits of being more readily available than bone marrow, and in an emergency situation, a transplant can be provided in just a few days. In addition, due to their immunological immaturity, hematopoietic stem cells from cord blood are generally better tolerated than those from bone marrow source while being safe and effective. In certain circumstances, cord blood can be transplanted successfully without a perfect HLA match between donor and recipient [36], thus opening the door for a wide range of patients to undergo transplant. The major disadvantage of unexpanded cord blood is the relatively low number of cells obtained compared to bone marrow. To overcome this limitation, pooled matched cord bloods from several donors are generally used for some transplant [37] while efforts continue to expand ex vivo cord blood cells [43–45].

9.3.2 Adult Stem Cells

Adult stem cells exist throughout the body and are found inside different types of body tissue, though in very low quantity. For example, they have been identified in tissues such as the brain, bone marrow, blood, blood vessels, fat, skeletal muscles, skin, teeth, and liver [46–53]. They usually remain in a quiescent state for years until activated by disease or tissue injury. The current review will focus on four cell types that are more advanced in development: bone marrow-derived mesenchymal stem cells (BM-MSCs), adipose-derived stem cells (ASCs), dental stem cells (DSCs), and endothelial stem cells. While adult stem cells were initially believed to have limited ability to differentiate based on their tissue of origin, there is now new evidence that they can differentiate into multiple cell types. They are also not tumorigenic although their relatively lower quantity remains a limitation.

9.3.2.1 Bone Marrow-Derived Mesenchymal Stem Cells (BM-MSCs)   Mesenchymal stem cells (MSCs) can be derived from many different sources, but bone marrow is the one first identified and most widely studied [54–67]. MSCs are a group of heterogeneous multipotent cells that have the capacity of self-renewal and differentiation along with a wide tissue distribution [58, 59]. MSCs participate in organ homeostasis and wound healing. In addition, MSCs also regulate successful aging either as precursors for certain lineages or through participation in "niches" (clusters of cells) that regulate stem cell proliferation and differentiation [60]. The multilineage differentiation potential with the ability to differentiate into cells of mesodermal origin (e.g., osteoblasts, chondrocytes, and adipocytes) is a hallmark of MSCs [61–63]. MSCs have been shown to differentiate into other cells such as cardiomyocytes and endothelial, smooth muscle, and neural cells [64]. With low acute toxicity, ease of preparation, ex vivo expansion, and "immunologic privilege" (neither induces nor is subjected to immune reaction), MSCs are good candidate for tissue regeneration. Their initial use was as graft-enhancing agent in patients with malignancies receiving HSCT to combat acute GVHD. MSCs have also been studied in cardiovascular and autoimmune diseases, bone/cartilage disorders, stroke, and spinal cord injuries [60,65,66]. However, MSCs have been shown to enhance tumor growth or undergo spontaneous malignant transformation; more data and caution is needed in their application [67].

9.3.2.2 Adipose-Derived Stem Cells (ASCs)   This is another source of MSCs that is normally extracted via liposuction. There is an abundant source of undifferentiated progenitor cells in various adipose tissues throughout the human body. These cells,

called ASCs or adipose-derived adult stromal cells, display cell surface marker profiles and differentiation characteristics very similar to other multipotent adult stem cells, such as BM-MSCs. However, ASCs are shown to have a wider therapeutic capacity and are easier to produce [68–75]. Under controlled culture conditions, these cells have been shown to differentiate into other cell types besides adipocytes such as chondrocytes, osteoblasts, neuronal cells, or muscle cells [69]. One major advantage is their abundance in normal human fat and their ease of availability through standard liposuction surgery. With such benefits, there are now many clinical trials testing ASCs in many conditions such as GVHD, stroke, spinal cord injury, intervertebral disk repair, rheumatoid arthritis, diabetes, and wound healing and repair, especially in the repair of perianal fistula associated with Crohn's disease [70–75]. ASCs are also being tested in less common conditions such as treatment of traumatic calvarial defect and urinary incontinence [74, 75].

9.3.2.3 Dental Stem Cells (DSCs) This is another extremely rich source for adult MSCs with easy harvest from extracted wisdom or deciduous teeth [76–85]. There are several types including dental follicle stem cells (DFSCs), apical papilla stem cells (APSCs), dental pulp stem cells (DPSCs), and periodontal ligament stem cells (PDLSCs) [79]. However, the former two types seem to exhibit higher proliferative potential and clonogenicity than the latter two. Based on data from growing curves and population doubling times, DSCs appear to have higher proliferative abilities than BM-MSCs [76,79]. These stem cells eventually form various parts of the dental tissues including enamel, dentin, periodontal ligament, and dental pulp as well as blood vessels and nervous tissues. DSCs also secrete trophic factors to enhance pulp vascularization and provide nutrients for the dentinogenic process. In addition, a new type of DSCs was also identified. They are called immature dental pulp stem cells (IDPSCs), which are derived from dental pulp of exfoliated deciduous teeth [80]. While all DSCs have been shown to differentiate into multiple cell lineages (osteogenic/odontogenic, adipogenic, and neurogenic) under the proper culture conditions, IDPSCs were found to express several cell markers such as Oct-4 and Nanog, which make them pluripotent. Besides dental diseases, DSCs have been studied in many conditions including cardiovascular problems, liver diseases, neurologic disorders, and ophthalmologic use [81–84]. A recent report showed induction of hepatocyte formation from dental pulp cells isolated from extracted full-grown wisdom and exfoliated deciduous teeth [85]. Because of their extreme ease for collection, they are becoming popular as an alternate source of stem cells for personal cell banking and research.

9.3.2.4 Endothelial Stem Cells Endothelial stem cells (also called endothelial progenitor cells) are multipotent cells usually derived from the bone marrow and identified by cell markers CD133 and VEGFR2 [86–96]. Nonmarrow source of endothelial stem cells has also been identified circulating in blood with the ability to differentiate into endothelial cells for blood vessels formation in a process called vasculogenesis [88]. However, recruitment and incorporation of these cells require a complicated sequence of events including adhesion, migration, chemoattraction, and the ultimate differentiation into endothelial cells. They have been shown to mobilize after myocardial infarction, which helps to restore the lining of blood vessels that are damaged during the event [89–91]. These cells have also been studied in leg ischemia caused by peripheral artery disease [92]. For therapeutic use, these cells are usually ex vivo expanded and administered to augment neovascularization of tissue after ischemia and re-endothelialization after endothelial injury. Major applications are for a broad range of cardiovascular diseases encompassing vasculature in the kidney, lung, and brain besides the heart [93–95]. Along with skeletal progenitor cells, they can also be used in bone regeneration and healing [96]. However, a recent setback is their link to pathologic angiogenesis such as retinopathy, endometriosis, and tumor growth, and caution is needed when using these cells [97–99].

9.3.3 Engineered Stem Cells

iPSCs can be derived from human skin cells after a series of complicated steps including gene insertion and manipulation. In many aspects, iPSCs are very similar to natural pluripotent stem cells (ESCs) and express comparable stem cell genes and proteins, chromatin methylation patterns, doubling time, and potency and differentiability [100]. Unlike adult stem cells, they have the advantage of abundant supply from an autologous host. Interestingly, while the current understanding of stem cell biology assumes unidirectional differentiation of stem cells into nonstem progeny, Weinberg et al. demonstrated a population of human mammary epithelial cells spontaneously dedifferentiating into stemlike cells [101]. The conversion was found to occur in both transformed and nontransformed cells isolated from cell lines or primary tissue, without any genetic manipulation. This observed plasticity offers new opportunity in producing patient-specific adult stem cells via spontaneous conversion, which has important implication in regenerative medicine.

9.3.3.1 Induced Pluripotent Stem Cells (iPSCs)   iPSCs are generated by reprogramming somatic cells (usually fibroblasts) via forced expression of a combination of transcription factors, such as Oct-3/4, Nanog, Klf4, c-Myc, Lin28, and Sox2 [102–118]. iPSCs can be maintained indefinitely in culture. They can be induced to undergo differentiation to give rise to any cell types like ESCs. In addition, frozen blood samples can also be used as a source of IPSCs, opening additional avenues to produce these cells. These characteristics make iPSCs potentially an unlimited supply of cells for cell replacement therapies, but without the ethical controversy like ESCs. One report estimates that as few as 15 iPSC clones would be sufficient to address the needs of all Caucasian patients with rare blood phenotypes/genotypes in the entire France [104]. Moreover, a single iPSC clone would meet 73% of the needs in alloimmunized patients with sickle cell disease. iPSCs represent renewable, potentially unlimited cell sources, in contrast to the hematopoietic stem cells originating from bone marrow, cord blood, or peripheral blood, which require more frequent sourcing of donor tissues to meet increasing treatment demands. A significant appeal for autologous iPSCs is the identical antigenic characteristics between the cells and the recipient. Several laboratories have established various culture techniques, for

example, by coculturing stem cells with stromal layers to induce the desired differentiation or by growing stem cells in suspension to specific differentiation [105]. In a span of a few years, the explosive growth in the number of iPSC lines has now far exceeded the number of ESC lines [106]. With their pluripotency and high potential for autologous use, iPSCs have tremendous potential in regenerative medicine as well as for drug discovery testing [107–111]. However, since recent work demonstrated potential aberrant epigenomic reprogramming and tumorigenicity in iPSCs [112], newer techniques are being developed to minimize this occurrence and mitigate concern [113–118].

9.4   STEM CELL TECHNOLOGY AND PREPARATION

9.4.1   Potency: Understanding Cellular Mechanism of Action

Cells are hypothesized to exert efficacy in one of two ways, either by replacing lost or dysfunctional cells in the recipient with donor (or self) stem cells or via secretion of trophic factors to influence the repair or development of the cells in the targeted tissue/organ. The former approach may be feasible when the disease is hinged on replacement of one cell type [119–128]. A good example is Parkinson's whereby disease progression is associated with temporal loss of dopaminergic neurons in the substantia nigra. While small molecules, such as levodopa and dopamine agonists, are commonly used to alleviate early motor symptoms, they lose effectiveness over time and can produce dyskinesia at a later stage of the disease. A huge unmet need for Parkinson's symptomatic relief and disease modification exists, and the concept of replacing dead neurons with live dopaminergic cells emerges with the advent of cell therapy research.

Parkinson's cell therapy research ensued with different stem and progenitor cells, some further differentiated for dopaminergic potential. This led to modality-specific assumptions and questions, for example, the need for long-term *in vivo* survival of transplanted cells for continuous dopamine secretion (as it is not known how long donor cells can last in the recipient's brain). Empirical observations on the effects of cell concentration, administration route, and graft survival *in vivo* led to refinement of cell preparation and administration techniques, as well as the coimplantation of cells on cellular matrices to enhance survival. A handful of cell types have been advanced through animal testing into clinic for Parkinson's disease. During the 1980s, transplantation of fetal brain dopaminergic neurons into the brains of Parkinson's patients was first attempted [119–121]. By the mid-2000s, Spheramine™ (cultured human retinal pigment epithelial cells on microcarriers) was tested in a phase II trial in 71 Parkinson's patients [122]. While efficacy was not observed with Spheramine™, major and long-lasting improvements were seen in some patients with fetal dopaminergic neurons [123]. In postmortem analysis, some of the transplanted fetal cells showed host-to-graft symptoms such as Lewy body formation [124, 125]. Another interesting observation was the unintentional coimplantation of serotonin neurons together with dopaminergic neurons. This became a calling point for techniques to propagate a high-purity single cell population as serotonin neurons have an undesired role in progression to dyskinesia.

The success of a cell replacement strategy is generally predicated on replacing a single cell type. However, many diseases are linked with multiple cell types, and transplantation of multiple cell types would pose a formidable challenge. An example is stroke, in which neuronal, astrocytic, glial, and endothelial cells are impacted. In this situation, the focus on cellular replacement has gradually moved toward trophic factor secretion by donor cells [126,127]. Preclinical testings support the hypothesis that newly introduced cells make and release trophic factors, and donor cells may be able to affect multiple tissues, especially when administered via a parenteral mean (intravenously or intra-arterially). The secreted trophic factors are thought to affect directly the local environment. For example, secreted fibroblast growth factor (FGF) is thought to enhance angiogenesis in the ischemic stroke brain. Trophic factors can also act indirectly by modulating secondary tissues such as the spleen, which in turn could downregulate the inflammatory cascade in the brain [128].

Whether donor cells act by direct replacement or via trophic factor secretion, they present a new paradigm of treatment where multiple mechanisms of action are possible, a leap from the single mode of action offered by conventional pharmacologic treatment. Harnessing the full potential of stem cell therapy requires continued refinement in product purification and qualification, mechanistic understanding, choice of cell delivery, and use of appropriate substrate. For stroke therapy, specific cell administration guidance is available [126]. For instance, with intra-arterial cell delivery, care must be taken to prevent microembolism and cerebral infarcts.

9.4.2   Cell Production: Isolation and Expansion

Therapeutic stem cells can be derived from many tissues of the body. The choice of tissue depends on ease of access and functional requirements [129–133]. For example, stem cells from the skin are greater in number and relatively easier to acquire compared to stem cells in the subventricular zone of the brain. On the other hand, skin stem cells may not be an efficacious option for neurological applications.

Once the tissue is selected, the path to cell isolation is often similar. Following aseptic acquisition and storage, the tissue is typically minced and/or enzymatically digested to yield heterogeneous cell populations. This is often followed by homogeneous cell enrichment through selective cell culture media and conditions that promote targeted cell survival or by cell separation. The latter may be achieved by various approaches. Some examples include using antibodies to label the differentiated cells, followed by ferromagnetic nanoparticle (immunomagnetic separation), fluorescence-activated cell sorting, or incubation with complement proteins for select cell lysis. Once the stem cells have been isolated, they are expanded for a number of population doublings and then either cryopreserved or prepped for administration.

In autologous therapies, the patient's cells are cultured and expanded in T-flasks or similar vessels at a GTP/GMP facility at or near the patient care center. When the targeted number of cells is reached, the cell batch is ready for harvest. With adherent cells, they are enzymatically removed from their substrate by trypsin digestion or other approaches. The cells are washed and reconstituted in the final formulation at an appropriate cell density for administration to the patient. Quality control is essential, with stringent product release criteria, including high cell viability and potency and lack of viral, bacterial, and mycoplasma contamination. However, these release tests are time and labor intensive, increasing the costs of production, thus making stem cell therapy relatively expensive at the moment. Another consideration for autologous therapy is the potentially limited window of administration. Donor cells require several days to months for processing, culture expansion, and release; therefore, autologous therapy may not be a desired option for indications requiring acute or subacute intervention (e.g., an acute myocardial infarction or stroke).

For allogeneic product development, donor cells are typically expanded as long as feasible to maximize the number of doses produced per donor tissue. Once cells have been isolated from tissue, they are cultured and tested for growth kinetics, morphology, yield and quality before primary banking into a master cell bank (MCB). One vial from the MCB can be expanded to yield a working cell bank (WCB), and a single WCB vial can be expanded to ultimately yield a batch of final product.

The stepwise expansion of cells requires increasingly larger vessels. The use of 2D and 3D cultures is a hot topic for bioprocessing engineers interested in maximum expansion while maintaining product quality and lowering costs of production. 2D cultures usually refer to single-use, disposable cell factories and cell cubes, where planar surfaces are stacked in rows. In general, 2D cultures are suitable for indications that require fewer cells per dose and/or indications with limited patient population demand. When patient numbers and dosage demands are high, engineers turn to 3D systems such as bioreactors. Pluristem, for instance, seeds placenta-derived cells on FibraCel disks in a 75 l bioreactor, estimated to be equivalent to 20,000 tissue culture flasks (T-175 ml) [129]. The bioreactor is intended as part of a production train capable of propagating cells sufficient to supply approximately 40,000 patient doses from one donor placenta.

While different stem cells can be isolated from a variety of tissues, the majority share a common requirement during culture—the need for serum in the growth media. With increasing global regulatory demand toward animal-free products, this has stimulated research into serum-free media, recombinant enzymes, and substrates. Once cells are harvested, they are processed and concentrated into the final formulation usually containing a cryopreservant such as DMSO before freezing. Cells can be stored for years in liquid nitrogen and maintain high viability upon rapid thaw. High viability is especially important as cells are live products and compromises to their viability could potentially impact efficacy and safety.

For iPSCs, large-scale generations of targeted differentiated cells from iPSCs are a prerequisite to clinical usage [130]. For example, the magnitude and cost of widespread application in replacing all blood transfusions with *in vitro* generated RBCs from iPSCs are prohibitively expensive. To make transfusion product cost-effective, considerable amount of progress in the expansion, maturation, and terminal differentiation/enucleation of erythrocytes will be needed [131]. A novel approach in culturing iPSCs in suspension may help in the scaling up production of iPSCs to address such need [130]. Furthermore, utilization of synthetic 3D structures mimicking bone marrow structure and the manipulation of transcriptional environment (e.g., downregulation of microRNAs-126/125*, ectopic expression of engineered Nup98-HoxA10 fusion protein, etc.) may also help to improve production efficiency [132,133].

9.4.3   Strategies for Long-Term Cell Survival

When the mode of action is cellular replacement, longevity of transplanted cells *in vivo* is desired [134–141]. With the majority of replacement cells being adherent, the use of a scaffold as substrate onto which donor cells can attach seems appropriate. In practice, however, empirical evidence points to poor survival of transplanted cells. For example, with Spheramine™ cell therapy for Parkinson's disease, less than 1% of transplanted cells on gelatin microcarriers were viable at 6 months post transplantation in the brain of a Parkinson's recipient, and overall efficacy was not demonstrated in the phase II trial [122]. On the other hand, in a primate model of Parkinson's, efficacy with the same cells and scaffold was demonstrated alongside cell survival at 18 months [134]. It seems reasonable to correlate cellular survival with long-term efficacy. New strategies and better scaffolds to enhance donor cell survival *in vivo* are necessary. Success will obviate need for readministration of cell product, often achieved by surgical means accompanied by costly medical and/or hospitalization, and potential complications with immunologic response to new transplants.

Scaffolds can be made from resorbable and nonresorbable synthetic polymers (e.g., polylactic acid (PLA), polyglycolic acid (PGA), polyethylene glycol (PEG)) or physiological materials (e.g., collagen, gelatin, hyaluronic acid, alginate, fibrin, decellularized extracellular matrix) [135,136]. Molecularly designed biomaterials for use as cell scaffolds can control many of the factors that guide differentiation and function of stem or progenitor cells. For example, injectable hydrogels could be designed to manipulate the biochemical/biophysical microenvironments for the transplanted stem cells.

Biomaterials have the ability to promote angiogenesis, improve engraftment and differentiation of stem cells, as well as hasten electromechanical integration of the transplanted cells [137,138]. For stem cell therapy of ischemic diseases, vasculogenic progenitors have been combined with injectable polylactic-co-glycolic acid-based scaffolds releasing single factors or combinations of angiopoietin-1, vascular endothelial growth factor (VEGF), and hepatocyte growth factor (HGF). The results also support that dual and triple combinations of scaffold-released growth factors are superior to single release [139].

Biomaterials can also be used to deliver genes, small RNAs, and proteins together with the stem cells. Integration of molecularly designed biomaterials and stem cell biology may be helpful for stable tissue regeneration and long-term utility. New technologies in local drug delivery to support the transplanted cells to achieve a tighter control of the microenvironment of these cells will be desired. To ensure successful tissue regeneration, delivery of multiple growth and differentiation factors along with different cell types in a temporally and spatially controlled fashion will also be critical.

Scaffolds can also extend to whole organs [135,136,140]. In an effort to create a bioartificial heart, for instance, cadaveric rodent hearts were first decellularized and subsequently repopulated as a scaffold with neonatal cardiocytes or aortic endothelial cells derived from stem cells [141]. The scaffold was further cultured under simulated physiological conditions for organ maturation. The construct was able to generate pump function (equivalent to about 2% of adult or 25% of 16-week fetal heart function) in a modified working heart preparation. There are several critical successful factors for this approach: optimal characterization of the extracellular matrix as a scaffold, methods for decellularization of vascular organs, potential cells to reseed such a scaffold, and proper techniques for the recellularization process.

9.5   REGULATORY CONSIDERATIONS AND PRODUCT TESTING

The key global regulatory agencies (e.g., FDA, EMA, and TGA) require stringent stem cell therapy product characterization and testing before release [142–149]. In the United States, the FDA's Center for Biologics Evaluation and Research (CBER) has issued several nonbinding guidance documents [143–147]. In brief, a product may be released after testing results confirm the following.

9.5.1   Cell Identity

Quantitative testing by phenotypic and/or biochemical assays should be used to confirm cell identity and assess heterogeneity (*21 CFR 610.14*). For example, flow cytometry can be used to assess heterogeneity of the cell prep and uniformity in expression of known cell surface markers.

9.5.2   Potency

Potency is defined as "the specific ability or capacity of the product, as indicated by appropriate laboratory tests or by adequately controlled clinical data obtained through the administration of the product in the manner intended, to effect a given result" (*21 CFR 600.3(s)*). For stem cell therapy, the relevant function of donor cells, if known, and/or their proteins relevant to efficacy should be defined and quantified as a measure of potency (*21.CFR 610.10*). With cell replacement strategy such as the use of pancreatic stem/progenitor cells to treat diabetes, potency may be confirmed by testing for insulin production [148]. In cases when donor cell efficacy depends on multiple mechanisms of action, one or more potency assays may be required.

9.5.3   Viability

A hallmark of product consistency is the percentage of cells that is expected to be viable at the time of administration to patients. Cell viability should be determined and an acceptable viability threshold established. Trypan blue exclusion assays are common and can also be substituted with more sensitive and reproducible methods

[149]. Unfortunately, the majority of somatic cell therapies are derived with adherent cells, which have the ability to form clumps. While clumps may be dispersed by pipetting during viability testing and escape notice, the remaining product to be administered may still contain cell aggregates, which in turn may impact product stability, viability, potency, and safety.

9.5.4 Adventitious Agent Testing

Tests should demonstrate that the cells are not contaminated with adventitious agents such as bacteria, fungi, mycoplasma, and viruses (*21 CFR 610.12*). For example, HIV, HBV, and EBV testing is most appropriate for products derived from human donors.

9.5.5 Purity

Purity (*21 CFR 610.13*) or validation of endotoxin testing by limulus amebocyte lysate (LAL) or other acceptable assays should be established.

9.5.6 General Safety Test

The general safety test (*21 CFR 610.11*) must be performed on the final product. When appropriate, modified procedures may be developed according to *21 CFR 610.9*.

9.5.7 Frozen Cell Banks

When cell populations frozen are thawed, expanded, and then administered to patients, lot release testing on the thawed cells is needed. Karyology is especially important for cell lines that exhibit instability with increasing population doublings.

In addition to the aforementioned requirements, current regulatory push for animal-free products has also stimulated research into serum-free media, recombinant enzymes, and substrates.

9.6 APPLICATION FOR MEDICAL PRACTICE

9.6.1 Potential of Stem Cells in Medical Treatment

Disease conditions where stem cell treatment could be used are emerging. There are numerous preclinical, animal, and case studies that show successes (or failures) of various types of stem cells in many disease conditions [14,107,109,150–165]. Diseases range from various cancers, Parkinson's disease, spinal cord injuries, amyotrophic lateral sclerosis (ALS), multiple sclerosis, and muscle damage, which are commonly mentioned in the literature [150–157]. Nonetheless, one potential concern in using transplanted stem cells is the fear that uncontrolled division of an immortal cell could turn into tumors [112]. While such potential could exist, there is currently not sufficient experience to accept or refute such concern.

Meanwhile, HSCT has been well established for several decades in treating many hematologic and immune disorders [4]. Nevertheless, early bone marrow-derived HSCT faced two major challenges: an invasive marrow harvesting procedure and finding a matched marrow donor. Fortunately, bone marrow stem cells were later discovered to detach continuously from the marrow into circulation, and an alternate technique was subsequently developed to harvest stem cells from the peripheral blood, using G-CSF or plerixafor as stem cell mobilizers [158]. Compared to marrow harvest, peripheral blood stem cells also have the benefit of quicker hematopoietic reconstitution. This approach has now replaced direct marrow harvest for autologous and the majority of the allogeneic transplants [4].

Furthermore, concurrent improvement in DNA typing to identify HLA alleles has led to better donor matching. With international registries now listing several million potential donors, acceptable donors are identified approximately 50% of the time. Unfortunately, the process of identifying unrelated matched donors followed by pro-curement of stem cells usually takes several months, and the delay sometimes limits its application [159]. To circumvent delay, autologous transplant can also be used and has the benefit of not needing a donor and/or the complication of GVHD. Unfortunately, the lack of graft-versus-tumor activity for autologous HSCT limits its effectiveness and thus usually has a higher chance of relapse [4].

Over the past decade, considerable efforts have been made to translate umbilical cord, cord blood, or placental stem cells from bench to bedside. An increasing global trend has parents contracting the collection and storage of postpartum cord blood for future autologous transplant purposes. In general, the majority of umbilical cords, cord blood, and placenta are discarded post delivery. With proper consent, these tis-sues can be collected and typed and be used for transplant or research [160, 161]. In fact, if an allotransplant is urgent and a suitable donor cannot be found, cord/pla-cental blood stem cells are becoming a popular alternative [162]. It requires less stringent HLA matching than HSCT because mismatched cord blood cells are less likely to cause GVHD without losing the graft-versus-tumor effect [36,163,164].

However, cord/placental source suffers from lower quantity of stem cells, and hematologic and immunologic reconstitution in the recipients is generally slower. Sometimes, the use of additional grafts from different donors may improve engraft-ment [160]. In addition, ex vivo expansion of cord blood stem cells is now under active research [161]. The less stringent HLA requirements for cord blood transplant allow the flexibility of using a smaller donor pool to serve most recipients. For many childhood hematologic disorders, UCB is now more often used than conventional source. For adult recipients, UCB is used when matched donors cannot be identified in a timely manner. An organization called Bone Marrow Donors Worldwide collects and lists the HLA types for all cord blood and adult registries. Cord blood banks in 21 countries currently store about 170,000 units [4].

Tissue engineering and tissue/organ regeneration is one of the most important potential applications of stem cell research. Transplanted organs are commonly in short supply with demand exceeding supply. Stem cells can potentially be used to grow a particular type of tissue or organ such as cartilage, cornea, pancreas, liver, or heart. For example, stem cells found below the skin have been used to form new skin tissue that can be grafted onto burn patients [153]. For patients with Parkinson's or Alzheimer's, replacing damaged brain tissue may lead to improvement, if not cure, for these conditions [157]. While the road to regenerate a completely replaceable body organ is likely long and tortuous, advances in technology that combine stem cells and novel scaffolds will fill the sizeable gap in organ reconstruction in the future.

Besides cell therapy and organ regeneration, stem cells can also be used to test new drugs, especially for safety [108]. It is well known that drugs commonly used in medical practice are often associated with potential organ toxicities. By using specific tissues engineered from stem cells, such toxicities may be detected during early drug development. A very toxic drug may then be dropped, or proper measures could be applied to minimize such risks before formal testing in humans.

While extensive research has been conducted either in academia or the medical community, large-scale commercialization of the use of stem cells (except HSCT) with regulatory approval is limited. Table 9.2 illustrates multiple major ongoing randomized studies (www.clinicaltrials.gov) using non-ESCs in the treatment of various diseases by a commercial sponsor [165].

9.6.2 Stem Cell Banking

In 1992, the University of Arizona began the world's first family cord blood sample banking facility [166–179]. In 1993, the university established the Designated Transplant Program (DTP) that provided free banking of cord blood for individuals and families with medical needs. Then, in 1998, it became the first accredited cord blood bank by the American Association of Blood Banks (AABB). Commercial stem cell banks quickly follow and there are many of them now throughout the world. In North America, Cryo-Cell International (Florida), NeoStem (New York), BioEden (Texas), and Medistem Lab (Arizona) are some bigger players involved in stem cell banking [167,168]. Besides UCB, these companies also offer to extract and store stem cells from adult blood, adipose tissues, and children's teeth. The fee for cell collection and processing ranges from several hundreds to thousands of dollars along with a lower annual charge for storage.

Collecting cord blood is relatively simple and takes just a few minutes after birth. After the umbilical cord has been severed, the remaining blood from the cord is harvested. The cord blood is then shipped to a cord blood bank and maintained in cryogenic liquid nitrogen storage tanks for long-term preservation. In contrast, to harvest adult blood hematopoietic stem cells for storage, the customer will need to take two injections of G-CSF, before undergoing apheresis for 4 h for stem cells harvested, with the rest of blood return to the body. Some common side effects for the procedure are bone and lung discomforts from the G-CSF and fatigue and itching from the extraction procedure itself. Use of G-CSF also increases the costs of adult blood stem cell banking. Meanwhile, adipose stem cells are harvested from liposuction and teeth resulted from natural detachment. However, among all types, cord blood banking seems to be more

TABLE 9.2 Major ongoing randomized stem cell trials conducted by a commercial sponsor

Indication	Sponsor	Trial name	Sample size	Product tested	Stem cell source	Comparator
Cardiac ischemia	Miltenyi	PERFECT	142	BMSC	CD133+ autologous bone marrow stem cells	Placebo
Cardiac ischemia	Cytori	ATHENA	45	ADRCs	ASCs	Placebo
Cardiomyopathy	Aastrom	AB155-1202-1	108	Ixmyelocel-T	BM-MSCs and other cells	Vehicle
Cardiomyopathy/ CAD	Miltenyi	0801-133	23	BMSC	CD133+ autologous bone marrow stem cells	Vehicle
Claudication	Aldagen	CCTRN581	80	ALD-301	BM-derived stem cells (bright cells)	Vehicle
Crohn's disease	Osiris	Study 603	270	Prochymal	Human adult MSCs	Placebo
Crohn's disease	Celgene	CCT-PDA001-CD003	27	PDA001	Human placenta-derived cells	Placebo
Diabetic nephropathy	Mesoblast	MSB-DN001	30	MPCs	Mesenchymal precursor cells	Placebo
Diabetes (type II)	Mesoblast	MSB-DM003	60	MPCs	Mesenchymal precursor cells	Placebo
Ischemic stroke	Aldagen	IS-101	100	ALD-401	BM-derived stem cells	Sham procedure
Ischemic stroke	Athersys	B01-02	140	MultiStem	Multipotent BM derived progenitor cells	Placebo
Myocardial infarction	Amorcyte	PreSERVE-AMI	160	AMR-001	BM-derived autologous stem cells	Placebo
Myocardial infarction	Capricor	ALLSTAR	274	CAP-1002	Allogeneic cardiosphere-derived stem cells	Placebo
Myocardial infarction	Cytori	ADVANCE	216	ADRCs	ASCs	Placebo
Myocardial infarction	Mesoblast	ANG.AMI-IC001	225	MPCs	Mesenchymal precursor cells	Placebo
Osteoarthritis	Mesoblast	MSB-CAR001	24	MSB-CAR001	Human adult MSCs	Vehicle
Osteoporotic fracture	Cytori	ROBUST	290	ADRCs	ASCs	Placebo
Renal injury	AlloCure	ACT-AKI	200	AC607	Allogeneic BM-MSCs	Vehicle
Ulcerative colitis	Athersys	B3041001	128	MultiStem	Multipotent BM-derived progenitor cells	Placebo

common than other kinds. By 2008, more than 400,000 cord blood stem cell units were banked at the Cord Blood Registry for use of over 120,000 clients including hospitals and other medical centers [166–168]. Cord blood has also been researched for use outside oncologic or immunologic diseases such as anoxic brain injury, traumatic brain injury, and cerebral palsy [36–45]. As of 2008, there were already more than 12,000 cord blood stem cell transplants performed worldwide [168].

Stem cell banks have now been in existence for more than couple decades [169–175]. Since the first stem cell bank establishment in the 1990s, they are increasingly recognized as a major resource of biological materials for both basic and translational research besides serving a storage function for motivated people for potential future use [176–178]. By providing transnational access to quality controlled and properly sourced stem cell lines, stem cell banks work together to foster international collaboration and innovation. Nonetheless, many local or regional stem cell banks operate under different policy, regulatory, and commercial frameworks; thus, transnational sharing of stem cell materials and data has many challenges. There are now numerous initiatives that have arisen to help harmonize and standardize stem cell banking and research to address such challenges [179].

9.6.3 Commercialization

There are many stem cell companies specializing in developing stem cell therapy for the treatment of human illness [180–187]. Unfortunately, there were a lot of overpromises about stem cell therapy from the industry early on. When the industry later failed to live up to these initial promises, enthusiasm from the medical community dwindled. Nevertheless, as new controlled clinical data begin to emerge, stem cell therapy is gaining attention and credibility again [182–185]. Alliance for Regenerative Medicine (ARM) is a professional organization that is made up of about 70 pharmaceutical and biotech companies specializing in stem cell therapy and regenerative medicine [180]. Besides industrial members, ARM also comprises multiple patient advocacy groups, research organizations, and members from the investing community. Major patient advocacy members include the ALS Association, Juvenile Diabetes Research Foundation, Parkinson's Action Network, Unite 2 Fight Paralysis, and many others. Together, they promote research and awareness of stem cell and regenerative medicine.

Unlike ESCs, adult stem cells generally have more limited capacity to differentiate but appear able to reduce inflammation and promote blood vessel formation effectively. These cells can also respond to body injury in a flexible and dynamic manner, offering advantages over traditional drugs. These cells seem to be preprogrammed to act more in tissue repair and less in forming a tissue or an entire organ. These are the type of stem cell treatments, delivered by infusion, injection, or catheters, that are being developed most commonly today by the commercial sponsors. Some companies, such as Celgene, Pluristem, Athersys, and Mesoblast, are, however, developing allogeneic stem cell products. Pfizer has a regenerative medicine unit as well as a partnership with Athersys. Baxter International is developing stem cell therapies for heart disease.

Aastrom Biosciences presented encouraging results from a phase II trial of its cell treatment for patients with critical limb ischemia at the American Heart Association's annual meeting [186]. Another phase II trial from Mesoblast (Australia) showed its stem cell product reduced the rate of heart attacks and the need for angioplasty by 78% [187]. Cephalon (acquired by Teva) took a 20% stake in Mesoblast, which has an estimated market value of about $2 billion [181]. Shire established a new regenerative medicine business and started it with a $750 million acquisition of Advanced BioHealing, a maker of skin substitute product for treating diabetic foot ulcers [181]. As more investments and positive news become available, industrialization of stem cells will continue to emerge.

9.7 CONCLUSIONS AND OUTLOOK FOR THE FUTURE

There remain many barriers to stem cell therapy. The best cell type, optimal delivery approach, and most applicable medical conditions for stem cell therapy will need to be elucidated. Continued improved techniques to derive, culture, and differentiate stem cells into the desired different cell types are certainly needed. Meanwhile, standardized control test, development of control comparator cell lines (to evaluate newly generated cell lines), and new and validated testing assays will be important. Improved pluripotency maintenance, better means for gene correction/insertion, and integration-free iPSC reprogramming will be useful. Besides challenges from the legal, patent, and regulatory fronts, lack of cost-efficient large-scale productions is also an area where more work is required.

Scientists, regulators, and policy makers will certainly need to work effectively together to advance breakthroughs from laboratories into actual medical practice. However, despite great progresses in basic research, clinical applications have fallen short. For clinical trials already conducted, follow-ups for long-term outcomes are often lacking. It is good to see that NIH has established an intramural center (NIH Center for Regenerative Medicine) to advance stem cell therapies, especially iPSC research [188]. Meanwhile, as expertise and experiences of regulatory authorities increase with these therapies, hopefully more efforts to harmonize international regulatory standards in approving stem cell therapy will evolve. Fortunately, as stem cell research offers new options to millions of patients with debilitating diseases or incurable conditions, public support remains high, and it is foreseeable that stem cell therapy will continue to move toward the therapeutic mainstream.

This chapter summarized the history and sources of stem cells, along with a general overview of the stem cell industry, including those involved in the development of stem cell therapeutics and stem cell banking. It also summarized the various potential diseases of which active clinical studies are conducted. Stem cell research can potentially help to treat a wide range of medical problems. It can help to replace or repair damaged organs, reduced risk of transplantation, and/or provide better treatment of these diseases. For example, use of stem cell therapy in Parkinson's disease, Alzheimer's disease, cardiovascular disorders, stroke, diabetes, birth defects, spinal cord injuries, etc. has all been widely published. Unfortunately, except for

HSCT, stem cell therapy has not yet passed the regulatory hurdles of widespread adoption. With more research, the days of using stem cells as an alternative solution to current medical and surgical options will emerge.

REFERENCES

1. McCulloch E, Till J. Repression of colony-forming ability of C57BL hematopoietic cells transplanted into non-isologous hosts. *J Cell Comp Physiol* 1963;61:301–308.

2. Kenny AB, Hitzig WH. Bone marrow transplantation for severe combined immuno-deficiency disease. Reported from 1968 to 1977. *Eur J Pediatr* 1979;131(3):155–177.

3. Maris M, Storb R. The transplantation of hematopoietic stem cells after non-myeloabla-tive conditioning: a cellular therapeutic approach to hematologic and genetic diseases. *Immunol Res* 2003;28(1):13–24.

4. Copelan E. Hematopoietic stem-cell transplantation. *N Engl J Med* 2006;354 (17):1813–1826.

5. Leavitt A, Hamlett L. Homologous recombination in human embryonic stem cells: a tool for advancing cell therapy and understanding and treating human disease. *Clin Transl Sci* 2011;4(4):298–305.

6. Yabut O, Bernstein H. The promise of human embryonic stem cells in aging-associated diseases. *Aging* 2011;3(5):494–508.

7. Power C, Rasko J. Will cell reprogramming resolve the embryonic stem cell controversy? A narrative review. *Ann Intern Med* 2011;155(2):114–121.

8. Robertson JA. Embryo stem cell research: ten years of controversy. *J Law Med Ethics* 2010;38(2):191–203.

9. Sohni A, Verfaillie C. Multipotent adult progenitor cells. *Best Pract Res Clin Haematol* 2011;24(1):3–11.

10. Chen P, Yen M, Liu K, Sytwu H, Yen B. Immunomodulatory properties of human adult and fetal multipotent mesenchymal stem cells. *J Biomed Sci* 2011;18:49.

11. Zhao J, Xu Q. Emerging restorative treatments for Parkinson's disease: manipulation and inducement of dopaminergic neurons from adult stem cells. *CNS Neurol Disord Drug Targets* 2011;10(4):509–516.

12. Yu F, Morshead C. Adult stem cells and bioengineering strategies for the treatment of cerebral ischemic stroke. *Curr Stem Cell Res Ther* 2011;6(3):190–207.

13. Harari-Steinberg O, Pleniceanu O, Dekel B. Selecting the optimal cell for kidney regen-eration: fetal, adult or reprogrammed stem cells. *Organogenesis* 2011;7(2):123–134.

14. Ebben J, Zorniak M, Clark P, Kuo J. Introduction to induced pluripotent stem cells: advancing the potential for personalized medicine. *World Neurosurg* 2011;76(3–4):270–275.

15. Stadtfeld M, Hochedlinger K. Induced pluripotency: history, mechanisms, and applica-tions. *Genes Dev* 2010;24(20):2239–2263.

16. Wilson J. Medicine: a history lesson for stem cells. *Science* 2009;324(5928):727–728.

17. Ma T, Xie M, Laurent T, Ding S. Progress in the reprogramming of somatic cells. *Circ Res* 2013;112(3):562–574.

18. Brunt KR, Weisel RD, Li RK. Stem cells and regenerative medicine—future perspectives. *Can J Physiol Pharmacol* 2012;90(3):327–335.

19. Stem Cell Research Timeline. Available at http://www.stemcellhistory.com/. Accessed May 31, 2013.

20. Stem Cell News. Available at http://www.sciencedaily.com/news/health_medicine/stem_cells/. Accessed May 31, 2013.

21. Stem Cell Research News. Available at http://www.medicalnewstoday.com/sections/stem_cell/. Accessed May 31, 2013.

22. Stem Cell: Key Research Events. Available at http://en.wikipedia.org/wiki/Stem_cell. Accessed May 31, 2013.

23. Shaw S, David A, De Coppi P. Clinical applications of prenatal and postnatal therapy using stem cells retrieved from amniotic fluid. *Curr Opin Obstet Gynecol* 2011;23(2):109–116.

24. Dobreva M, Pereira P, Deprest J, Zwijsen A. On the origin of amniotic stem cells: of mice and men. *Int J Dev Biol* 2010;54(5):761–777.

25. Walther G, Gekas J, Bertrand O. Amniotic stem cells for cellular cardiomyoplasty: promises and premises. *Catheter Cardiovasc Interv* 2009;73(7):917–924.

26. Parolini O, Soncini M, Evangelista M, Schmidt D. Amniotic membrane and amniotic fluid-derived cells: potential tools for regenerative medicine? *Regen Med* 2009;4(2):275–291.

27. Siegel N, Rosner M, Hanneder M, Valli A, Hengstschläger M. Stem cells in amniotic fluid as new tools to study human genetic diseases. *Stem Cell Rev* 2007;3(4):256–264.

28. Siegel N, Rosner M, Hanneder M, Freilinger A, Hengstschläger M. Human amniotic fluid stem cells: a new perspective. *Amino Acids* 2008;35(2):291–293.

29. Delo D, De Coppi P, Bartsch G, Atala A. Amniotic fluid and placental stem cells. *Methods Enzymol* 2006;419:426–438.

30. Fauza D. Amniotic fluid and placental stem cells. *Best Pract Res Clin Obstet Gynaecol* 2004;18(6):877–891.

31. Gosden CM. Amniotic fluid cell types and culture. *Br Med Bull* 1983;39:348–354.

32. Prusa AR, Marton E, Rosner M, Bernaschek G, Hengstschläger M. Oct-4-expressing cells in human amniotic fluid: a new source for stem cell research? *Hum Reprod* 2003; 18(7):1489–1493.

33. Orciani M, Emanuelli M, Martino C, Pugnaloni A, Tranquilli A, Di Primio R. Potential role of culture mediums for successful isolation and neuronal differentiation of amniotic fluid stem cells. *Int J Immunopathol Pharmacol* 2008;21:595–602.

34. Antonucci L, Lezzi L, Morizio E, Mastrangelo F, Pantalone A, Mattioli-Belmonte M, Gigante A, Salini V, Calabrese G, Tete S, Palka G, Stuppia L. Isolation of osteogenic progenitors from human amniotic fluid using a single step culture protocol. *BMC Biotechnol* 2009;9:9.

35. Perin L, Giuliani S, Jin D, Sedrakyan S, Carraro G, Habibian R, Warburton D, Atala A, De Filippo R. Renal differentiation of amniotic fluid stem cells. *Cell Prolif* 2007;40:936–948.

36. Anasetti C, Aversa F, Brunstein C. Back to the future: mismatched unrelated donor, haploidentical related donor, or unrelated umbilical cord blood transplantation? *Biol Blood Marrow Transplant* 2012;18(1 Suppl):S161–S165.

37. Sideri A, Neokleous N, De La Grange P, Guerton B, Le Bousse KM, Uzan G, Peste-Tsilimidos C, Gluckman E. An overview of the progress on double umbilical cord blood transplantation. *Haematologica* 2011;96(8):1213–1220.

38. Liao Y, Geyer M, Yang A, Cairo M. Cord blood transplantation and stem cell regenerative potential. *Exp Hematol* 2011;39(4):393–412.

39. McKenna D, Brunstein C. Umbilical cord blood: current status and future directions. *Vox Sang* 2011;100(1):150–162.

40. Zhong X, Zhang B, Asadollahi R, Low S, Holzgreve W. Umbilical cord blood stem cells: what to expect. *Ann N Y Acad Sci* 2010;1205:17–22.

41. Broxmeyer H. Insights into the biology of cord blood stem/progenitor cells. *Cell Prolif* 2011;44(Suppl 1):55–59.

42. Carvalho M, Teixeira F, Reis R, Sousa N, Salgado A. Mesenchymal stem cells in the umbilical cord: phenotypic characterization, secretome and applications in central nervous system regenerative medicine. *Curr Stem Cell Res Ther* 2011;6(3):221–228.

43. Robinson SN, Simmons PJ, Yang H, Alousi AM, Marcos de Lima J, Shpall EJ. Mesenchymal stem cells in ex vivo cord blood expansion. *Best Pract Res Clin Haematol* 2011;24(1):83–92.

44. Forraz N, McGuckin C. The umbilical cord: a rich and ethical stem cell source to advance regenerative medicine. *Cell Prolif* 2011;44(Suppl 1):60–69.

45. Arien-Zakay H, Lazarovici P, Nagler A. Tissue regeneration potential in human umbilical cord blood. *Best Pract Res Clin Haematol* 2010;23(2):291–303.

46. Jackson W, Nesti L, Tuan R. Potential therapeutic applications of muscle-derived mesenchymal stem and progenitor cells. *Expert Opin Biol Ther* 2010;10(2):505–517.

47. Zuk P, Zhu M, Mizuno H, Huang J, Futrell J, Katz A, Benhaim P, Lorenz H, Hedrick M. Multilineage cells from human adipose tissue: implications for cell-based therapies. *Tissue Eng* 2001;7(2):211–228.

48. Buffo A, Rite I, Tripathi P, Lepier A, Colak D, Horn AP, Mori T, Götz M. Origin and progeny of reactive gliosis: a source of multipotent cells in the injured brain. *Proc Natl Acad Sci USA* 2008;105(9):3581–3586.

49. Huang G, Gronthos S, Shi S. Mesenchymal stem cells derived from dental tissues vs. those from other sources: their biology and role in regenerative medicine. *J Dental Res* 2009;88(9):792–806.

50. Jones E, English A, Henshaw K, Kinsey S, Markham A, Emery P, McGonagle D. Enumeration and phenotypic characterisation of synovial fluid multipotential mesenchymal progenitor cells in inflammatory and degenerative arthritis. *Arthritis Rheum* 2004;50(3):817–827.

51. Janjanin S, Djouad F, Shanti R, Baksh D, Gollapudi K, Prgomet D, Rackwitz L, Joshi A, Tuan R. Human palatine tonsil: a new potential tissue source of multipotent mesenchymal progenitor cells. *Arthritis Res Ther* 2008;10(4):R83.

52. Shih Y, Kuo T, Yang A, Lee O, Lee C. Isolation and characterization of stem cells from the human parathyroid gland. *Cell Prolif* 2009;42(4):461–470.

53. Jazedje T, Perin P, Czeresnia C, Maluf M, Halpern S, Secco M, Bueno D, Vieira N, Zucconi E, Zatz M. Human fallopian tube: a new source of multipotent adult mesenchymal stem cells discarded in surgical procedures. *J Transl Med* 2009;7:46.

54. Pontikoglou C, Deschaseaux F, Sensebé L, Papadaki H. Bone marrow mesenchymal stem cells: biological properties and their role in hematopoiesis and hematopoietic stem cell transplantation. *Stem Cell Rev* 2011;7(3):569–589.

55. Pontikoglou C, Delorme B, Charbord P. Human bone marrow native mesenchymal stem cells. *Regen Med* 2008;3(5):731–741.

56. Jones E, McGonagle D. Human bone marrow mesenchymal stem cells *in vivo*. *Rheumatology* 2008;47(2):126–131.

57. Prockop D. Marrow stromal cells as stem cells for nonhematopoietic tissues. *Science* 1997;276(5309):71–74.

58. Dominici M, Le Blanc K, Mueller I, Slaper-Cortenbach I, Marini F, Krause D, Deans R, Keating A, Prockop D, Horwitz E. Minimal criteria for defining multipotent mesenchymal stromal cells. The international society for cellular therapy position statement. *Cytotherapy* 2006;8(4):315–317.

59. Caplan A. The mesengenic process. *Clin Plast Surg* 1994;21(3):429–435.

60. Williams A, Hare J. Mesenchymal stem cells: biology, pathophysiology, translational findings, and therapeutic implications for cardiac disease. *Circ Res* 2011;109(8):923–940.

61. Bruder S, Jaiswal N, Haynesworth S. Growth kinetics, self-renewal, and the osteogenic potential of purified human mesenchymal stem cells during extensive subcultivation and following cryopreservation. *J Cell Biochem* 1997;64(2):278–294.

62. Haynesworth S, Goshima J, Goldberg V, Caplan A. Characterization of cells with osteogenic potential from human marrow. *Bone* 1992;13(1):81–88.

63. Pittenger M, Mackay A, Beck S, Jaiswal R, Douglas R, Mosca J, Moorman M, Simonetti D, Craig S, Marshak D. Multilineage potential of adult human mesenchymal stem cells. *Science* 1999;284(5411):143–147.

64. Jiang Y, Jahagirdar B, Reinhardt R, Schwartz R, Keene C, Ortiz-Gonzalez X, Reyes M, Lenvik T, Lund T, Blackstad M, Du J, Aldrich S, Lisberg A, Low W, Largaespada D, Verfaillie C. Pluripotency of mesenchymal stem cells derived from adult marrow. *Nature* 2002;418(6893):41–49.

65. Gupta E, Nayyer N, Chin S, Cheok C, Cheong S. Clinical safety and efficacy of autologous bone marrow mesenchymal stem cell injection for the treatment of severe osteoarthritis. *Int J Rheum Dis* 2010;13(Suppl 1):40–43.

66. Chin S, Poey A, Wong C, Chang S, Teh W, Mohr T, Cheong S. Cryopreserved mesenchymal stromal cell treatment is safe and feasible for severe dilated ischemic cardiomyopathy. *Cytotherapy* 2010;12(1):31–37.

67. Wong RS. Mesenchymal stem cells: angels or demons? *J Biomed Biotechnol* 2011;2011:459–510.

68. Gimble J, Katz A, Bunnell B. Adipose-derived stem cells for regenerative medicine. *Circ Res* 2007;100(9):1249–1260.

69. Hass R, Kasper C, Bohm S, Jacobs R. Different populations and sources of human mesenchymal stem cells (MSC): a comparison of adult and neonatal tissue-derived MSC. *Cell Commun Signal* 2011;9:12.

70. Zuk P. Adipose tissue-derived cell: looking back and looking ahead. *Mol Bio Cell* 2010;21 (11):1783–1787.

71. Garcia-Olmo D, Herreros D, Pascual L, Pascual J, Del-Valle E, Zorrilla J, De-La-Quintana P, Garcia-Arranz M, Pascual M. Expanded adipose-derived stem cells for the treatment of complex perianal fistula: a phase II clinical trial. *Dis Colon Rectum* 2009;52(1):79–86.

72. Fang B, Song Y, Zhao R, Han Q, Lin Q. Using human adipose tissue-derived mesenchymal stem cells as salvage therapy for hepatic graft-versus-host disease resembling acute hepatitis. *Transplant Proc* 2007;39(5):1710–1713.

73. Fang B, Song Y, Lin Q, Zhang Y, Cao Y, Zhao R, Ma Y. Human adipose tissue-derived mesenchymal stromal cells as salvage therapy for treatment of severe refractory acute graft-vs.-host disease in two children. *Pediatr Transplant* 2007;11(7):814–817.

74. Lendeckel S, Jödicke A, Christophis P, Heidinger K, Wolff J, Fraser J, Hedrick M, Berthold L, Howaldt H. Autologous stem cells (adipose) and fibrin glue used to treat

widespread traumatic calvarial defects: case report. *J Craniomaxillofac Surg* 2004;32 (6):370–373.

75. Yamamoto T, Gotoh M, Hattori R, Toriyama K, Kamei Y, Iwaguro H, Matsukawa Y, Funahashi Y. Periurethral injection of autologous adipose-derived stem cells for the treatment of stress urinary incontinence in patients undergoing radical prostatectomy: report of two initial cases. *Int J Urol* 2010;17(1):75–82.

76. Estrela C, Alencar AH, Kitten GT, Vencio EF, Gava E. Mesenchymal stem cells in the dental tissues: perspectives for tissue regeneration. *Braz Dent J* 2011;22(2):91–98.

77. Shi S, Bartold P, Miura M, Seo B, Robey P, Gronthos S. The efficacy of mesenchymal stem cells to regenerate and repair dental structures. *Orthod Craniofac Res* 2005;8:191–199.

78. Morscheck C, Gotz W, Schierholz J, Zeilhofer F, Kuhn U, Mohl C, Sippel C, Hoffmann K. Isolation of precursor cells (PCs) from human dental follicle of wisdom teeth. *Matrix Biol* 2005;24:155–165.

79. Nakahara T. Potential feasibility of dental stem cells for regenerative therapies: stem cell transplantation and whole-tooth engineering. *Odontology* 2011;99(2):105–111.

80. Kerkis I, Kerkis A, Dozortsev D, Stukart-Parsons G, Gomes Massironi S, Pereira L, Caplan A, Cerruti H. Isolation and characterization of a population of immature dental pulp stem cells expressing OCT-4 and other embryonic stem cell markers. *Cells Tissues Organs* 2006;184:105–116.

81. Gomes JA, Geraldes Monteiro B, Melo GB, Smith RL, Cavenaghi Pereira da Silva M, Lizier NF, Kerkis A, Cerruti H, Kerkis I. Corneal reconstruction with tissue-engineered cell sheets composed of human immature dental pulp stem cells. *Invest Ophthalmol Vis Sci* 2010;51:1408–1414.

82. Iohara K, Zheng L, Wake H, Ito M, Nabekura J, Wakita H, Nakamura H, Into T, Matsushita K, Nakashima M. A novel stem cell source for vasculogenesis in ischemia: subfraction of side population cells from dental pulp. *Stem Cells* 2008;26:2408–2418.

83. Ikeda E, Yagi K, Kojima M, Yagyuu T, Ohshima A, Sobajima S, Tadokoro M, Katsube Y, Isoda K, Kondoh M, Kawase M, Go M, Adachi H, Yokota Y, Kirita T, Ohgushi H. Multipotent cells from the human third molar: feasibility of cell-based therapy for liver disease. *Differentiation* 2008;76:495–505.

84. Huang A, Snyder B, Cheng P, Chan A. Putative dental pulp-derived stem/stromal cells promote proliferation and differentiation of endogenous neural cells in the hippocampus of mice. *Stem Cells* 2008;26:2654–2663.

85. Ishkitiev N, Yaegaki K, Calenic B, Nakahara T, Ishikawa H, Mitiev V, Haapasalo M. Deciduous and permanent dental pulp mesenchymal cells acquire hepatic morphologic and functional features in vitro. *J Endod* 2010;36:469–474.

86. Urbich C, Dimmeler S. Endothelial progenitor cells functional characterization. *Trends Cardiovasc Med* 2004;14(8):318–322.

87. Friedrich E, Walenta K, Scharlau J, Nickenig G, Werner N. CD34-/CD133+/VEGFR-2+ endothelial progenitor cell subpopulation with potent vasoregenerative capacities. *Circ Res* 2006;98(3):20–25.

88. George A, Bangalore-Prakash P, Rajoria S, Suriano R, Shanmugam A, Mittelman A, Tiwari R. Endothelial progenitor cell biology in disease and tissue regeneration. *J Hematol Oncol* 2011;4:24.

89. Jia L, Takahashi M, Yoshioka T, Morimoto H, Ise H, Ikeda U. Therapeutic potential of endothelial progenitor cells for cardiovascular diseases. *Curr Vasc Pharmacol* 2006;4 (1):59–65.

90. Werner N, Kosiol S, Schiegl T, Ahlers P, Walenta K, Link A, Böhm M, Nickenig G. Circulating endothelial progenitor cells and cardiovascular outcomes. *N Engl J Med* 2005;353(10):999–1007.

91. Alaiti M, Ishikawa M, Costa M. Bone marrow and circulating stem/progenitor cells for regenerative cardiovascular therapy. *Transl Res* 2010;156(3):112–129.

92. Coppolino G, Buemi A, Bolignano D, Lacquaniti A, La Spada M, Stilo F, De Caridi G, Benedetto F, Loddo S, Buemi M, Spinelli F. Perioperative iloprost and endothelial progenitor cells in uremic patients with severe limb ischemia undergoing peripheral revascularization. *J Surg Res* 2009;157(1):129–135.

93. Lavoie JR, Stewart DJ. Genetically modified endothelial progenitor cells in the therapy of cardiovascular disease and pulmonary hypertension. *Curr Vasc Pharmacol* 2012; 10(3):289–299.

94. Rabelink T, de Boer H, van Zonneveld A. Endothelial activation and circulating markers of endothelial activation in kidney disease. *Nat Rev Nephrol* 2010;6(7):404–414.

95. Wei H, Jiang R, Liu L, Zhang J. Circulating endothelial progenitor cells in traumatic brain injury: an emerging therapeutic target? *Chin J Traumatol* 2010;13(5):316–318.

96. Matsumoto T, Kuroda R, Mifune Y, Kawamoto A, Shoji T, Miwa M, Asahara T, Kurosaka M. Circulating endothelial/skeletal progenitor cells for bone regeneration and healing. *Bone* 2008;43(3):434–439.

97. Li Calzi S, Neu M, Shaw L, Grant M. Endothelial progenitor dysfunction in the pathogenesis of diabetic retinopathy: treatment concept to correct diabetes-associated deficits. *EPMA J* 2010;1(1):88–100.

98. Laschke M, Giebels C, Nickels R, Scheuer C, Menger M. Endothelial progenitor cells contribute to the vascularization of endometriotic lesions. *Am J Pathol* 2011;178(1):442–450.

99. Dome B, Timar J, Ladanyi A, Paku S, Renyi-Vamos F, Klepetko W, Lang G, Dome P, Bogos K, Tovari J. Circulating endothelial cells, bone marrow-derived endothelial progenitor cells and proangiogenic hematopoietic cells in cancer: from biology to therapy. *Crit Rev Oncol Hematol* 2009;69(2):108–124.

100. Takahashi K, Yamanaka S. Induction of pluripotent stem cells from mouse embryonic and adult fibroblast cultures by defined factors. *Cell* 2006;126(4):663–676.

101. Chaffer C, Brueckmann I, Scheel C, Kaestli A, Wiggins P, Rodrigues L, Brooks M, Reinhardt F, Su Y, Polyak K, Arendt L, Kuperwasser C, Bierie B, Weinberg R. Normal and neoplastic nonstem cells can spontaneously convert to a stem-like state. *Proc Natl Acad Sci USA* 2011;108(19):7950–7955.

102. Takahashi K, Okita K, Nakagawa M, Yamanaka S. Induction of pluripotent stem cells from fibroblast cultures. *Nat Protoc* 2007;2(12):3081–3089.

103. Yu J, Vodyanik M, Smuga-Otto K, Antosiewicz-Bourget J, Frane J, Tian S, Nie J, Jonsdottir G, Ruotti V, Stewart R, Slukvin II, Thomson J. Induced pluripotent stem cell lines derived from human somatic cells. *Science* 2007;318(5858):1917–1920.

104. Peyrard T, Bardiaux L, Krause C, Kobari L, Lapillonne H, Andreu G, Douay L. Banking of pluripotent adult stem cells as an unlimited source for red blood cell production: potential applications for alloimmunized patients and rare blood challenges. *Transf Med Rev* 2011;25(3):206–216.

105. Chang K, Bonig H, Papayannopoulou T. Generation and characterization of erythroid cells from human embryonic stem cells and induced pluripotent stem cells: an overview. *Stem Cells Int* 2011;2011:791604.

106. Stem Cell Basics—What Are Induced Pluripotent Stem Cells? Available at http://stemcells. nih.gov/info/basics/basics10.asp. Accessed May 31, 2013.

107. Wu S, Hochedlinger K. Harnessing the potential of induced pluripotent stem cells for regenerative medicine. *Nat Cell Biol* 2011;13(5):497–505.

108. Grskovic M, Javaherian A, Strulovici B, Daley G. Induced pluripotent stem cells— opportunities for disease modelling and drug discovery. *Nat Rev Drug Discov* 2011;10 (12):915–929.

109. Nelson T, Martinez-Fernandez A, Terzic A. Induced pluripotent stem cells: developmental biology to regenerative medicine. *Nat Rev Cardiol* 2010;7(12):700–710.

110. Kane NM, Xiao Q, Baker AH, Luo Z, Xu Q, Emanueli C. Pluripotent stem cell differentiation into vascular cells: a novel technology with promises for vascular regeneration. *Pharmacol Ther* 2011;129(1):29–49.

111. Marchetto M, Brennand K, Boyer L, Gage F. Induced pluripotent stem cells (IPSCs) and neurological disease modeling: progress and promises. *Hum Mol Genet* 2011;20 (R2):R109–R115.

112. Okita K, Yamanaka S. Induced pluripotent stem cells: opportunities and challenges. *Philos Trans R Soc Lond B: Biol Sci* 2011;366(1575):2198–2207.

113. Tavernier G, Wolfrum K, Demeester J, De Smedt S, Adjaye J, Rejman J. Activation of pluripotency-associated genes in mouse embryonic fibroblasts by non-viral transfection with in vitro-derived mRNAs encoding Oct4, Sox2, Klf4 and cMyc. *Biomaterials* 2012;33(2):412–417.

114. Ruan J, Shen J, Wang Z, Ji J, Song H, Wang K, Liu B, Li J, Cui D. Efficient preparation and labeling of human induced pluripotent stem cells by nanotechnology. *Int J Nanomed* 2011;6:425–435.

115. Hockemeyer D, Jaenisch R. Gene targeting in human pluripotent cells. *Cold Spring Harb Symp Quant Biol* 2010;75:201–209.

116. Tichy E. Mechanisms maintaining genomic integrity in embryonic stem cells and induced pluripotent stem cells. *Exp Biol Med (Maywood)* 2011;236(9):987–996.

117. Chambers S, Studer L. Cell fate plug and play: direct reprogramming and induced pluripotency. *Cell* 2011;145(6):827–830.

118. Zhang F, Citra F, Wang DA. Prospects of induced pluripotent stem cell technology in regenerative medicine. *Tissue Eng Part B: Rev* 2011;17(2):115–124.

119. Braak H, Del Tredici K. Assessing fetal nerve cell grafts in Parkinson's disease. *Nat Med* 2008;14(5):483–485.

120. McKay R, Kittappa R. Will stem cell biology generate new therapies for Parkinson's disease? *Neuron* 2008;58(5):659–661.

121. Folkerth R, Durso R. Survival and proliferation of nonneural tissues with obstruction of cerebral ventricles in a Parkinsonian patient treated with fetal allografts. *Neurology* 1996;46(5):1219–1225.

122. Farag E, Vinters H, Bronstein J. Pathologic findings in retinal pigment epithelial cell implantation for Parkinson disease. *Neurology* 2009;73(14):1095–1102.

123. Mendez I, Viñuela A, Astradsson A, Mukhida K, Hallett P, Robertson H, Tierney T, Holness R, Dagher A, Trojanowski JQ, Isacson O. Dopamine neurons implanted into people with Parkinson's disease survive without pathology for 14 years. *Nat Med* 2008;14(5):507–509.

124. Li J, Englund E, Holton J, Soulet D, Hagell P, Lees A, Lashley T, Quinn N, Rehncrona S, Björklund A, Widner H, Revesz T, Lindvall O, Brundin P. Lewy bodies in grafted neurons in subjects with Parkinson's disease suggest host-to-graft disease propagation. *Nat Med* 2008;14(5):501–503.

125. Kordower J, Chu Y, Hauser R, Freeman T, Olanow C. Lewy body-like pathology in long-term embryonic nigral transplants in Parkinson's disease. *Nat Med* 2008;14 (5):504–506.

126. Wechsler L, Steindler D, Borlongan C, Chopp M, Savitz S, Deans R, Caplan L, Hess D, Mays R, Boltze J, Boncoraglio G, Borlongan C, Caplan L, Carmichael S, Chopp M, Davidoff A, Deans R, Fisher M, Hess D, Kondziolka D, Mays R, Norrving B, Parati E, Parent J, Reynolds B, Gonzalez-Rothi L, Savitz S, Sanberg P, Schneider D, Sinden J, Snyder E, Steinberg G, Steindler D, Wechsler L, Weiss M, Weiss S, Victor S, Zheng T. Stem cell therapies as an emerging paradigm in stroke (STEPS): bridging basic and clinical science for cellular and neurogenic factor therapy in treating stroke. *Stroke* 2009;40(2):510–515.

127. Savitz S, Chopp M, Deans R, Carmichael S, Phinney D, Wechsler L. Stem cell therapy as an emerging paradigm for stroke (STEPS) II. *Stroke* 2011;42(3):825–829.

128. Ajmo C, Vernon D, Collier L, Hall A, Garbuzova-Davis S, Willing A, Pennypacker K. The spleen contributes to stroke-induced neurodegeneration. *J Neurosci Res* 2008;86 (10):2227–2234.

129. Tripoint Global Research – Pluristem Therapeutics, Inc (PSTI). November 29, 2010. Available at http://www.pluristem.com/images/Analyst_Reports/PSTI_TRIPOINT_ Report_112910.pdf. Accessed May 31, 2013.

130. Amit M, Chebath J, Margulets V, Laevsky I, Miropolsky Y, Shariki K, Peri M, Blais I, Slutsky G, Revel M, Itskovitz-Eldor J. Suspension culture of undifferentiated human embryonic and induced pluripotent stem cells. *Stem Cell Rev* 2010;6(2):248–259.

131. Anstee D. Production of erythroid cells from human Embryonic Stem Cells (hESC) and human induced Pluripotent Stem cells (hiPS). *Transfus Clin Biol* 2010;17(3):104–109.

132. Ji J, Risueño R, Hong S, Allan D, Rosten P, Humphries K, Bhatia M. Brief report: ectopic expression of NUP98-HOXA10 augments erythroid differentiation of human embryonic stem cells. *Stem Cells* 2011;29(4):736–741.

133. Huang X, Gschweng E, Van Handel B, Cheng D, Mikkola H, Witte O. Regulated expression of MicroRNAs-126/126* inhibits erythropoiesis from human embryonic stem cells. *Blood* 2011;117(7):2157–2165.

134. Doudet D, Cornfeldt M, Honey C, Schweikert A, Allen R. PET Imaging of implanted human retinal pigment epithelial cells in the MPTP-induced primate model of Parkinson's disease. *Exp Neurol* 2004;189(2):361–368.

135. Segers V, Lee R. Biomaterials to enhance stem cell function in the heart. *Circ Res* 2011;109(8):910–22.

136. Garbayo E, Delcroix G, Schiller P, Montero-Menei C. 2011. Advances in the combined use of adult cell therapy and scaffolds for brain tissue engineering. Available at http://www. intechopen.com/source/pdfs/18204/InTech-Advances_in_the_combined_use_of_adult_ cell_therapy_and_scaffolds_for_brain_tissue_engineering.pdf. Accessed May 31, 2013.

137. Chen A, Chen X, Choo A, Reuveny S, Oh S. Critical microcarrier properties affecting the expansion of undifferentiated human embryonic stem cells. *Stem Cell Res* 2011;7 (2):97–111.

138. Leung H, Chen A, Choo A, Reuveny S, Oh S. Agitation can induce differentiation of hESC in microcarrier cultures. *Tissue Eng* 2011;17(2):165–172.

139. Saif J, Schwarz T, Chau D, Henstock J, Sami P, Leicht S, Hermann P, Alcala S, Mulero F, Shakesheff K, Heeschen C, Aicher A. Combination of injectable multiple growth factor-releasing scaffolds and cell therapy as an advanced modality to enhance tissue neovascularization. *Arterioscler Thromb Vasc Biol* 2010;30(10):1897–1904.

140. Badylak S, Taylor D, Uygun K. Whole-organ tissue engineering: decellularization and recellularization of three-dimensional matrix scaffolds. *Annu Rev Biomed Eng* 2011;13:27–53.

141. Ott H, Matthiesen T, Goh S, Black L, Kren S, Netoff T, Taylor D. Perfusion-decellularized matrix: using nature's platform to engineer a bioartificial heart. *Nat Med* 2008;14(2):213–221.

142. Gimble J, Bunnell B, Chiu E, Guilak F. Taking stem cells beyond discovery: a milestone in the reporting of regulatory requirements for cell therapy. *Stem Cells Dev* 2011;20(8):1295–1296.

143. Guidance for Industry: Guidance for Human Somatic Cell Therapy and Gene Therapy. Available at http://www.fda.gov/downloads/BiologicsBloodVaccines/GuidanceCompliance RegulatoryInformation/Guidances/CellularandGeneTherapy/ucm081670.pdf. Accessed May 31, 2013.

144. Draft Guidance for Industry: Potency Tests for Cellular and Gene Therapy Products. Available at http://www.fda.gov/downloads/BiologicsBloodVaccines/GuidanceCompliance RegulatoryInformation/Guidances/CellularandGeneTherapy/ucm078687.pdf. Accessed May 31, 2013.

145. FDA Draft Guidance for Reviewers: Instructions and Template for Chemistry, Manufacturing, and Control (CMC) Reviewers of Human Somatic Cell Therapy Investigational New Drug Applications (INDs). 2003. Available at http://www.fda.gov/OHRMS/DOCKETS/98fr/03d0349gdl.pdf. Accessed May 31, 2013.

146. Guidance for Industry: Eligibility Determination for Donors of Human Cells, Tissues, and Cellular and Tissue-Based Products (HCT/Ps). 2007. Available at http://www.fda.gov/downloads/BiologicsBloodVaccines/GuidanceComplianceRegulatoryInformation/Guidances/Tissue/ucm091345.pdf. Accessed May 31, 2013.

147. Points to Consider in the Characterization of Cell Lines Used to Produce Biologicals, CBER, FDA. 1993. Available at http://www.fda.gov/downloads/biologicsbloodvaccines/safetyavailability/ucm162863.pdf. Accessed May 31, 2013.

148. Noguchi H. Pancreatic stem/progenitor cells for the treatment of diabetes. *Rev Diabet Stud* 2010;7(2):105–111.

149. Hong K, McCaman M. High throughput cytometric analysis of whole cells: implication for cell therapy product characterization. *Cell Gene Ther* 2004;1:33–42.

150. Burra P, Bizzaro D, Ciccocioppo R, Marra F, Piscaglia A, Porretti L, Gasbarrini A, Russo F. Therapeutic application of stem cells in gastroenterology: an up-date. *World J Gastroenterol* 2011;17(34):3870–3880.

151. Wesson R, Cameron A. Stem cells in acute liver failure. *Adv Surg* 2011;45:117–130.

152. Tzouvelekis A, Antoniadis A, Bouros D. Stem cell therapy in pulmonary fibrosis. *Curr Opin Pulm Med* 2011;17(5):368–373.

153. Staniszewska M, Słuczanowska-Głąbowska S, Drukała J. Stem cells and skin regeneration. *Folia Histochem Cytobiol* 2011;49(3):375–380.

154. Bernardi S, Severini G, Zauli G, Secchiero P. Cell-based therapies for diabetic complications. *Exp Diabetes Res* 2012;2012:872504.

155. Penn M, Dong F, Klein S, Mayorga M. Stem cells for myocardial regeneration. *Clin Pharmacol Ther* 2011;90(4):499–501.

156. Heile A, Brinker T. Clinical translation of stem cell therapy in traumatic brain injury: the potential of encapsulated mesenchymal cell biodelivery of glucagon-like peptide-1. *Dialogues Clin Neurosci* 2011;13(3):279–286.

157. Lunn J, Sakowski S, Hur J, Feldman E. Stem cell technology for neurodegenerative diseases. *Ann Neurol* 2011;70(3):353–361.

158. Keating G. Plerixafor: a review of its use in stem-cell mobilization in patients with lymphoma or multiple myeloma. *Drugs* 2011;71(12):1623–1647.

159. Bone Marrow Transplant: Despite Recruitment Successes, National Programs may be Underutilized. General Accounting Office (GAO-03-182). 2002. Available at http://www.gao.gov/new.items/d03182.pdf. Accessed May 31, 2013.

160. Barker J, Weisdorf D, DeFor T, Blazar B, McGlave P, Miller J, Verfaillie C, Wagner J. Transplantation of 2 partially HLA matched umbilical cord blood units to enhance engraftment in adults with hematologic malignancy. *Blood* 2005;105:1343–1347.

161. Jaroscak J, Goltry K, Smith A, Waters-Pick B, Martin P, Driscoll T, Howrey R, Chao N, Douville J, Burhop S, Fu P, Kurtzberg J. Augmentation of umbilical cord blood (UCB) transplantation with ex vivo-expanded UCB cells: results of a phase 1 trial using the AastromReplicell system. *Blood* 2003;101:5061–5067.

162. Chun YS, Byun K, Lee B. Induced pluripotent stem cells and personalized medicine: current progress and future perspectives. *Anat Cell Biol* 2011;44(4):245–255.

163. Gluckman E, Broxmeyer H, Auerbach A, Friedman H, Douglas G, Devergie A, Esperou H, Thierry D, Socie G, Lehn P, Cooper S, English D, Kurtzberg J, Bard J, Boyse E. Hematopoietic reconstitution in a patient with Fanconi's anemia by means of umbilical-cord blood from an HLA-identical sibling. *N Engl J Med* 1989;321:1174–1178.

164. Wagner J, Barker J, DeFor T, Baker K, Blazar B, Eide C, Goldman A, Kersey J, Krivit W, MacMillan M, Orchard P, Peters C, Weisdorf D, Ramsay N, Davies S. Transplantation of unrelated donor umbilical cord blood in 102 patients with malignant and nonmalignant diseases: influence of CD34 cell dose and HLA disparity on treatment-related mortality and survival. *Blood* 2002;100:1611–1618.

165. ClinicalTrials.gov: A service of the US National Institute of Health. Available at http://www.clinicaltrials.gov. Accessed May 31, 2013.

166. Stem Cell Global Foundation: A Brief History of Cord Blood Banking. Available at http://www.stemcellgf.org/cord-blood-banking-history-future.aspx. Accessed May 31, 2013.

167. National Stem Cell Bank (NSCB). Available at http://www.nationalstemcellbank.org/. Accessed May 31, 2013.

168. Cord Blood Registry: Family Banking and Public Donation. Available at http://www.cordblood.com/cord_blood_banking_with_cbr/banking/family_public_banking.asp. Accessed May 31, 2013.

169. Pollack A. Questioning the allure of putting cells in the bank. *New York Times*. January 29, 2008. Available at http://www.nytimes.com/2008/01/29/health/29stem.html?pagewanted=all&_r=0. Accessed May 31, 2013.

170. Alkindi S, Dennison D. Umbilical cord blood banking and transplantation: a short review. *Sultan Qaboos Univ Med J* 2011;11(4):455–461.

171. Cooper K, Viswanathan C. Establishment of a mesenchymal stem cell bank. *Stem Cells Int* 2011;2011:905621.

172. Ausubel L, Lopez P, Couture L. GMP scale-up and banking of pluripotent stem cells for cellular therapy applications. *Methods Mol Biol* 2011;767:147–159.

173. Cavallo C, Cuomo C, Fantini S, Ricci F, Tazzari P, Lucarelli E, Donati D, Facchini A, Lisignoli G, Fornasari P, Grigolo B, Moroni L. Comparison of alternative mesenchymal stem cell sources for cell banking and musculoskeletal advanced therapies. *J Cell Biochem* 2011;112(5):1418–1430.

174. De Rosa A, De Francesco F, Tirino V, Ferraro G, Desiderio V, Paino F, Pirozzi G, D'Andrea F, Papaccio G. A new method for cryopreserving adipose-derived stem cells: an attractive and suitable large-scale and long-term cell banking technology. *Tissue Eng Part C: Methods* 2009;15(4):659–667.

175. Tirino V, Paino F, d'Aquino R, Desiderio V, De Rosa A, Papaccio G. Methods for the identification, characterization and banking of human DPSCs: current strategies and perspectives. *Stem Cell Rev* 2011;7(3):608–615.

176. Knoppers B, Isasi R. Stem cell banking: between traceability and identifiability. *Genome Med* 2010;2(10):73.

177. Thirumala S, Goebel W, Woods E. Clinical grade adult stem cell banking. *Organogenesis* 2009;5(3):143–154.

178. Healy L, Young L, Stacey G. Stem cell banks: preserving cell lines, maintaining genetic integrity, and advancing research. *Methods Mol Biol* 2011;767:15–27.

179. Isasi R, Knoppers B. From banking to international governance: fostering innovation in stem cell research. *Stem Cells Int* 2011;2011:498132.

180. Alliance for Regenerative Medicine: Members. Available at http://alliancerm.org/member-profiles. Accessed May 31, 2013.

181. Toni Clarke T, Beasley D. Insight: stem cell therapy poised to come in from the cold. *Reuters News*. December 5, 2011. Available at http://uk.reuters.com/article/2011/12/05/us-insight-stem-cell-therapy-idUKTRE7B30FH20111205. Accessed May 31, 2013.

182. Sipp D. The unregulated commercialization of stem cell treatments: a global perspective. *Front Med* 2011;5(4):348–355.

183. Courtney A, de Sousa P, George C, Laurie G, Tait J. Balancing open source stem cell science with commercialization. *Nat Biotechnol* 2011;29(2):115–116.

184. Smith D. Commercialization challenges associated with induced pluripotent stem cell-based products. *Regen Med* 2010;5(4):593–603.

185. Resnik D. The commercialization of human stem cells: ethical and policy issues. *Health Care Anal* 2002;10(2):127–154.

186. Schmid S. Aastrom reports positive phase 2 trial results. *Xconomy*. November 15, 2011. Available at http://www.xconomy.com/detroit/2011/11/15/aastrom-reports-positive-phase-2-trial-results/. Accessed May 31, 2013.

187. Anonymous. Mesoblast presented positive results from phase 2 trial of adult stem cell therapy at the American Heart Association annual meeting. Available at http://www.drugs.com/clinical_trials/mesoblast-limited-asx-msb-presented-positive-results-phase-2-trial-adult-stem-cell-therapy-american-12700.html. Accessed May 31, 2013.

188. NIH Center for Regenerative Medicine (NIH-CRM). Available at http://commonfund.nih.gov/stemcells/overview.aspx. Accessed May 31, 2013.

SECTION 5

ANALYTICAL SUPPORT NEEDED FOR THE RESEARCH AND DEVELOPMENT

10

SPECIFICATION SETTING AND STABILITY STUDIES IN THE DEVELOPMENT OF THERAPEUTIC DELIVERY SOLUTION

CHUNG CHOW CHAN

10.1 INTRODUCTION

During the development of therapeutic solution, specification setting and stability studies are two important activities that are needed to ensure quality of the products for clinical trial (CT) work. This chapter will focus on discussion using **drug substance** (DS) and **drug product** (DP) as the theme. The principles will apply equally well to devices and cellular therapy products.

The manufacturing and use of a DP, including its components, entail some degree of risk. The risk to its quality is one component of the overall risk. Product quality should be maintained throughout the product life cycle such that the attributes that are important to the quality of the DP remain consistent with those used in the clinical studies. These quality attributes are combined into a specification that forms part of a total control strategy for the DS and DP to ensure product quality and consistency.

The purpose of stability testing is to provide evidence on how the quality of a DS or DP varies with time under the influence of a variety of environmental factors such as temperature, humidity, and light and to establish a retest period for the DS or a **shelf life** for the DP and recommended storage conditions.

The International Conference on Harmonisation of Technical Requirements for Registration of Pharmaceuticals for Human Use (ICH) brings together the regulatory

Therapeutic Delivery Solutions, First Edition. Edited by Chung Chow Chan, Kwok Chow, Bill McKay, and Michelle Fung.
© 2014 John Wiley & Sons, Inc. Published 2014 by John Wiley & Sons, Inc.

authorities and pharmaceutical industry of Europe, Japan, and the United States. It provides harmonized guidances for setting up global specifications for new DSs and new DPs as well as stability studies. The following section discusses more details of these guidances from ICH which is applicable to development work.

10.2 SPECIFICATIONS FOR TEST PROCEDURES AND ACCEPTANCE CRITERIA FOR DRUG SUBSTANCES AND DRUG PRODUCTS

Specifications are one part of a total control strategy for the DS and DP to ensure product quality and consistency. Other parts of this strategy include thorough product characterization during development, upon which specifications are based, and adherence to good manufacturing practices (GMP) such as suitable facilities, a validated manufacturing process, validated test procedure, raw material testing, in-process testing, and stability testing. Specifications are necessary to assure the strength, identity, safety, purity, and quality of the DSs and DPs [1].

During the drug development, specification setting is continuous. The philosophy and criterion regarding specification setting change due to greater availability of data and increased expectations from regulatory authorities as the DP development proceeds. The general phases of a specification follow the development of the DS and DP that will go through first human dose, later phases of development (i.e., phase II), definitive stability studies, establishing the process validation protocol, establishing registration, and final product monograph.

A specification is defined as a list of tests, references to analytical procedures, and appropriate **acceptance criteria**, which are numerical limits, ranges, or other criteria for the tests described. It establishes the set of criteria to which a DS or DP should conform to be considered acceptable for its intended use. Specifications are critical quality standards that are proposed and justified by the developer and approved by regulatory authorities as conditions of approval. Setting and refining the DS characteristics from the beginning and throughout the formulation program help focus the project team to develop a quality product quickly in a cost-effective manner. DP characteristics defined and refined during the drug development program guide the development program and ensure the product is developed to meet the clinical, marketing, manufacturing, and patient compliance goals on budget.

An example specification for DS is listed in Table 10.1. Critical attributes used in this DS example for the specification are appearance, identity, assay, impurities, solvent residue, XRD, DSC, TGA, titration, and particle size.

An example specification of ABC capsule is listed in Table 10.2. Critical attributes used in this example for the specification are appearance, identity, assay, impurities, dissolution, content uniformity, and water.

Specifications are chosen to confirm the quality of the DS and DP rather than to establish full characterization and should focus on those characteristics found to be useful in ensuring the safety and efficacy of the DS and DP at release and during the different clinical development stages and at commercialization. Global agencies expect there will be changes to specification as the DP goes through the different

TABLE 10.1   Example specification of ABC DS

SPECIFICATIONS						
DRUG SUBSTANCE						
Document Code:	**XXX**	Revision Number:	**DRAFT**	Page Number:		**x of x**
Effective Date:						
Product:		**ABC Drug Substance**				
Test	**Method**	**Acceptance Criteria**				
IDENTIFICATION						
Appearance	M-XXX	White to off-white powder				
¹H-NMR	M-XXX	Complies with structure				
Identification B (UV scan)	M-XXX	Absorbance maxima and minima of the sample match the absorbance maxima and minima of the reference standard				
Identification A (HPLC retention time)	M-XXX	The retention time of the main peak obtained for the test solution in the HPLC assay testing matches the retention time obtained for the main peak of the standard solution				
XRD	M-XXX	Report results				
IMPURITIES						
Related Substances by HPLC	M-XXX					
Specified Impurities: A (% area) B (% area) C (% area) D (% area)		Report results				
Unspecified impurities (% area)		Report results				
Total impurities (% area)		Report results				
Solvent Residue	XXX	Toluene: 800 ppm				
		THF: 700 ppm				
DSC	XXX	Report results				
TGA	XXX	Report results				

(*Continued*)

TABLE 10.1   (Cont'd)

MISCELLEANEOUS		
Titration	M-XXX	90.0–105.0%
Assay by HPLC	M-XXX	Report results
Particle size	M-XXX	Report results

TABLE 10.2   Example specification of ABC capsule

SPECIFICATIONS FINISHED DRUG PRODUCT					
Document Code:	**XXX**	Revision Number:	**DRAFT**	Page Number:	**x of x**
Effective Date:					
Product:	ABC Capsule xx mg				

Test	Acceptance Criteria	Method
Appearance	Opaque white capsule, containing a white to light-brown solid	Visual
Identity by retention time	The RT of the ABC peak in the chromatogram of a sample preparation corresponds to that of the working standard preparation	HPLC
Assay	90–110% of label claim	HPLC Mxxx
Impurities Individual Unknown Total	NMT 1.0% NMT 5.0%	HPLC MXXX
Dissolution (@50 rpm) 10 Minutes 15 Minutes 30 Minutes 45 Minutes Infinity (60 Minutes @250 rpm)	USP<711>	Report Report Report Q = 75% Report

TABLE 10.2   (Cont'd)

Content Uniformity
  Mean, % Label Claim
  Max, % Label Claim
  Min, % Label Claim
  %RSD
	HPLC Mxxx
	USP<905>
		Acceptance Value
		(AV) <15.0

Water	USP<921>	Report

TABLE 10.3   Example changes during development of XYZ capsule

Test	Phase II specification	Registration stability specification	Registration specification

Assay	90.0–110.0% of
	  label claim
		90.0–110.0% of
		  label claim
			NLT 90.0% and NMT
			  110.0% of label claim

Total related substances	NMT 2.0%	NMT 1.0%	NMT 0.3%
Largest individual related substance	NMT 1.0%	NMT 0.5%	NMT 0.2%

phases of clinical development. An example for XYZ capsule is listed in Table 10.3 to show typical changes through development of a DP when more data are collected.

The following sections discuss the detail for setting specification of the common quality attributes and testing procedures for regulatory submission.

10.2.1   In-Process Tests

In-process tests are tests that may be performed during the manufacture of either the DS or DP, rather than as part of the formal set of tests that are conducted prior to release. In-process tests that are only used for the purpose of adjusting process parameters within an operating range, for example, hardness and friability of tablet cores that will be coated and individual tablet weights, are not included in the specification.

10.2.2   Pharmacopoeial Tests and Acceptance Criteria

References to certain test procedures are found in pharmacopoeias in the United States, EU, and Japan. Pharmacopoeial procedures should be utilized wherever possible. Where harmonization has been achieved, an appropriate reference to the harmonized procedure and acceptance criteria is considered acceptable for a specification in all three regions.

10.2.3   Impact of Drug Substance on Drug Product Specifications

In general, it should not be necessary to test the DP for quality attributes uniquely associated with the DS. It is normally not considered necessary to test the DP for synthesis impurities that are controlled in the DS and are not degradation products.

10.2.4 Reference Standard

A reference standard, or reference material, is a substance prepared for use as the standard in an assay, identification, or purity test. It should have a quality appropriate to its use. It is often characterized and evaluated for its intended purpose by additional procedures other than those used in routine testing. For new DS reference standards intended for use in assays, the impurities should be adequately identified and/or controlled, and purity should be measured by a quantitative procedure.

10.2.5 Justification of Specifications

Justification should be presented for each test procedure and each acceptance criterion. The justification should refer to relevant development data, pharmacopoeial standards, test data for DS and DP used in toxicology and clinical studies, and results from accelerated and long-term stability studies, where appropriate. Additionally, a reasonable range of expected analytical and manufacturing variability should be considered. It is important to consider all of this information.

For final product specification, test results from CT, stability, and scale-up/validation batches, with emphasis on the primary stability batches, should be considered in setting and justifying specifications. If multiple manufacturing sites are used, it is valuable to consider data from these sites in establishing the initial tests and acceptance criteria. This is particularly true when there is limited initial experience with the manufacture of the DS or DP at any particular site. If data from a single representative manufacturing site are used in setting tests and acceptance criteria, product manufactured at all sites should still comply with these criteria.

Presentation of test results in graphic format is helpful in justifying individual acceptance criteria, particularly for assay values and impurity levels. Data from development work should be included in such a presentation, along with stability data available for new DS or new DP batches manufactured by the proposed commercial processes. Justification for proposing exclusion of a test from the specification should be based on development data and appropriate on process validation data.

10.2.6 Universal Tests/Criteria for Drug Substances and Drug Products

Test methods used for DSs and DPs can be classified as **universal tests** or specific tests. Table 10.4 and Table 10.5 list all the common methods that will be applicable to DSs and DPs.

TABLE 10.4 Test methods used in DSs

Universal tests	Specific tests
Description	Physicochemical properties
Identification	Particle size
Assay	Polymorphic forms
Impurities	Chirality
	Moisture content
	Inorganic impurities
	Microbial limit

TABLE 10.5 Test methods used in DPs

Universal tests	Specific tests
Description	Dissolution
Identification	Disintegration
Assay	Hardness/friability
Impurities	Uniformity of dosage units
	Moisture content
	Microbial limit
	pH
	Antimicrobial preservative content
	Antimicrobial preservative effectiveness
	Antioxidant
	Extractables/leachables
	Alcohol content
	Particle size distribution
	Redispersibility
	Rheological properties
	Reconstitution time
	Sterility
	Endotoxins
	Particulate matter
	Functional testing
	Osmolarity
	Particle size distribution
	Aerodynamic size distribution
	Minimum fill
	Pressure test
	Foreign particulate matter
	Spray pattern
	Net fill weight
	Leak rate

10.2.6.1 Universal Tests for New Drug Substances The following tests and acceptance criteria are generally applicable to all new DSs [2].

Description/appearance Description/appearance is a qualitative statement about the physical state (e.g., solid, liquid) and color of the new DS. If any of these characteristics change during storage, this change should be investigated and appropriate action taken.

Identification Identification testing should optimally be able to discriminate between compounds of closely related structure that are likely to be present. Identification tests should be specific for the new DS, for example, infrared (IR) spectroscopy. Identification solely by a single chromatographic retention time, for example, is not regarded as being specific. However, the use of two chromatographic procedures, where the separation is based on different principles, or a combination of tests into a single procedure, such as HPLC/ultraviolet (UV) diode array, HPLC/ MS, or GC/MS, is acceptable. If the new DS is a salt, identification testing should be specific for the individual ions. An identification test that is specific for the salt itself should suffice. New DS that is optically active needs specific identification testing or performance of a **chiral** assay.

Assay   A specific, stability-indicating procedure should be included to determine the content of the new DS. It is possible to employ the same procedure (e.g., HPLC) for both assay of the new DS and quantitation of impurities.

In cases where use of a nonspecific assay is justified, other supporting analytical procedures should be used to achieve overall specificity. For example, where titration is adopted to assay the DS, the combination of the assay and a suitable test for impurities should be used.

Impurities   Organic and inorganic impurities and residual solvents are included in this category.

10.2.6.2   *Universal Tests for New Drug Products*   The following tests and acceptance criteria are generally applicable to all new DPs.

Description/appearance   A qualitative description of the **dosage form** should be provided (e.g., size, shape, and color). If any of these characteristics change during manufacture or storage, this change should be investigated and appropriate action taken. The acceptance criteria should include the final acceptable appearance. If color changes during storage, a quantitative procedure may be appropriate.

Identification   Identification testing should establish the identity of the DS(s) in the DP and should be able to discriminate between compounds of closely related structure that are likely to be present. Identity tests should be specific for the DS, for example, IR spectroscopy. Identification solely by a single chromatographic retention time, for example, is not regarded as being specific. However, the use of two chromatographic procedures, where the separation is based on different principles, or a combination of tests into a single procedure, such as HPLC/UV diode array, HPLC/ MS, or GC/MS, is acceptable.

Assay   A specific, stability-indicating assay to determine strength (content) should be included for all DPs. It is possible to employ the same procedure (e.g., HPLC) for both assay of the DS and quantitation of impurities. Results of content uniformity testing for DPs can be used for quantitation of DP strength, if the methods used for content uniformity are also appropriate for assays.

In cases where use of a nonspecific assay is justified, other **supporting analytical procedures** should be used to achieve overall specificity. For example, where titration is adopted to assay the DS for release, the combination of the assay and a suitable test for impurities can be used. A specific procedure should be used when there is evidence of **excipient** interference with the nonspecific assay.

Impurities   Organic and inorganic impurities (degradation products) and residual solvents are included in this category. Organic impurities arising from degradation of the DS and impurities that arise during the manufacturing process for the DP should be monitored in the new DP. Acceptance limits should be stated for individual specified degradation products, which may include both identified and unidentified degradation products as appropriate, and total degradation products. Process impurities from the DS synthesis are controlled during DS testing and therefore are not included in the total impurity limit. However, when a synthesis impurity is also a degradation product, its level should be monitored and included in the total degradation product limit. When it has been conclusively demonstrated via appropriate analytical methodology, that the DS does not degrade in the specific formulation, and under the specific storage conditions proposed in the new drug application, degradation product testing may be reduced or eliminated upon approval by the regulatory authorities.

10.2.7   Specific Tests/Criteria for Drug Substances and Drug Products

In addition to the universal tests, the following tests should be considered for DSs and/or DPs. Individual tests/criteria should be included in the specification when the tests have an impact on the quality of the DS and DP. Tests other than those listed in the following sections may be needed in particular situations or as new information becomes available.

10.2.7.1   New Drug Substances Specific Tests

Physicochemical properties   These are properties such as pH of an aqueous solution, melting point/range, and refractive index. The procedures used for the measurement of these properties are usually unique and do not need much elaboration, for example, capillary melting point. The tests performed in this category should be determined by the physical nature of the DS and by its intended use.

Particle size   For some DSs intended for use in solid or suspension DPs, particle size can have a significant effect on dissolution rates, bioavailability, and/or stability. In such instances, testing for particle size distribution should be carried out using an appropriate procedure, and acceptance criteria should be provided.

Polymorphic forms   Some DSs exist in different crystalline forms that differ in their physical properties. Polymorphism may also include solvation or hydration products (also known as pseudopolymorphs) and amorphous forms. Differences in these forms could affect the quality or performance of the new DPs. In cases where differences exist that have been shown to affect DP performance, bioavailability, or stability, then the appropriate solid state should be specified. Physicochemical measurements and techniques are commonly used to determine whether multiple forms exist. Examples of these procedures are as follows: melting point (including hot-stage microscopy), solid-state IR, X-ray powder diffraction, thermal analysis procedures (like DSC, TGA, and DTA), Raman spectroscopy, optical microscopy, and solid-state NMR. It is generally technically very difficult to measure polymorphic changes in DPs. A surrogate test (e.g., dissolution) can be used to monitor product performance, and polymorph content should only be used as a test and acceptance criterion of last resort.

Drug substance chirality   Where a DS is predominantly one enantiomer, the opposite enantiomer is normally excluded from the qualification and identification thresholds. However, for chiral DS that is developed as a single enantiomer, control of the other enantiomer should be considered in the same manner as for other impurities. However, technical limitations may preclude the same limits of quantification or qualification from being applied. Assurance of control also could be given by appropriate testing of a starting material or intermediate, with suitable justification.

Drug substance assay   An enantioselective determination of the DS should be part of the specification. It is acceptable for this to be achieved either through use of a chiral assay procedure or by the combination of an achiral assay together with appropriate methods of controlling the enantiomeric impurity.

Drug substance identity   For a DS developed as a single enantiomer, the identity test(s) should be capable of distinguishing both **enantiomers** and the racemic mixture.

DP chirality   Control of the other enantiomer in a DP is considered necessary unless racemization has been shown to be insignificant during manufacture of the dosage form and on storage.

Drug product chiral assay   An achiral assay may be sufficient where racemization has been shown to be insignificant during manufacture of the dosage form and on storage. Otherwise, a chiral assay should be used, or alternatively, the combination of an achiral assay plus a validated procedure to control the presence of the opposite enantiomer may be used.

Drug product chiral identity   A stereospecific identity test is not needed in the DP release specification. When racemization is insignificant during manufacture of the dosage form and on storage, stereospecific identity testing is more appropriately addressed as part of the DS specification. When racemization in the dosage form is a concern, chiral assay or enantiomeric impurity testing of the DP will serve to verify identity.

Water content   Water content test is important where the DS is known to be hygroscopic or degraded by moisture or when the DS is known to be a stoichiometric hydrate. The acceptance criteria are justified with data on the effects of hydration or moisture absorption. A loss on drying procedure may be considered adequate; however, a detection procedure that is specific for water (e.g., Karl Fischer titration) is preferred.

Inorganic impurities The need for tests and acceptance criteria for inorganic impurities (e.g., catalysts) should be studied during development and based on knowledge of the manufacturing process. Procedures and acceptance criteria for sulfated ash/residue on ignition should follow pharmacopoeial precedents; other inorganic impurities may be determined by other appropriate procedures, for example, atomic absorption spectroscopy.

Microbial limits There may be a need to specify the total count of aerobic microorganisms, the total count of yeasts and molds, and the absence of specific objectionable bacteria (e.g., *Staphylococcus aureus*, *Escherichia coli*, *Salmonella*, *Pseudomonas aeruginosa*). These objectionable bacteria limits can suitably be determined using pharmacopoeial procedures. The type of microbial test(s) and acceptance criteria should be based on the nature of the DS, method of manufacture, and the intended use of the DP. For example, sterility testing may be appropriate for DSs manufactured as sterile, and endotoxin testing may be appropriate for DSs used to formulate an injectable DP.

10.2.7.2 New Drug Products Specific Tests Additional tests and acceptance criteria should be included for particular DPs. The following selection presents a representative sample of both the DPs and the types of tests and acceptance criteria that may be appropriate for solid oral DPs, liquid oral DPs, and parenterals (small and large volume).

Tablets (coated and uncoated) and hard capsules One or more of these tests may also be applicable to soft capsules and granules.

Dissolution The specification for solid oral dosage forms normally includes a test to measure release of DS from the DP. Single-point measurements are normally suitable for immediate-release dosage forms. For modified-release dosage forms, appropriate test conditions and sampling procedures should be established. Multiple time point sampling should be performed for **extended-release** dosage forms, and two-stage testing (using different media in succession or in parallel) is appropriate for **delayed-release** dosage forms.

For immediate-release DP where changes in dissolution rate have been demonstrated to significantly affect bioavailability, it is required to develop test conditions that can distinguish batches with unacceptable bioavailability.

Where dissolution significantly affects bioavailability, the acceptance criteria should be set to reject batches with unacceptable bioavailability. Otherwise, test conditions and acceptance criteria should be established, which pass clinically acceptable batches.

For extended-release DP, *in vitro/in vivo* correlation may be used to establish acceptance criteria when human bioavailability data are available for formulations exhibiting different release rates. Where such data are not available, and drug release cannot be shown to be independent of *in vitro* test conditions, then acceptance criteria should be established on the basis of available batch data.

Normally, the permitted variability in mean release rate at any given time point should not exceed a total numerical difference of ±10% of the labeled content of DS (i.e., a total variability of 20%; a requirement of 50±10% thus means an acceptable range from 40% to 60%), unless a wider range is supported by a bioequivalence study.

Disintegration   For rapidly dissolving (dissolution >80% in 15 min at pH 1.2, 4.0, and 6.8) products containing drugs that are highly soluble throughout the physiological range (dose/solubility volume <250 ml from pH 1.2 to 6.8), disintegration may be substituted for dissolution. Disintegration testing is most appropriate when a relationship to dissolution has been established or when disintegration is shown to be more discriminating than dissolution. In such cases, dissolution testing may not be necessary.

Hardness/friability   It is normal to perform hardness and/or friability testing as an in-process control and not necessary to include these attributes in the specification. If the characteristics of hardness and friability have a critical impact on DP quality (e.g., chewable tablets), acceptance criteria should be included in the specification.

Uniformity of dosage units   This term includes both the mass of the dosage form and the content of the DS in the dosage form. A pharmacopoeial procedure should be used. In general, the specification should include one or the other but not both. If appropriate, these tests may be performed in process; the acceptance criteria should be included in the specification. When weight variation is applied for DPs exceeding the threshold value to allow testing uniformity by weight variation, the laboratory should verify that the homogeneity of the product is adequate.

Water content   A test for water content should be included when appropriate. The acceptance criteria should be justified with data on the effects of hydration or water absorption on the DP. In some cases, a loss on drying procedure may be adequate; however, a detection procedure that is specific for water (e.g., Karl Fischer titration) is preferred.

Microbial limits   Microbial limit testing is an attribute of GMP as well as of quality assurance. In general, it is needed to test microbial limit in the DP unless its components are tested before manufacture and the manufacturing process is known, through validation studies, not to carry a significant risk of microbial contamination or proliferation.

Acceptance criteria should be set for the total count of aerobic microorganisms, the total count of yeasts and molds, and the absence of specific objectionable bacteria (e.g., *S. aureus, E. coli, Salmonella, P. aeruginosa*). These limits should be determined by suitable procedures, using pharmacopoeial procedures, and at a sampling frequency or time point in manufacture that is justified by data and experience. The type of microbial test(s) and acceptance criteria should be based on the nature of the

DS, the method of manufacture, and the intended use of the DP. With acceptable scientific justification, it should be possible to propose no microbial limit testing for solid oral dosage forms.

Oral liquids   One or more of the following specific tests will normally be applicable to oral liquids and to powders intended for reconstitution as oral liquids.

Uniformity of dosage units   This term includes both the mass of the dosage form and the content of the DS in the dosage form. A pharmacopoeial procedure should be used. The specification should include one or the other but not both. When weight variation is applied for the DP exceeding the threshold value to allow testing uniformity by weight variation, the laboratory should verify that the homogeneity of the product is adequate.

If appropriate, the tests may be performed in process; however, the acceptance criteria should be included in the specification. This concept may be applied to both single-dose and multiple-dose packages.

The dosage unit is considered to be the typical dose taken by the patient. If the actual unit dose, as taken by the patient, is controlled, it may either be measured directly or calculated, based on the total measured weight or volume of drug divided by the total number of doses expected. If dispensing equipment (such as medicine droppers or dropper tips for bottles) is an integral part of the packaging, this equipment should be used to measure the dose. Otherwise, a standard volume measure should be used.

For powders for reconstitution, uniformity of mass testing is generally considered acceptable.

pH   Acceptance criteria for pH should be provided where applicable and the proposed range justified.

Microbial limits   Microbial limit testing is an attribute of GMP as well as of quality assurance. In general, it is needed to test microbial limits in the DP unless its components are tested before manufacture and the manufacturing process is known, through validation studies, not to carry a significant risk of microbial contamination or proliferation. With acceptable scientific justification, it may be possible to propose no microbial limit testing for powders intended for reconstitution as oral liquids.

Acceptance criteria should be set for the total count of aerobic microorganisms, the total count of yeasts and molds, and the absence of specific objectionable bacteria (e.g., *S. aureus, E. coli, Salmonella, P. aeruginosa*). These limits should be determined by suitable procedures, using pharmacopoeial procedures, and at a sampling frequency or time point in manufacture that is justified by data and experience.

Antimicrobial preservative content   For oral liquids needing an antimicrobial preservative, acceptance criteria for preservative content should be established. Acceptance criteria for preservative content should be based upon the levels of antimicrobial preservative necessary to maintain microbiological quality of the product at all stages throughout its proposed usage and shelf life. The lowest specified concentration of antimicrobial preservative should be demonstrated to be effective in controlling microorganisms by using a pharmacopoeial antimicrobial preservative effectiveness test.

Testing for antimicrobial preservative content should be performed at release. Under certain circumstances, in-process testing may suffice in lieu of release testing. When antimicrobial preservative content testing is performed as an in-process test, the acceptance criteria should remain part of the specification.

Antimicrobial preservative effectiveness should be demonstrated during development, during scale-up, and throughout the shelf life, although chemical testing for preservative content is the attribute normally included in the specification.

Antioxidant preservative content　Release testing for antioxidant content should be performed. Under certain circumstances, where justified by developmental and stability data, shelf life testing may be unnecessary, and in-process testing may suffice in lieu of release testing where permitted. When antioxidant content testing is performed as an in-process test, the acceptance criteria should remain part of the specification. If only release testing is performed, this decision should be reinvestigated whenever either the manufacturing procedure or the **container/closure system** changes.

Extractables　Where development and stability data show evidence that extractables from the container/closure systems are consistently below the levels that are demonstrated to be acceptable and safe, elimination of this test can be accepted. This should be reinvestigated if the container/closure system or formulation changes.

Where data demonstrate the need, tests and acceptance criteria for extractables from the container/closure system components (e.g., rubber stopper, cap liner, plastic bottle, etc.) are considered appropriate for oral solutions packaged in nonglass systems or in glass containers with nonglass closures. The container/closure components should be listed, and data collected for these components as early in the development process as possible.

Alcohol content　Where it is declared quantitatively on the label in accordance with pertinent regulations, the alcohol content should be specified. It may be assayed or calculated.

Dissolution　It is appropriate (e.g., insoluble DS) to include dissolution testing and acceptance criteria for oral suspensions and dry powder products for resuspension. Dissolution testing should be performed at release. This test may be performed as an in-process test when justified. The testing apparatus, media, and conditions should be pharmacopoeial, if possible, or otherwise justified. Dissolution procedures using either pharmacopoeial or nonpharmacopoeial apparatus and conditions should be validated.

Single-point measurements are normally suitable for immediate-release dosage forms. Multiple-point sampling, at appropriate intervals, should be performed for modified-release dosage forms. Acceptance criteria should be set based on the observed range of variation and should take into account the dissolution profiles of the batches that showed acceptable performance *in vivo*.

Particle size distribution   Quantitative acceptance criteria and a procedure for determination of particle size distribution may be appropriate for oral suspensions. Developmental data should be considered when determining the need for either a dissolution procedure or a particle size distribution procedure for these formulations.

Particle size distribution testing should be performed at release. It may be performed as an in-process test when justified. If these products have been demonstrated to have consistently rapid drug-release characteristics, exclusion of a particle size distribution test from the specification may be proposed.

Particle size distribution testing may also be proposed in place of dissolution testing with the proper justification. The acceptance criteria should include acceptable particle size distribution in terms of the percent of total particles in given size ranges. The mean, upper, and/or lower particle size limits should be well defined.

Acceptance criteria should be set based on the observed range of variation and should take into account the dissolution profiles of the batches that showed acceptable performance *in vivo*, as well as the intended use of the product. The potential for particle growth should be investigated, and the acceptance criteria should take the results of these studies into account.

Redispersibility   For oral suspensions that settle on storage (produce sediment), acceptance criteria for redispersibility are appropriate. Shaking may be an appropriate procedure.

The procedure (mechanical or manual) should be indicated. Time required to achieve resuspension by the indicated procedure should be clearly defined.

Rheological properties   For relatively viscous solutions or suspensions, it is appropriate to include rheological properties (viscosity/specific gravity) in the specification. The test and acceptance criteria should be stated.

Reconstitution time   Acceptance criteria for reconstitution time should be provided for dry powder products that require reconstitution. The choice of diluent should be justified.

Water content   For oral products requiring reconstitution, a test and acceptance criterion for water content should be proposed when appropriate. Loss on drying is sufficient if the effect of absorbed moisture versus water of hydration has been adequately characterized. In certain cases, a more specific procedure (e.g., Karl Fischer titration) may be preferable.

Parenteral DPs   The following tests may be applicable to parenteral DPs.

Uniformity of dosage units   This term includes both the mass of the dosage form and the content of the active substance in the dosage form. A pharmacopoeial procedure should be used. The specification should include one or the other but not both and is applicable to powders for reconstitution.

These tests may be performed in process and the acceptance criteria should be included in the specification. This test may be applied to both single-dose and multiple-dose packages.

For powders for reconstitution, uniformity of mass testing is generally acceptable.

pH   Acceptance criteria for pH should be provided where applicable and the proposed range justified.

Sterility   All parenteral products should have a test procedure and acceptance criterion for evaluation of sterility.

Endotoxins/pyrogens   A test procedure and acceptance criterion for endotoxins, using a procedure such as the limulus amebocyte lysate test, should be included in the specification. Pyrogenicity testing may be proposed as an alternative to endotoxin testing where justified.

Particulate matter   Parenteral products should have appropriate acceptance criteria for particulate matter. This will normally include acceptance criteria for visible particulates and/or clarity of solution as well as for subvisible particulates as appropriate.

Water content   For nonaqueous parenterals and for parenteral products for reconstitution, a test procedure and acceptance criterion for water content should be proposed when appropriate. Loss on drying is generally considered sufficient for parenteral products, if the effect of absorbed moisture versus water of hydration has been adequately characterized during development. In certain cases, a more specific procedure (e.g., Karl Fischer titration) may be preferred.

Antimicrobial preservative content   For parenteral products needing an antimicrobial preservative, acceptance criteria for preservative content should be established. Acceptance criteria for preservative content should be based upon the levels of antimicrobial preservative necessary to maintain microbiological quality of the product at all stages throughout its proposed usage and shelf life. The lowest specified concentration of antimicrobial preservative should be demonstrated to be effective in controlling microorganisms by using a pharmacopoeial antimicrobial preservative effectiveness test.

Testing for antimicrobial preservative content should normally be performed at release. Under certain circumstances, in-process testing may suffice in lieu of release testing where permitted. When antimicrobial preservative content testing is performed as an in-process test, the acceptance criteria should remain part of the specification.

Antimicrobial preservative effectiveness should be demonstrated during development, during scale-up, and throughout the shelf life (e.g., in stability testing), although chemical testing for preservative content is the attribute normally included in the specification.

Antioxidant preservative content   Release testing for antioxidant content should normally be performed. Under certain circumstances, where justified by developmental and stability data, shelf life testing may be unnecessary, and in-process testing may suffice in lieu of release testing. When antioxidant content testing is performed as an in-process test, the acceptance criteria should remain part of the specification. If only release testing is performed, this decision should be reinvestigated whenever either the manufacturing procedure or the container/closure system changes.

Extractables   Control of extractables from container/closure systems is considered significantly more important for parenteral products than for oral liquids. However, where development and stability data show evidence that extractables are consistently below the levels that are demonstrated to be acceptable and safe, elimination of this test can normally be accepted. This should be reinvestigated if the container/closure system or formulation changes.

Where data demonstrate the need, acceptance criteria for extractables from the container/closure components are considered appropriate for parenteral products packaged in nonglass systems or in glass containers with elastomeric closures. This testing may be performed at release only, where justified by data obtained during development. The container/closure system components (e.g., rubber stopper) should be listed, and data collected for these components as early in the development process as possible.

Functionality testing of delivery systems   Parenteral formulations packaged in prefilled syringes, autoinjector cartridges, or the equivalent should have test procedures and acceptance criteria related to the functionality of the delivery system. These may include control of syringeability, pressure, and seal integrity (leakage) and/or parameters such as tip cap removal force, piston release force, piston travel force, and power injector function force. Under certain circumstances, these tests may be performed in process. Data generated during product development may be sufficient to justify skip lot testing or elimination of some or all attributes from the specification.

Osmolarity   When the tonicity of a product is declared in its labeling, appropriate control of its osmolarity should be performed. Data generated during development and validation may be sufficient to justify performance of this procedure as an in-process control, skip lot testing, or direct calculation of this attribute.

Particle size distribution   Quantitative acceptance criteria and a procedure for determination of particle size distribution may be appropriate for injectable suspensions. Developmental data should be considered when determining the need for either a dissolution procedure or a particle size distribution procedure.

Particle size distribution testing should be performed at release. It may be performed as an in-process test when justified by product development data. If the product has been demonstrated during development to have consistently rapid drug-release characteristics, exclusion of particle size controls from the specification may be proposed.

Particle size distribution testing may also be proposed in place of dissolution testing, when development studies demonstrate that particle size is the primary factor influencing dissolution; justification should be provided. The acceptance criteria should include acceptable particle size distribution in terms of the percent of total particles in given size ranges. The mean, upper, and/or lower particle size limits should be well defined.

Acceptance criteria should be set based on the observed range of variation and should take into account the dissolution profiles of the batches that showed acceptable performance *in vivo* and the intended use of the product. The potential for particle growth should be investigated during product development; the acceptance criteria should take the results of these studies into account.

Redispersibility   For injectable suspensions that settle on storage (produce sediment), acceptance criteria for redispersibility may be appropriate. Shaking may be an appropriate procedure. The procedure (mechanical or manual) should be indicated. Time required to achieve resuspension by the indicated procedure should be clearly defined. Data generated during product development may be sufficient to justify skip lot testing, or elimination of this attribute from the specification may be proposed.

Reconstitution time   Acceptance criteria for reconstitution time should be provided for all parenteral products that require reconstitution. The choice of diluent should be justified. Data generated during product development and process validation may be sufficient to justify skip lot testing or elimination of this attribute from the specification for rapidly dissolving products.

10.3   STABILITY TESTING FOR NEW DRUG SUBSTANCES AND DRUG PRODUCTS

The purpose of stability testing is to provide evidence on how the quality of a DS or DP varies with time under the influence of a variety of environmental factors such as temperature, humidity, and light and to establish a retest period for the DS or a shelf life for the DP and recommended storage conditions.

In the ICH definition, the choice of test conditions is based on the analysis of the effects of climatic conditions in the three regions of the EC, Japan, and the United States. The mean kinetic temperature in any part of the world can be divided into four **climatic zones**, I–IV. The EC, Japan, and the United States fall into climatic zones I and II. It had been agreed that stability information generated in any one of the three

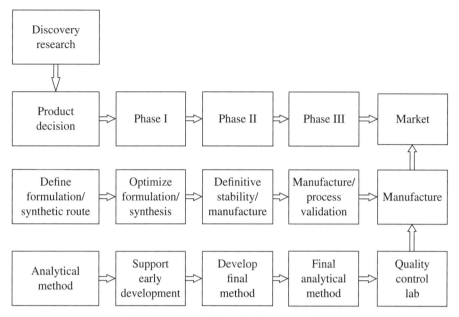

FIGURE 10.1   Drug development cycle.

regions of the EC, Japan, and the United States would be mutually acceptable to the other two regions.

The drug development cycle is summarized in Figure 10.1.

As the DS and DP go through the different phases of development, different stability studies will be conducted to collect data that will be used to support each phase of development. The different studies can be summarized below:

1. Excipient Compatibility Study
2. Prototype Formulation Stability Studies
3. DS and DP Stress Testing Study
4. CT Stability and Toxicology Stability Studies
5. Preliminary and Confirmatory Photostability Study
6. Cycle Stability Study
7. Registration Stability Studies

10.3.1   Excipient Compatibility Study

This study involves simple binary mixtures in vials. The samples are exposed to extreme conditions of temperature and humidity. Quality attributes studied in this study normally include assay, related substances, moisture, and other appropriate analytical properties. Analytical data critical to determine suitability of the formulation to meet desired final product attributes are studied. This study is usually started at initiation of formulation development and lasts 1–3 months.

10.3.2 Prototype Formulation Stability Studies

This is usually an open-dish study. Quality attributes studied in this study normally include content uniformity, assay, related substances, moisture, and other appropriate analytical properties. The analytical data is critical to determine suitability of the formulation to meet desired final product attributes. The study is normally initiated with commercial formulation development and takes 3–6 months. Stability conditions used will be dependent on the molecule.

10.3.3 Drug Substance and Drug Product Stress Testing Study

Results from these studies form an integral part of the information provided to regulatory authorities. Stress testing of the DS and DP can help identify the likely degradation products, which can in turn help establish the degradation pathways and the intrinsic stability of the molecule. It validates the stability-indicating power of the analytical procedures used. The nature of the stress testing will depend on the individual DS and the type of DP involved.

Stress testing is likely to be carried out on a single batch of the DS. It should include the effect of temperatures (in 10°C increments (e.g., 50°C, 60°C, etc.) above that for **accelerated testing**), humidity (e.g., 75% relative humidity (RH) or greater) oxidation, and photolysis on the DS. The testing should also evaluate the susceptibility of the DS to hydrolysis across a wide range of pH values in solution or suspension. Photostability testing should be an integral part of stress testing.

Degradation products identified under stress conditions are useful in establishing degradation pathways and developing and validating suitable analytical procedures. However, it may not be necessary to examine specifically for certain degradation products if it has been demonstrated that they are not formed under accelerated or long-term storage conditions.

DP stress study determines the potential degradation profile of the DP to assist in the development of analytical methods, define container/closure system, and determine specifications. As for the DS, the DP is exposed to extreme conditions of temperature, humidity, and light. This study will help to develop and validate discriminative stress test methods. The study will use qualitative/quantitative formulation and placebo formulation samples and takes about 2 months.

10.3.4 CT Stability and Toxicology Stability Studies

These studies will be ongoing activities throughout the development of a therapeutic solution when the molecule is going through different stages of CT. The samples include CT samples, swab stability samples, and toxicology dosage forms stability samples.

10.3.5   Preliminary and Confirmatory Photostability Study

This study is conducted to establish that light exposure does not result in unacceptable changes to DS and DP. Photostability information serves as a guideline to determine the potential degradation profile of the DS and DP to assist in the development of analytical methods, define container/closure system, and determine specifications. The samples are exposed to extreme conditions of light. The preliminary photostability study is completed during early development. The **confirmatory study** is completed at the start of registration/**formal stability study** and takes about 1 month. More detailed discussion is presented in later sections.

10.3.6   Cycle Stability Studies

Cycle stability studies are conducted to further support the stability of the DS and DP, for example, in-use stability and excursion stability. The excursion study will be used to support the supply chain for potential temperature deviation during transportation and during product distribution. An example for an excursion study is to cycle between −10°C and +50°C, with 2 days in each condition and employing four cycles of temperature changes.

10.3.7   Registration/Formal Stability Studies

Formal stability studies will be discussed in more detail in Sections 10.4.1 and 10.4.2. Generally, climatic zones I and II will be studied. Data generated in any one of the EC, Japan, and the United States is mutually acceptable and if it is in accordance with national/regional requirements of the country.

10.4   INITIATION OF STABILITY STUDIES

When a stability study is defined, a formal protocol should be written to define the scope of the study. The study should be formally written, reviewed, and approved. Approval should be in the form of signed document by management and the quality unit. The document also includes testing that needs to be conducted, acceptance criteria of the test methods, the storage conditions, and the schedules for the sample pull and testing.

An example stability protocol is given in Table 10.6. The depth and details of the stability study protocol will vary depending on the scope and stage of the therapeutic solution development. Detailed requirements for DS and DP marketing approval are discussed in the following sections. The principles of the stability requirement are also applicable to the device and cellular therapeutic solution development.

TABLE 10.6   Example stability protocol template for ABC capsules

EXAMPLE STABILITY PROTOCOL ABC Capsules	

Prepared by:

_____   _____

ABC1   Date

Reviewed by:

_____   _____

ABC2   Date

Approvals:

_____   _____

ABC3   Date

_____   _____

Quality Assurance   Date

PURPOSE OF STUDY

This protocol outlines the stability activities to be performed for ABC Capsules, X mg, Y mg, and Z mg. Samples will be stored for up to 12 months at 25°C/60%RH, 30°C/65%RH, and 40°C/75%RH.

A. PRODUCT DESCRIPTION
Provide detail description of formulation

TABLE 10.6.1   Example table summary of ABC capsule formulations

	X mg	Y mg	Z mg
	Formula	Formula	Formula
Ingredients	%	%	%
API			
Excipients			
Fill weight			
Capsules size			
Color			
Method of preparation			

B. STUDY DESIGN
The required number of bottle(s) from the specified chamber for the indicated month will be removed for testing at the time intervals according to Table 10.6.2. The testing to be performed at each time pull is detailed in the testing parameters section of this protocol (Table 10.6.3). Samples to be pulled and tested are marked with "X."

TABLE 10.6.2　Pull points for ABC capsules

Conditions[a] (°C/%RH)	Pull points (months)						Total containers stored
	T_0	1	2	3	6	12	
25°C/60%RH	X	X		X	X	X	
30°C/65%RH		X		X	X	X	
40°C/75%RH	X	X		X	X		

[a]Temperatures should be held ±2°C of the specified temperature and the relative humidity (RH) should be ±5% of the specified RH.

TESTING PARAMETERS AND ACCEPTANCE CRITERIA

Table 10.6.3 presents the details of parameters to be tested, the method to be utilized, and the relevant acceptance criteria.

TABLE 10.6.3　Testing guidelines

Parameters	Method tracker	Acceptance criteria
Appearance	Visual	Report observations
Assay	M-xxx	90–110% label claim
Related substances	M-xxx	Report results for impurities ≥ 0.05% by % area,
		RRT and identity (if known)
		Total impurities ≤ 2.0%
Dissolution	M-xxx	Report the results
Moisture content	Karl Fisher	Report the results

C. SAMPLE STORAGE

Upon testing completion, the sample will be retained for 60 days and then scheduled for disposal in accordance to relevant SOPs (include SOP reference).

D. REPORTS

A stability report will be generated at the end of the study and will include all stability information, including but not limited to packaging information, conditions, method(s), and results generated during testing. Any Out of Specification or Atypical results will be investigated and conducted in accordance with SOP XXX (Investigating Out of Specifications and Atypical Test Results).

TABLE 10.7 Usual storage conditions used in stability studies

25°C±2°C/60% RH±5% RH
30°C±2°C/65% RH±5% RH
40°C±2°C/75% RH±5% RH
5°C±2°C
−20°C±5°C
Below −20°C

10.4.1 Formal Stability of Drug Substance

Formal stability data of DS is required to assess the stability of the DS at the time of regulatory submission [3]. Table 10.7 lists the usual storage conditions that are used for both DS and DP stability studies.

10.4.1.1 Selection of Batches Data from formal stability studies should be provided on at least three primary batches of the DS. The batches should be manufactured to a minimum of pilot scale by the same synthetic route as, and using a method of manufacture and procedure that simulates the final process to be used for production batches. The overall quality of the batches of DS placed on formal stability studies should be representative of the quality of the material to be made on a production scale.

10.4.1.2 Container/Closure System The stability studies should be conducted on the DS packaged in a container/closure system that is the same as or simulates the packaging proposed for storage and distribution.

10.4.1.3 Specification Specification which is a list of tests, reference to analytical procedures, and proposed acceptance criteria, is addressed in previous sections. Stability studies should include testing of those attributes of the drug substance that are susceptible to change during storage and are likely to influence quality, safety, and/or efficacy. The testing should cover the physical, chemical, biological, and microbiological attributes. Validated stability-indicating analytical procedures should be applied. Whether and to what extent replication should be performed will depend on the results from validation studies. Specification with the appropriate acceptance criteria should be used with the stability studies.

10.4.1.4 Testing Frequency For long-term studies, frequency of testing should be sufficient to establish the stability profile of the DS. For DSs with a proposed retest period of at least 12 months, the frequency of testing at the long-term storage condition should normally be every 3 months over the first year, every 6 months over the second year, and annually thereafter through the proposed retest period.

At the accelerated storage condition, a minimum of three time points, including the initial and final time points (e.g., 0, 3, and 6 months), from a 6-month study is recommended. Where an expectation (based on development experience) exists that results from accelerated studies are likely to approach significant change criteria, increased testing should be conducted either by adding samples at the final time point or by including a fourth time point in the study design.

When testing at the intermediate storage condition is called for as a result of significant change at the accelerated storage condition, a minimum of four time points, including the initial and final time points (e.g., 0, 6, 9, 12 months), from a 12-month study is recommended.

10.4.1.5 Storage Conditions   A DS should be evaluated under storage conditions that test its thermal stability and its sensitivity to moisture. The storage conditions and the lengths of studies chosen should be sufficient to cover storage, shipment, and subsequent use.

The long-term testing should cover a minimum of 12 months duration on at least three primary batches at the time of submission to regulatory authorities and should be continued for a period of time sufficient to cover the proposed retest period. Additional data accumulated during the assessment period of the registration application should be submitted to the authorities if requested. Data from the accelerated storage condition and from the intermediate storage condition can be used to evaluate the effect of short-term excursions outside the label storage conditions (such as might occur during shipping).

Long-term, accelerated, and, where appropriate, intermediate storage conditions for DSs are detailed in the following sections. The general case applies if the DS is not specifically covered by a subsequent section. Alternative storage conditions can be used if justified.

10.4.1.6 General Case for Controlled Room Temperature Storage

Study	Storage condition	Minimum time period covered by data at submission
Long term[a]	25°C ± 2°C/60% RH ± 5% RH or 30°C ± 2°C/65% RH ± 5% RH	12 months
Intermediate[b]	30°C ± 2°C/65% RH ± 5% RH	6 months
Accelerated	40°C ± 2°C/75% RH ± 5% RH	6 months

[a] The laboratory will decide whether long-term stability studies are performed at 25 ± 2°C/60% RH ± 5% RH or 30°C ± 2°C/65% RH ± 5% RH.
[b] If 30°C ± 2°C/65% RH ± 5% RH is the long-term condition, there is no intermediate condition.

If long-term studies are conducted at 25°C ± 2°C/60%RH ± 5%RH and "significant change" occurs at any time during 6 months testing at the accelerated storage condition, additional testing at the intermediate storage condition should be conducted and evaluated against significant change criteria. Testing at the intermediate storage condition should include all tests, unless otherwise justified. The initial application should include a minimum of 6 months data from a 12-month study at the intermediate storage condition.

"Significant change" for a DS is defined as failure to meet its specification.

10.4.1.7   Drug Substances Intended for Storage in a Refrigerator

Study	Storage condition	Minimum time period covered by data at submission
Long term	5°C ± 3°C	12 months
Accelerated	25°C ± 2°C/60% RH ± 5% RH	6 months

Data from refrigerated storage should be assessed according to the Evaluation section, except where explicitly noted in the following.

If significant change occurs between 3 and 6 months testing at the accelerated storage condition, the proposed retest period should be based on the real-time data available at the long-term storage condition.

If significant change occurs within the first 3 months testing at the accelerated storage condition, a discussion should be provided to address the effect of short-term excursions outside the label storage condition, for example, during shipping or handling. This discussion can be supported by further testing on a single batch of the DS for a period shorter than 3 months but with more frequent testing than usual. It is unnecessary to continue to test a DS through 6 months when a significant change has occurred within the first 3 months.

10.4.1.8   Drug Substances for Storage in a Freezer

Study	Storage condition	Minimum time period covered by data at submission
Long term	–20°C ± 5°C	12 months

For DSs intended for storage in a freezer, the retest period should be based on the real-time data obtained at the long-term storage condition. In the absence of an accelerated storage condition for DSs intended to be stored in a freezer, testing on a single batch at an elevated temperature (e.g., 5°C ± 3°C or 25°C ± 2°C) for an appropriate time period should be conducted to address the effect of short-term excursions outside the proposed label storage condition, for example, during shipping or handling.

10.4.1.9   Drug Substances for Storage below –20°C   DSs intended for storage below –20°C should be treated on a case-by-case basis with specific justification.

10.4.1.10   Stability Commitment   When available long-term stability data on primary batches do not cover the proposed retest period granted at the time of regulatory approval, a commitment should be made to continue the stability studies post approval in order to firmly establish the retest period.

Where the submission includes long-term stability data on three production batches covering the proposed retest period, a postapproval commitment is unnecessary. Otherwise, one of the following commitments should be made:

1. If the submission includes data from stability studies on at least three production batches, a commitment should be made to continue these studies through the proposed retest period.
2. If the submission includes data from stability studies on fewer than three production batches, a commitment should be made to continue these studies through the proposed retest period and to place additional production batches, to a total of at least three, on long-term stability studies through the proposed retest period.
3. If the submission does not include stability data on production batches, a commitment should be made to place the first three production batches on long-term stability studies through the proposed retest period.

The stability protocol used for long-term studies for the stability commitment should be the same as that for the primary batches, unless otherwise scientifically justified.

10.4.1.11 Evaluation The purpose of the stability study is to establish, based on testing a minimum of three batches of the DS and evaluating the stability information (including results of the physical, chemical, biological, and microbiological tests), a retest period applicable to all future batches of the DS manufactured under similar circumstances. The degree of variability of individual batches affects the confidence that a future production batch will remain within specification throughout the assigned retest period. Any evaluation should cover not only the assay but also the levels of degradation products and other appropriate attributes.

When the data show so little degradation and so little variability that it is apparent from looking at the data that the requested retest period will be granted, it is normally unnecessary to go through the formal statistical analysis. Under this circumstance, providing a justification for the omission should be sufficient.

An approach for analyzing the data on a quantitative attribute that is expected to change with time is to determine the time at which the 95% one-sided confidence limit for the mean curve intersects the acceptance criterion. If analysis shows that the batch-to-batch variability is small, it is advantageous to combine the data into one overall estimate. This can be done by first applying appropriate statistical tests (e.g., p values for level of significance of rejection of more than 0.25) to the slopes of the regression lines and zero time intercepts for the individual batches. If it is inappropriate to combine data from several batches, the overall retest period should be based on the minimum time a batch can be expected to remain within acceptance criteria.

The nature of any degradation relationship will determine whether the data should be transformed for linear regression analysis. Usually, the relationship can be represented by a linear, quadratic, or cubic function on an arithmetic or logarithmic scale.

Statistical methods should be employed to test the goodness of fit of the data on all batches and combined batches to the assumed degradation line or curve.

10.4.2 Formal Stability Study of Drug Product

The design of the formal stability studies for the DP should be based on knowledge of the behavior and properties of the DS and from stability studies on the DS and on experience gained from clinical formulation studies. The likely changes on storage and the rationale for the selection of attributes to be tested in the formal stability studies should be stated.

10.4.2.1 Selection of Batches   Data from stability studies should be provided on at least three primary batches of the DP. The primary batches should be of the same formulation and packaged in the same container/closure system as proposed for marketing. The manufacturing process used for primary batches should simulate that to be applied to production batches and should provide product of the same quality and meeting the same specification as that intended for marketing. Two of the three batches should be at least pilot scale batches and the third one can be smaller, if justified. Where possible, batches of the DP should be manufactured by using different batches of the DS. Stability studies should be performed on each individual strength and container size of the DP unless **bracketing** or matrixing is applied.

10.4.2.2 Container/Closure System   Stability testing should be conducted on the dosage form packaged in the container/closure system proposed for marketing (including any secondary packaging and container label). Any available studies carried out on the DP outside its immediate container or in other packaging materials can form a useful part of the stress testing of the dosage form or can be considered as supporting information, respectively.

10.4.2.3 Specification   Stability studies should include testing of those attributes of the DP that are susceptible to change during storage and are likely to influence quality, safety, and/or efficacy. The testing should cover the physical, chemical, biological, and microbiological attributes, preservative content (e.g., antioxidant, antimicrobial preservative), and functionality tests (e.g., for a dose delivery system). Analytical procedures should be fully validated and stability indicating. Whether and to what extent replication should be performed will depend on the results of validation studies. Specification with the appropriate acceptance criteria should be used with the stability studies.

Shelf life acceptance criteria should be derived from consideration of all available stability information. It may be appropriate to have justifiable differences between the shelf life and release acceptance criteria based on the stability evaluation and the changes observed on storage. Any differences between the release and shelf life acceptance criteria for antimicrobial preservative content should be supported by a validated correlation of chemical content and preservative effectiveness demonstrated during drug development on the product in its final formulation (except for preservative concentration) intended for marketing. A single primary stability batch of the DP should be tested for antimicrobial preservative effectiveness (in addition to preservative content) at the proposed shelf life for verification purposes, regardless of whether there is a difference between the release and shelf life acceptance criteria for preservative content.

10.4.2.4 Testing Frequency For long-term studies, frequency of testing should be sufficient to establish the stability profile of the DP. For products with a proposed shelf life of at least 12 months, the frequency of testing at the long-term storage condition should normally be every 3 months over the first year, every 6 months over the second year, and annually thereafter through the proposed shelf life.

At the accelerated storage condition, a minimum of three time points, including the initial and final time points (e.g., 0, 3, and 6 months), from a 6-month study is recommended. Where an expectation (based on development experience) exists that results from accelerated testing are likely to approach significant change criteria, increased testing should be conducted either by adding samples at the final time point or by including a fourth time point in the study design.

When testing at the intermediate storage condition is called for as a result of significant change at the accelerated storage condition, a minimum of four time points, including the initial and final time points (e.g., 0, 6, 9, 12 months), from a 12-month study is recommended.

Reduced designs, that is, matrixing or bracketing, where the testing frequency is reduced or certain factor combinations are not tested at all, can be applied, if justified.

10.4.2.5 Storage Conditions A DP should be evaluated under storage conditions that test its thermal stability and its sensitivity to moisture or potential for solvent loss. The storage conditions and the lengths of studies chosen should be sufficient to cover storage, shipment, and subsequent use.

Stability testing of the DP after constitution or dilution, if applicable, should be conducted to provide information for the labeling on the preparation, storage condition, and in-use period of the constituted or diluted product. This testing should be performed on the constituted or diluted product through the proposed in-use period on primary batches as part of the formal stability studies at initial and final time points and, if full shelf life long-term data will not be available before regulatory submission, at 12 months or the last time point for which data will be available. In general, this testing need not be repeated on **commitment batches**.

The long-term testing should cover a minimum of 12 months duration on at least three primary batches at the time of submission and should be continued for a period of time sufficient to cover the proposed shelf life. Additional data accumulated during the assessment period of the registration application should be submitted to the authorities if requested. Data from the accelerated storage condition and, if appropriate, from the intermediate storage condition can be used to evaluate the effect of short-term excursions outside the label storage conditions (such as might occur during shipping).

Long-term, accelerated, and, where appropriate, intermediate storage conditions for DPs are detailed in the following sections. Alternative storage conditions can be used, if justified.

10.4.2.6 General Case for Controlled Room Temperature Storage

The storage condition and table used in the general case of DS (Section 10.4.1.6) can be applied to the DP.

If long-term studies are conducted at $25°C \pm 2°C/60\%RH \pm 5\%RH$ and "significant change" occurs at any time during 6 months testing at the accelerated storage condition, additional testing at the intermediate storage condition should be conducted and evaluated against significant change criteria. The initial application should include a minimum of 6 months data from a 12-month study at the intermediate storage condition.

"Significant change" for a DP is defined as:

1. A 5% change in assay from its initial value or failure to meet the acceptance criteria for potency when using biological or immunological procedures;
2. Any degradation products exceeding its acceptance criterion;
3. Failure to meet the acceptance criteria for appearance, physical attributes, and functionality test (e.g., color, phase separation, resuspendability, caking, hardness, dose delivery per actuation); however, some changes in physical attributes (e.g., softening of suppositories, melting of creams) may be expected under accelerated conditions;
4. Failure to meet the acceptance criterion for pH; or
5. Failure to meet the acceptance criteria for dissolution for 12 dosage units.

10.4.2.7 Drug Products Packaged in Impermeable Containers

Sensitivity to moisture or potential for solvent loss is not a concern for DPs packaged in impermeable containers that provide a permanent barrier to passage of moisture or solvent. Thus, stability studies for products stored in impermeable containers can be conducted under any controlled or ambient humidity condition.

10.4.2.8 Drug Products Packaged in Semipermeable Containers

Aqueous-based products packaged in semipermeable containers should be evaluated for potential water loss in addition to physical, chemical, biological, and microbiological stability. This evaluation can be carried out under conditions of low RH, as discussed in the following paragraph. Ultimately, it should be demonstrated that aqueous-based DPs stored in semipermeable containers can withstand low-RH environments.

Other comparable approaches can be developed and reported for nonaqueous, solvent-based products.

Study	Storage condition	Minimum time period covered by data at submission
Long term[a]	25°C±2°C/40% RH±5% RH or 30°C±2°C/35% RH±5% RH	12 months
Intermediate[b]	30°C±2°C/65% RH±5% RH 40°C±2°C/not more than (NMT) 25% RH	6 months
Accelerated		6 months

[a] It is up to the laboratory to decide whether long-term stability studies are performed at 25±2°C/40% RH±5% RH or 30°C±2°C/35% RH±5% RH.
[b] If 30°C±2°C/35% RH±5% RH is the long-term condition, there is no intermediate condition.

For long-term studies conducted at 25°C±2°C/40% RH±5% RH, additional testing at the intermediate storage condition should be performed to evaluate the temperature effect at 30°C if significant change other than water loss occurs during the 6 months testing at the accelerated storage condition. A significant change in water loss alone at the accelerated storage condition does not necessitate testing at the intermediate storage condition. However, data should be provided to demonstrate that the DP will not have significant water loss throughout the proposed shelf life if stored at 25°C and the reference RH of 40% RH.

A 5% loss in water from its initial value is considered a significant change for a product packaged in a semipermeable container after an equivalent of 3 months storage at 40°C/NMT 25% RH. However, for small containers (1 ml or less) or unit-dose products, a water loss of 5% or more after an equivalent of 3 months storage at 40°C/NMT 25% RH may be appropriate, if justified.

10.4.2.9 *Drug Products Intended for Storage in a Refrigerator*

Study	Storage condition	Minimum time period covered by data at submission
Long term	5°C±3°C	12 months
Accelerated	25°C±2°C/60% RH±5% RH	6 months

If the DP is packaged in a semipermeable container, appropriate information should be provided to assess the extent of water loss.

Data from refrigerated storage should be assessed according to the evaluation section.

If significant change occurs between 3 and 6 months testing at the accelerated storage condition, the proposed shelf life should be based on the real-time data available from the long-term storage condition.

If significant change occurs within the first 3 months testing at the accelerated storage condition, a discussion should be provided to address the effect of short-term excursions outside the label storage condition, for example, during shipment and handling. This discussion can be supported, if appropriate, by further testing on a single batch of the DP for a period shorter than 3 months but with more frequent testing than usual. It is unnecessary to continue to test a product through 6 months when a significant change has occurred within the first 3 months.

10.4.2.10   *Drug Products Intended for Storage in a Freezer*

Study	Storage condition	Minimum time period covered by data at submission
Long term	−20°C ± 5°C	12 months

For DPs intended for storage in a freezer, the shelf life should be based on the real-time data obtained at the long-term storage condition. In the absence of an accelerated storage condition for DPs intended to be stored in a freezer, testing on a single batch at an elevated temperature (e.g., 5°C ± 3°C or 25°C ± 2°C) for an appropriate time period should be conducted to address the effect of short-term excursions outside the proposed label storage condition.

10.4.2.11   *Drug Products Intended for Storage below −20°C*   DPs intended for storage below −20°C should be treated on a case-by-case basis based on scientific data justification.

10.4.2.12   *Stability Commitment*   When available long-term stability data on primary batches do not cover the proposed shelf life granted at the time of regulatory approval, a commitment should be made to continue the stability studies post approval in order to firmly establish the shelf life.

Where the submission includes long-term stability data from three production batches covering the proposed shelf life, a postapproval commitment is unnecessary. Otherwise, one of the following commitments should be made:

1. If the submission includes data from stability studies on at least three production batches, a commitment should be made to continue the long-term studies through the proposed shelf life and the accelerated studies for 6 months.

2. If the submission includes data from stability studies on fewer than three production batches, a commitment should be made to continue the long-term

studies through the proposed shelf life and the accelerated studies for 6 months and to place additional production batches, to a total of at least three, on long-term stability studies through the proposed shelf life and on accelerated studies for 6 months.

3. If the submission does not include stability data on production batches, a commitment should be made to place the first three production batches on long-term stability studies through the proposed shelf life and on accelerated studies for 6 months.

The stability protocol used for studies on commitment batches should be the same as that for the primary batches, unless otherwise scientifically justified.

Where intermediate testing is called for by a significant change at the accelerated storage condition for the primary batches, testing on the commitment batches can be conducted at either the intermediate or the accelerated storage condition. However, if significant change occurs at the accelerated storage condition on the commitment batches, testing at the intermediate storage condition should also be conducted.

10.4.2.13 Evaluation

A systematic approach should be adopted in the presentation and evaluation of the stability information, which should include, as appropriate, results from the physical, chemical, biological, and microbiological tests, including particular attributes of the dosage form (e.g., dissolution rate for solid oral dosage forms). Any evaluation should consider not only the assay but also the degradation products and other appropriate attributes. Attention should be paid to reviewing the adequacy of the mass balance and different stability and degradation performance.

The purpose of the stability study is to establish, based on testing a minimum of three batches of the DP, a shelf life and label storage instructions applicable to all future batches of the DP manufactured and packaged under similar circumstances. The degree of variability of individual batches affects the confidence that a future production batch will remain within specification throughout its shelf life.

Where the data show so little degradation and so little variability that it is apparent from looking at the data that the requested shelf life will be granted, it is normally unnecessary to go through the formal statistical analysis; providing a justification for the omission should be sufficient.

An approach for analyzing data of a quantitative attribute that is expected to change with time is to determine the time at which the 95% one-sided confidence limit for the mean curve intersects the acceptance criterion. If analysis shows that the batch-to-batch variability is small, it is advantageous to combine the data into one overall estimate. This can be done by first applying appropriate statistical tests (e.g., p values for level of significance of rejection of more than 0.25) to the slopes of the regression lines and zero time intercepts for the individual batches. If it is inappropriate to combine data from several batches, the overall shelf life should be based on the minimum time a batch can be expected to remain within acceptance criteria.

The nature of the degradation relationship will determine whether the data should be transformed for linear regression analysis. Usually, the relationship can be represented by a linear, quadratic, or cubic function on an arithmetic or logarithmic scale. Statistical methods should be employed to test the goodness of fit on all batches and combined batches (where appropriate) to the assumed degradation line or curve.

10.5 PHOTOSTABILITY TESTING OF NEW DRUG SUBSTANCES AND DRUG PRODUCTS

The intrinsic photostability characteristics of new DSs and DPs should be evaluated to demonstrate that light exposure does not result in unacceptable change.

A systematic approach to photostability testing is recommended covering studies such as:

1. Tests on the DS
2. Tests on the exposed DP outside of the immediate pack
3. Tests on the DP in the immediate pack
4. Tests on the DP in the marketing pack

The extent of DP testing should be established by assessing whether or not acceptable change has occurred at the end of the light exposure testing. Acceptable change is change within justified limits.

10.5.1 Light Sources

The light sources described in the following section may be used for photostability testing. The laboratory should either maintain an appropriate control of temperature to minimize the effect of localized temperature changes or include a dark control in the same environment unless otherwise justified.

10.5.1.1 Option 1 Any light source that is designed to produce an output similar to the D65/ID65 emission standard such as an artificial daylight fluorescent lamp combining visible and UV outputs, xenon, or metal halide lamp. D65 is the internationally recognized standard for outdoor daylight as defined in ISO 10977 (1993). ID65 is the equivalent indoor indirect daylight standard. For a light source emitting significant radiation below 320 nm, an appropriate filter(s) may be fitted to eliminate such radiation.

10.5.1.2 Option 2 For option 2, the same sample should be exposed to both the cool white fluorescent and near-UV lamp:

1. A cool white fluorescent lamp designed to produce an output similar to that specified in ISO 10977(1993).
2. A near-UV fluorescent lamp having a spectral distribution from 320 to 400 nm with a maximum energy emission between 350 and 370 nm; a significant proportion of UV should be in both bands of 320–360 nm and 360–400 nm.

10.5.1.3 Procedure For confirmatory studies, samples should be exposed to light providing an overall illumination of not less than (NLT) 1.2 million lx h and an integrated near-UV energy of NLT 200 W h/m^2 to allow direct comparisons to be made between the DS and DP.

If protected samples (e.g., wrapped in aluminum foil) are used as dark controls to evaluate the contribution of thermally induced change to the total observed change, these should be placed alongside the authentic sample.

10.5.2 Drug Substance

For DSs, photostability testing should consist of two parts: **forced degradation testing** and confirmatory testing [4].

The purpose of forced degradation testing studies is to evaluate the overall photosensitivity of the material for method development purposes and/or degradation pathway elucidation. This testing may involve the DS alone and/or in simple solutions/suspensions to validate the analytical procedures. In these studies, the samples should be in chemically inert and transparent containers. In these forced degradation studies, a variety of exposure conditions may be used, depending on the photosensitivity of the DS involved and the intensity of the light sources used. For development and validation purposes, it is appropriate to limit exposure and end the studies if extensive decomposition occurs. For photostable materials, studies should be terminated after an appropriate exposure level has been used.

Under forcing conditions, decomposition products may be observed that are unlikely to be formed under the conditions used for confirmatory studies. This information may be useful in developing and validating suitable analytical methods. If in practice it has been demonstrated they are not formed in the confirmatory studies, these degradation products need not be further examined.

Normally, only one batch of DS is tested during the development phase, and then the photostability characteristics should be confirmed on a single primary batch of DS to determine if the DS is photostable or photolabile. If the results of the confirmatory study are equivocal, testing of up to two additional batches should be conducted.

10.5.2.1 Presentation of Samples Care should be taken to ensure that the physical characteristics of the samples under test are taken into account, and efforts should be made, such as cooling and/or placing the samples in sealed containers, to ensure that the effects of the changes in physical states such as sublimation, evaporation, or melting are minimized. All such precautions should be chosen to provide minimal interference with the exposure of samples under test. Possible interactions between the samples and any material used for containers or for general protection of the sample should also be considered and eliminated wherever not relevant to the test being carried out.

As a direct challenge for samples of solid DSs, an appropriate amount of sample should be taken and placed in a suitable glass or plastic dish and protected with a suitable transparent cover. Solid DSs should be spread across the container to give a thickness of typically NMT 3 mm. DSs that are liquids should be exposed in chemically inert and transparent containers.

10.5.2.2 Analysis of Samples At the end of the exposure period, the samples should be examined for any changes in physical properties (e.g., appearance, clarity, or color of solution) and for assay and degradants by a method suitably validated for products likely to arise from photochemical degradation processes.

Where solid DS samples are involved, sampling should ensure that a representative portion is used in individual tests. Similar sampling considerations, such as homogenization of the entire sample, apply to other materials that may not be homogeneous after exposure. The analysis of the exposed sample should be performed concomitantly with that of any protected samples used as dark controls if these are used in the test.

10.5.2.3 Evaluation of Photostability Results The forced degradation studies should be designed to provide suitable information to develop and validate test methods for the confirmatory studies. These test methods should be capable of resolving and detecting photolytic degradants that appear during the confirmatory studies. When evaluating the results of these studies, it is important to recognize that they form part of the stress testing and are not therefore designed to establish qualitative or quantitative limits for change.

The confirmatory studies should identify precautionary measures needed in manufacturing or in formulation of the DP and if light-resistant packaging is needed. When evaluating the results of confirmatory studies to determine whether change due to exposure to light is acceptable, it is important to consider the results from other formal stability studies in order to assure that the drug will be within justified limits at time of use.

10.5.3 Drug Product

Normally, the studies on DPs should be carried out in a sequential manner starting with testing the fully exposed product and then progressing as necessary to the product in the immediate pack and then in the marketing pack. Testing should progress until the results demonstrate that the DP is adequately protected from exposure to light.

Normally, only one batch of DP is tested during the development phase, and then the photostability characteristics should be confirmed on a single primary batch of DP to determine if the product is clearly photostable or photolabile. If the results of the confirmatory study are equivocal, testing of up to two additional batches should be conducted.

For some products where it has been demonstrated that the immediate pack is completely impenetrable to light, such as aluminum tubes or cans, testing should normally only be conducted on directly exposed DP.

It may be appropriate to test certain products such as infusion liquids, dermal creams, etc. to support their photostability in use. The extent of this testing should depend on and relate to the directions for use. The analytical procedures used should be suitably validated.

10.5.3.1 *Presentation of Samples*

Care should be taken to ensure that the physical characteristics of the samples under test are taken into account, and efforts, such as cooling and/or placing the samples in sealed containers, should be made to ensure that the effects of the changes in physical states are minimized, such as sublimation, evaporation, or melting. All such precautions should be chosen to provide a minimal interference with the irradiation of samples under test. Possible interactions between the samples and any material used for containers or for general protection of the sample should also be considered and eliminated wherever not relevant to the test being carried out.

Where practicable when testing samples of the DP outside of the primary pack, these should be presented in a way similar to the conditions mentioned for the DS. The samples should be positioned to provide maximum area of exposure to the light source. For example, tablets, capsules, etc. should be spread in a single layer.

If direct exposure is not practical (e.g., due to oxidation of a product), the sample should be placed in a suitable protective inert transparent container (e.g., quartz).

If testing of the DP in the immediate container or as marketed is needed, the samples should be placed horizontally or transversely with respect to the light source, whichever provides for the most uniform exposure of the samples. Some adjustment of testing conditions may have to be made when testing large-volume containers (e.g., dispensing packs).

10.5.3.2 *Analysis of Samples*

At the end of the exposure period, the samples should be examined for any changes in physical properties (e.g., appearance, clarity, or color of solution, dissolution/disintegration for dosage forms such as capsules, etc.) and for assay and degradants by a method suitably validated for products likely to arise from photochemical degradation processes.

When powder samples are involved, sampling should ensure that a representative portion is used in individual tests. For solid oral dosage form products, testing should be conducted on an appropriately sized composite of, for example, 20 tablets or capsules. Similar sampling considerations, such as homogenization or solubilization of the entire sample, apply to other materials that may not be homogeneous after exposure (e.g., creams, ointments, suspensions, etc.). The analysis of the exposed sample should be performed concomitantly with that of any protected samples used as dark controls if these are used in the test.

10.5.3.3 Evaluation of Photostability Results   Depending on the extent of change, special labeling or packaging may be needed to mitigate exposure to light. When evaluating the results of photostability studies to determine whether change due to exposure to light is acceptable, it is important to consider the results obtained from other formal stability studies in order to assure that the product will be within proposed specifications during the shelf life.

GLOSSARY

Accelerated testing: Studies designed to increase the rate of chemical degradation or physical change of a DS or DP by using exaggerated storage conditions as part of the formal stability studies. Data from these studies, in addition to long-term stability studies, can be used to assess longer-term chemical effects at nonaccelerated conditions and to evaluate the effect of short-term excursions outside the label storage conditions such as might occur during shipping. Results from accelerated testing studies are not always predictive of physical changes.

Acceptance criteria: Numerical limits, ranges, or other suitable measures for acceptance of the results of analytical procedures.

Bracketing: The design of a stability schedule such that only samples on the extremes of certain design factors, for example, strength and package size, are tested at all time points as in a full design. The design assumes that the stability of any intermediate levels is represented by the stability of the extremes tested. Where a range of strengths is to be tested, bracketing is applicable if the strengths are identical or very closely related in composition (e.g., for a tablet range made with different compression weights of a similar basic granulation or a capsule range made by filling different plug fill weights of the same basic composition into different size capsule shells). Bracketing can be applied to different container sizes or different fills in the same container/closure system.

Chiral: Not superimposable with its mirror image, as applied to molecules, conformations, and macroscopic objects, such as crystals. The term has been extended to samples of substances whose molecules are chiral, even if the macroscopic assembly of such molecules is racemic.

Climatic zones: The four zones in the world that are distinguished by their characteristic prevalent annual climatic conditions. This is based on the concept described by Grimm [5].

Commitment batches: Production batches of a DS or DP for which the stability studies are initiated or completed post approval through a commitment made in the registration application.

Confirmatory studies: Those undertaken to establish photostability characteristics under standardized conditions. These studies are used to identify precautionary measures needed in manufacturing or formulation and whether light-resistant packaging and/or special labeling is needed to mitigate exposure to light.

Container/closure system: The sum of packaging components that together contain and protect the dosage form. This includes primary packaging components and secondary packaging components, if the latter are intended to provide additional protection to the DP. A packaging system is equivalent to a container/closure system.

Delayed release: Release of a drug (or drugs) at a time other than immediately following oral administration.

Dosage form: A pharmaceutical product type (e.g., tablet, capsule, solution, cream) that contains a DS generally, but not necessarily, in association with excipients.

DP: The dosage form in the final immediate packaging intended for marketing.

DS: The unformulated DS that may subsequently be formulated with excipients to produce the dosage form.

Enantiomers: Compounds with the same molecular formula as the DS, which differ in the spatial arrangement of atoms within the molecule and are non-superimposable mirror images.

Excipient: Anything other than the DS in the dosage form.

Expiration date: The date placed on the container label of a DP designating the time prior to which a batch of the product is expected to remain within the approved shelf life specification if stored under defined conditions and after which it must not be used.

Extended release: Products that are formulated to make the drug available over an extended period after administration.

Forced degradation testing studies: Those studies undertaken to degrade the sample deliberately. These studies, which may be undertaken in the development phase normally on the DSs, are used to evaluate the overall photosensitivity of the material for method development purposes and/or degradation pathway elucidation.

Formal stability studies: Long-term and accelerated (and intermediate) studies undertaken on primary and/or commitment batches according to a prescribed stability protocol to establish or confirm the retest period of a DS or the shelf life of a DP.

Immediate (primary) pack: Constituent of the packaging that is in direct contact with the DS or DP, and includes any appropriate label.

Immediate release: Allows the drug to dissolve in the gastrointestinal contents, with no intention of delaying or prolonging the dissolution or absorption of the drug.

Impermeable containers: Containers that provide a permanent barrier to the passage of gases or solvents, for example, sealed aluminum tubes for semisolids and sealed glass ampoules for solutions.

Impurity: (1) Any component of the new DS that is not the chemical entity defined as the new DS. (2) Any component of the DP that is not the chemical entity defined as the DS or an excipient in the DP.

In-process tests: Tests that may be performed during the manufacture of either the DS or DP, rather than as part of the formal battery of tests that are conducted prior to release.

Intermediate testing: Studies conducted at 30°C/65% RH and designed to moderately increase the rate of chemical degradation or physical changes for a DS or DP intended to be stored long term at 25°C.

Long-term testing: Stability studies under the recommended storage condition for the retest period or shelf life proposed (or approved) for labeling.

Marketing pack: The combination of immediate pack and other secondary packaging such as a carton used for marketing the product.

Mass balance: The process of adding together the assay value and levels of degradation products to see how closely these add up to 100% of the initial value, with due consideration of the margin of analytical error.

Matrixing: The design of a stability schedule such that a selected subset of the total number of possible samples for all factor combinations is tested at a specified time point. At a subsequent time point, another subset of samples for all factor combinations is tested. The design assumes that the stability of each subset of samples tested represents the stability of all samples at a given time point. The differences in the samples for the same DP should be identified as, for example, covering different batches, different strengths, different sizes of the same container/closure system, and, possibly in some cases, different container/closure systems.

Mean kinetic temperature: A single derived temperature that, if maintained over a defined period of time, affords the same thermal challenge to a DS or DP as would be experienced over a range of both higher and lower temperatures for an equivalent defined period. The mean kinetic temperature is higher than the arithmetic mean temperature and takes into account the Arrhenius equation [6].

Modified release: Dosage forms whose drug-release characteristics of time course and/or location are chosen to accomplish therapeutic or convenience objectives not offered by conventional dosage forms such as a solution or an immediate-release dosage form. Modified-release solid oral dosage forms include both delayed- and extended-release DPs.

New drug product: A pharmaceutical product type, for example, tablet, capsule, solution, cream, etc., which has not previously been registered, and which contains a drug ingredient generally, but not necessarily, in association with excipients.

New drug substance: The designated therapeutic moiety, which has not previously been registered. This is also referred to as a new molecular entity or new chemical entity (NCE). It may be a complex, simple ester, or salt of a previously approved DS.

New molecular entity: An active pharmaceutical substance not previously contained in any DP registered with the national or regional authority concerned. A new salt, ester, or noncovalent-bond derivative of an approved DS is considered a new molecular entity for the purpose of stability testing under this guidance.

Pilot scale batch: A batch of a DS or DP manufactured by a procedure fully representative of and simulating that to be applied to a full production scale batch. For solid oral dosage forms, a pilot scale is generally, at a minimum, 1/10th that of a full production scale or 100,000 tablets or capsules, whichever is the larger.

Polymorphism: The occurrence of different crystalline forms of the same DS. This may include solvation or hydration products (also known as pseudopolymorphs) and amorphous forms.

Primary batch: A batch of a DS or DP used in a formal stability study, from which stability data are submitted in a registration application for the purpose of establishing a retest period or shelf life, respectively. A primary batch of a DS should be at least a pilot scale batch. For a DP, two of the three batches should be at least pilot scale batch, and the third batch can be smaller if it is representative with regard to the critical manufacturing steps. However, a primary batch may be a production batch.

Production batch: A batch of a DS or DP manufactured at production scale by using production equipment in a production facility as specified in the application.

Quality: The suitability of either a DS or DP for its intended use. This term includes such attributes as the identity, strength, and purity.

Rapidly dissolving products: An immediate-release solid oral DP is considered rapidly dissolving when NLT 80% of the label amount of the DS dissolves within 15 min in each of the following media: (1) pH 1.2, (2) pH 4.0, and (3) pH 6.8.

Retest date: The date after which samples of the DS should be examined to ensure that the material is still in compliance with the specification and thus suitable for use in the manufacture of a given DP.

Retest period: The period of time during which the DS is expected to remain within its specification and, therefore, can be used in the manufacture of a given DP, provided that the DS has been stored under the defined conditions. After this period, a batch of DS destined for use in the manufacture of a DP should be retested for compliance with the specification and then used immediately. A batch of DS can be retested multiple times and a different portion of the batch used after each retest, as long as it continues to comply with the specification. For most biotechnological/biological substances known to be labile, it is more appropriate to establish a shelf life than a retest period. The same may be true for certain antibiotics.

Semipermeable containers: Containers that allow the passage of solvent, usually water, while preventing solute loss. The mechanism for solvent transport occurs by absorption into one container surface, diffusion through the bulk of the container material, and desorption from the other surface. Transport is driven by a partial-pressure gradient. Examples of semipermeable containers include plastic bags and semirigid, low-density polyethylene (LDPE) pouches for large-volume parenterals (LVPs) and LDPE ampoules, bottles, and vials.

Shelf life (also referred to as expiration dating period): The time period during which a DP is expected to remain within the approved shelf life specification, provided that it is stored under the conditions defined on the container label.

Specific test: A test that is considered to be applicable to particular new DSs or particular new DPs depending on their specific properties and/or intended use.

Specification: A list of tests, references to analytical procedures, and appropriate acceptance criteria, which are numerical limits, ranges, or other criteria for the tests described. It establishes the set of criteria to which a DS or DP should conform to be considered acceptable for its intended use. "Conformance to specifications" means that the DS and/or DP, when tested according to the listed analytical procedures, will meet the listed acceptance criteria. Specifications are critical quality standards that are proposed and justified by the manufacturer and approved by regulatory authorities.

Specification—Shelf life: The combination of physical, chemical, biological, and microbiological tests and acceptance criteria that determine the suitability of a DS throughout its retest period or that a DP should meet throughout its shelf life.

Stress testing (DP): Studies undertaken to assess the effect of severe conditions on the DP. Such studies include photostability testing and specific testing on certain products (e.g., metered dose inhalers, creams, emulsions, refrigerated aqueous liquid products).

Stress testing (DS): Studies undertaken to elucidate the intrinsic stability of the DS. Such testing is part of the development strategy and is normally carried out under more severe conditions than those used for accelerated testing.

Supporting data: Data, other than those from formal stability studies, that support the analytical procedures, the proposed retest period or shelf life, and the label storage statements. Such data include (1) stability data on early synthetic route batches of DS, small-scale batches of materials, investigational formulations not proposed for marketing, related formulations, and product presented in containers and closures other than those proposed for marketing; (2) information regarding test results on containers; and (3) other scientific rationales.

Universal test: A test that is considered to be potentially applicable to all new DSs or all new DPs, for example, appearance, identification, assay, and impurity tests.

REFERENCES

1. CFR Title 21 section 211.165 Testing and release for distribution.
2. ICH Q6A: Specifications: Test Procedures and Acceptance Criteria for New Drug Substances and New Drug Products, October 1999.
3. ICH Q1A (R2): Stability Testing of New Drug Substances and Products, February 2003.
4. ICH Q1B: Photostability Testing of New Drug Substances and Drug Products, November 1996.
5. W Grimm Drugs Made in Germany 1985;28:196–202, and 1986;29:39–47.
6. Haynes JD. Worldwide virtual temperature for product stability testing. J Pharm Sci 1971;60:927–929.

11

LC-MS FOR PHARMACEUTICAL ANALYSIS

Herman Lam

11.1 INTRODUCTION

When considering the wide scope of analytical requirements for pharmaceutical analysis, there is little doubt that the liquid chromatography-mass spectrometry (LC-MS) is among the most valuable technique in terms of sensitivity, selectivity, dynamic range, reliability, throughput, and diversity in applications [1–10]. LC-MS has been used in all stages of drug development process including discovery, preclinical, clinical, and manufacturing. The LC-MS applications being explored in various stages of drug development are outlined in Table 11.1.

The prevalence of the LC-MS technique in pharmaceutical analysis originated from the combination of two very powerful techniques, liquid chromatography (LC) and mass spectrometry (MS), through the use of an atmospheric pressure ionization (API) interface. LC, which provides very good separation and selectivity in liquid phase, is more amendable to pharmaceutical molecules and macromolecules such as proteins. This front-end separation power is coupled with the tremendous sensitivity and selectivity of the MS detection to enable the analysis of very complex samples. Recent developments in tandem multidimensional LC coupled to tandem MS detection have further enhanced the capability.

Therapeutic Delivery Solutions, First Edition. Edited by Chung Chow Chan, Kwok Chow, Bill McKay, and Michelle Fung.
© 2014 John Wiley & Sons, Inc. Published 2014 by John Wiley & Sons, Inc.

TABLE 11.1   LC-MS applications in different stages of drug development

Development stage	LC-MS applications
Discovery	High-throughput screening
	Structural identification
	Pharmacokinetics
	Membrane permeability
	Drug–drug interaction
	Metabolite identification
	Metabolic stability
	Drug–protein interaction
	Protein PTM–Post Translational Modification
	Biomarkers
Preclinical	Impurity identification
	Degradant identification
	Metabolite identification
Clinical	Bioanalysis
	Impurity identification
	Degradant identification
	Metabolite identification
Manufacturing	Impurity identification
	Quality control

11.2   LC-MS INSTRUMENTATION

There are many possible combinations to configure an LC-MS system. The configuration depends on the application requirements and the budget available. The selection of LC, ionization interface, and mass detector combinations is shown in Table 11.2.

11.2.1   HPLC Front End

The performance of high-performance liquid chromatography (HPLC) instrumentation has come a long way. Advancement in the HPLC instrumentation in all modules (viz., the pump, detector, and injector) that comprises of an LC together with the new HPLC column technology has greatly improved the resolution power and the sensitivity of HPLC. The new LC systems are designed to deliver mobile phase at a much higher pressure than traditional LC and allow the use of smaller particle sorbents to achieve better resolution. Many reverse-phase column sorbents are now available in sub-2 micron size. The ultrahigh-performance liquid chromatography (UHPLC) is an example of utilizing higher pressure with smaller particle size sorbent material to achieve superior resolution, sensitivity, and selectivity to accomplish the analysis in a much shorter time than traditional LC. Capillary LC and nanoflow LCs where the flow rates in the microliter or nanoliter per minute ranges are available for applications that require ultrahigh-resolution power. The resolution enhancement and

TABLE 11.2 Common liquid chromatographic systems, ionization interfaces, and mass detectors for LC-MS and LC-MS/MS systems

LC (flow rate)	Interface	MS	MS/MS
Traditional LC (ml/min)	ESI	Q	QqQ
UPLC (ml/min)	APCI	IT	Quadrupole-ion trap
Capillary LC (μl/min)	APPI	LIT	Q-TOF
Nanoflow LC (nl/min)		Orbitrap MS	TOF/TOF
		TOF	Quadrupole-Orbitrap MS

peak capacity are very important to reduce the sample complexity for the reduction of the matrix effects and ion suppression in LC-MS applications.

The development of two-dimensional LC (2D-LC) that couples columns of different separation chemistry together is aimed to gain further improvement in resolution and peak capacity [11–13]. For example, a hydrophilic interaction chromatography (HILIC) column can be coupled with a reversed-phase column online to extend the peak capacity of the front-end separation of an LC-MS system [14]. Online 2D-LC separation, which can be automated and with minimum sample loss, is preferred to the offline separation where the eluates are collected and reinjected into a second column of different chemistry.

11.2.2 LC-MS Interface

Since MS can only be used to analyze charged or ionized species, it is necessary to convert the neutral analytes into ions. The development of different API techniques is the key driving factor for the widespread use of LC-MS. Without the availability of various API interfaces, the use of LC-MS may still be confined to research laboratories with limited real-world applications. There are two main types of API interfaces, the electrospray ionization (ESI) and atmospheric pressure chemical ionization (APCI). The ESI is used in the analysis of polar analytes, while the APCI is used in the analysis of nonpolar molecules. Both API devices used as an interface between the LC and the MS have to cope with the eluent from the LC, ionize the analytes within the eluate, and transfer the analyte ions under atmospheric pressure to the high vacuum inside the mass spectrometer.

11.2.2.1 Electrospray Ionization (ESI)   For the ESI, the eluent from the HPLC is channeled through a fine capillary. A high voltage (typically 3–6 kV) is applied to the capillary tip to produce a spray of very fine droplets containing ions (Fig. 11.1). Depending on the polarity of the applied potential, either positive ions or negative ions will be formed. The charged droplets are drawn toward the inlet of the mass spectrometer by applying a potential difference between the spray tip and the inlet. A stream of hot dry gas usually nitrogen is blowing across the charged droplets

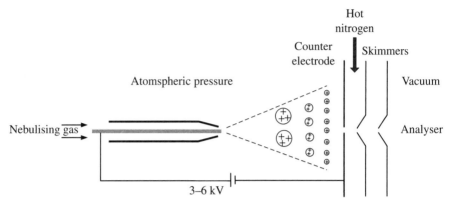

FIGURE 11.1 Schematic of an ESI interface.

to assist evaporation of the solvent in the droplets. Evaporation leads to solvent loses from the droplets and reduction in the size of the droplet. The size of the droplets reduces as they move toward the inlet. The charge density of the droplets increases with reducing droplet size. At a certain droplet size, the coulombic force overcomes the surface tension of the droplet [15, 16]. The droplets break into smaller droplets. This evaporation and droplet-breaking process continues until free ions are formed. The formation of the ions is under atmospheric pressure. The ions enter the mass analyzer through a small opening at the inlet and travel to the high vacuum section guided by electrical lens operating at cascading potential difference. The pressure at the inlet is reduced through several stages to reach the high vacuum condition inside the mass spectrometer.

ESI is considered a soft ionization technique where analytes usually remained intact during the ionization process [15]. A very useful feature of ESI is that multiply charged ions can be produced within large molecules. As the mass-to-charge (*m/z*) ratio is the physical property that is measured in MS, multiply charged ions have apparent *m/z* values that are a fraction of their actual masses. This enables the analyses of the multiply charged species of very large molecules such as proteins due to apparent reduction in the *m/z* ratio. The sensitivity of the ESI is dependent on the LC flow rate; the slower the flow rate, the more sensitive the analysis can be achieved. It is thus beneficial to use capillary LC or nanoflow LC with columns of smaller internal diameter in the LC front end of the LC-MS using an ESI interface.

The presence of a background matrix can affect the ionization efficiency in ESI by either suppressing the formation of ions or enhancing the ions formation. The matrix effect suppression can cause significant decrease in sensitivity and introduce large variation in quantitation.

11.2.2.2 Atmospheric Pressure Chemical Ionization (APCI)

In the APCI analysis, the eluent from the HPLC column is pushed through a heated nonconductive capillary tube surrounded by a coaxial jacket of nitrogen (Fig. 11.2). The spray of

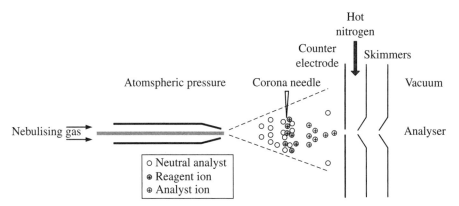

FIGURE 11.2 Schematic of an APCI interface.

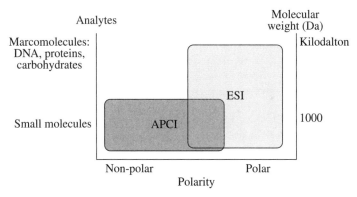

FIGURE 11.3 Schematic diagram showing the different operation ranges for APCI and ESI in terms of relative polarity, mass range, and target analytes.

fine droplets surrounded by hot gas is converted into gas phase. The gas phase analytes are then ionized chemically by transfer of charge between a reagent ion and a target molecule [16]. The mechanism of the ionization is believed to involve primary ions such as N_2^+ or O_2^+ formed by the corona discharge from a corona needle set to 2–3 kV. Subsequent charge transfer from the primary ions to solvent molecules took place to form solvated reagent ions, which then ionized the analyte molecules. APCI is also a soft ionization technique. Usually, $[M+H]^+$ ions are produced in the positive mode, and $[M-H]^-$ ions are produced in the negative mode. The ions enter the mass analyzer through a small opening at the inlet and travel to the high vacuum section guided by electrical potential difference.

APCI can be operated at a higher linear velocity of the mobile phase (often close to 1 ml/min). APCI is more suitable for analyzing less polar molecules and complements the operational polarity range for ESI. A schematic diagram showing the different operation ranges for APCI and ESI in terms of relative polarity, mass range, and target analytes is shown in Figure 11.3.

11.2.3 Mass Analyzers

Among the many different types of mass analyzers being used in MS, the analyzers based on quadrupole (Q), ion trap (IT), time of flight (TOF), and orbital trap (Orbitrap) are commonly encountered in LC-MS applications. These mass analyzers operate on different principles of ion separation, which will be discussed in the following sections. In order to provide a perspective of the capability of the different mass analyzers, it is useful to introduce several key performance characteristics such as mass range, mass accuracy, and mass resolution for consideration [17–19].

The mass range refers to the limit of m/z of the ions that the mass analyzer is capable of measuring. The mass accuracy is the difference between the measured mass and the theoretical mass. The mass accuracy can be expressed in absolute mass difference or in a part per million (ppm) in relationship to the mass of the ion being measured. A lower ppm value implies better mass accuracy. The mass resolution is the ability to differentiate neighboring peaks of ions with small difference in their mass (ΔM). Two peaks are considered resolved if the valley between them has 50% of the intensity of the smaller peak of the two. The resolution (R) between two peaks of masses M and $M + \Delta M$ is given by ($M/\Delta M$). The smaller the mass difference ΔM that can be differentiated, the better the resolution. Another way to estimate the resolution power is to use the full width at half maximum (FWHM) of a peak as the ΔM in the resolution Equation.

11.2.3.1 Quadruple (Q) The quadrupole is made up of four metallic rods arranged in parallel. Ideally, the rods should be hyperbolic (Fig. 11.4). One pair of opposing rods is connected electrically to a direct current (DC) voltage U and superimposed with an oscillating radio-frequency (RF) voltage $V\cos\omega t$ where ω is the angular frequency (in rad$^{s-1)}$ and t is time. The other pair of rods receives a DC voltage of U and superimposed with a RF voltage of the same magnitude but 180° out of phase. The configuration and the voltages applied to each pair of rods in the quadrupole are shown in Figure 11.4. During a m/z scan, the applied voltages U and V are increased in a linear manner. The ions that entered the quadrupole are subjected to a quadrupolar

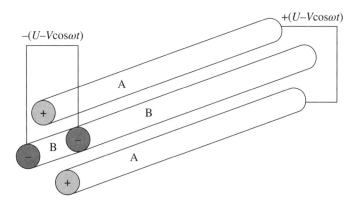

FIGURE 11.4 Schematic of the configuration of a quadrupole mass analyzer.

field. Only the ions of a narrow m/z range can have a stable trajectory to pass through the quadrupole at a particular voltage setting, while the others collide on the rod surface due to unstable trajectories. The ratio of U and V dictates the resolution. A higher ratio increases the resolution but decreases the number of ions that can have a stable trajectory reducing the sensitivity. Quadrupole is not a high-resolution mass analysis. The typical resolution is unit mass. The practical m/z range for a quadrupole is about 4000, which is sufficient for small molecules but not for big biomolecules such as proteins.

11.2.3.2 Ion Trap Traditional IT mass analyzers have a ring electrode and two endcap electrodes. The endcap electrodes have a small opening to allow introduction of ions into the trap and ejection ions to the detector (Fig. 11.5). The ions introduced to the space between the electrodes are trapped by an oscillating electric field by applying a potential $\Phi_0 = U - V\cos\omega t$ to the ring electrode. The trapped ions with a broad m/z range precess in trajectories within the space defined by the radius R_0 of the ring electrode and the distance (Z_0) between the endcap electrodes. As ions repel each other within the small confine of the trap, which leads to trajectory deterioration, helium at a pressure of 10^{-3} Torr is introduced to remove excess energy by collision to confine the ions.

Different modes of operation can be used to analyze the ions in the trap. In the mass-selective instability mode, ramming the RF potential leads to ions of increasing m/z to be ejected from the trap successively from the endcap to the detector. In the resonant excitation mode, a bipolar supplementary RF potential applied to the endcap excites the ions and ejects the ions. An ion with a particular m/z can be selected to remain in the trap by ejecting ions of m/z higher and lower than the m/z of interest. In the mass-selective mode, potentials to the ring and endcap electrode are analogous to the opposing electrodes in a quadrupole mass filter. A mass spectrum can be obtained by scanning the U and V components of the applied potentials or to trap selected ions with specific m/z. The major advantages of IT are abilities to accumulate ions of interests to enhance the sensitivity and to enable multiple fragmentations in a cascade reaction resulting in product ions in one experiment (MS^n), which can be very informative in structure identification. However, there is a limit in the number of ions

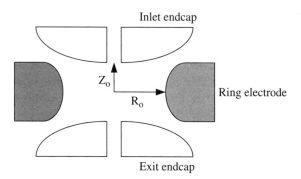

FIGURE 11.5   Schematic of the configuration of the electrodes in a 3D IT mass analyzer.

in the trap. If there are too many ions in the trap, a space charge effect resulting from the slight modification of the electrical field experienced by the inner ions due to shielding by the outer ions can lead to reduction in resolution power.

The linear ion trap (LIT) is a variant of the traditional three-dimensional ion trap (3D IT). A 2D LIT is a quadrupole with endcaps appended at the entrance and exit, so it can work as a quadrupole or as an IT, depending on whether potential is applied on the endcaps or not. Compared to the 3D IT, the LIT has better trapping efficiency, larger ion storage capability, and enhanced ion ejection efficiency and sensitivity.

11.2.3.3   Time of Flight (TOF)   For the TOF mass analyzer, ions are separated on the basis of their velocity difference inside a field-free flight tube (Fig. 11.6). Packages of ions are being drawn into the acceleration region of the analyzer by a potential difference to impart the ions with the same kinetic energy. Ions then travel into the flight tube of length L with different velocities depending on their masses (m) and the charges (ze) they carried. The relationship between the velocity of ion, mass, and charge is given by Equation 11.1:

$$v = \sqrt{\frac{2zeV}{m}} \qquad (11.1)$$

where V is the acceleration potential.

The time (t) it takes for the ions to reach the detector at a distance L away is related to m/z given by Equation 11.2:

$$t^2 = \frac{m}{z}\left(\frac{L^2}{2eV}\right) \qquad (11.2)$$

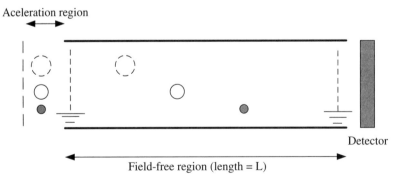

Aceleration region

Detector

Field-free region (length = L)

FIGURE 11.6   Schematic diagram showing the basic operation principle of TOF mass analyzer.

TOF mass analyzers offer many attractive features. The mass range is potentially unlimited though there is a practical limit. Good sensitivity is achieved by having high transmission efficiency and using multichannel detection. The acquisition rate is very fast to enable the detection of transient species. Good mass accuracy is down to less than 5 ppm. Mass resolution, which used to be a weakness for TOF, has been tremendously improved by the use of delayed ion extraction and the energy-correction device (reflectron) to correct slight variation in velocity of the ions due to the dispersion of initial kinetic energy.

11.2.3.4 Orbital Trap (Orbitrap) The Orbitrap is a new type of Fourier transform mass analyzer introduced in 2004. The Orbitrap is very compact in size but big in performances. Ions are introduced into the Orbitrap (a static field), cycling around the central electrode rings [20] while moving back and forth along the axis of the central electrode. The ions with a specific m/z ratio move in the static field oscillate along the central electrode at a specific frequency. The frequency of these harmonic oscillations is inversely proportional to the square root of the m/z ratio. The circulating ions induce a current that can be detected and converted into mass spectra by using Fourier transform algorithm. The Orbitrap MS exhibits a very impressive performance in mass accuracy (1–2 ppm), mass resolution (up to 200,000), and dynamic range.

Overall, there is a range of mass analyzers for LC-MS applications. The typical performance of each class of mass detector is given in Table 11.3.

11.2.4 Tandem Mass Spectrometry (MS/MS)

A lot of LC-MS applications involve the use of more than one MS analyzer working in tandem. The idea is to use the first mass analyzer to select a target (precursor) ion of interest for fragmentation studies. The collision-induced dissociation with an inert gas at reduced pressure is a popular way to introduce fragmentation to the target ion. The product ions resulting from the fragmentation are analyzed by a second MS analyzer. The unique relationship between the precursor ion and the product ions plays an important role in structural elucidation and identification of target compounds. There are many combinations of mass analyzers that can be used in LC-MS/MS. Some of the common tandem configurations are the triple quadrupoles (QqQ) and hybrid quadrupole/TOF (Q-TOF). The increasing popularity tandem MS can be linked to the emergence of QqQ. The QqQ systems are the workhorses of many bioanalysis in clinical trials. The Q-TOF systems that couple great sensitivity, mass

TABLE 11.3 Performance of mass detectors

Mass analyzer type	Mass range (μ)	Mass resolution (FWHM)	Mass accuracy (ppm)
Q	4,000	2,000	100
IT	6,000	4,000	100
TOF with reflectron	10,000	100,000	10
Orbitrap	50,000	200,000	<5

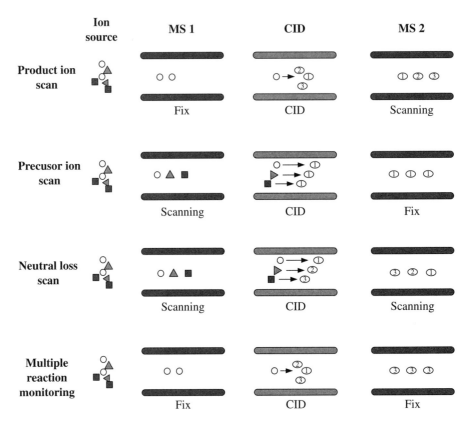

FIGURE 11.7   Scan mode for LC-MS/MS applications.

accuracy, and resolution with ion fragmentation capability play a crucial role in MS-based proteomics and metabolomics.

There are four possible scan modes that can be used in tandem MS/MS. Schematic illustration of the scan modes is shown in Figure 11.7 [2, 17].

11.2.4.1   Product Ion Scan   The first mass analyzer operates in a nonscanning setting to transmit only the target precursor ion with a fixed *m/z*. The precursor ion undergoes fragmentation to produce the product ions, which are analyzed by the second mass analyzer in scanning mode over a range of *m/z*. The spectral information from the fragments is useful in structural elucidation and peptide sequencing.

11.2.4.2   Precursor Ion Scan   The first mass analyzer operates in a scanning setting to transmit ions over a range of *m/z*. The second mass analyzer operates in a nonscanning setting to transmit only a selected product ion with a fixed *m/z*. All precursors that may produce the selected product ion fragment are identified. Precursor ion scan is useful to identify related compounds in a mixture. For example, a fragment of *m/z* 79 corresponding to the PO_3^- ion can be set to detect the presence of phosphopeptides in a mixture.

11.2.4.3 Neutral Loss Scan The first and the second mass analyzers are scanned in a synchronized fashion with a mass offset that corresponds to the mass of a specific neutral species. For example, the loss of 44 Da corresponding to the CO_2 is a common reaction for carboxylic acids. The scan in the second mass analyzer is offset by a m/z of 44; the neutral loss scan can reveal the analytes containing the carboxylic functionality.

11.2.4.4 Multiple Reaction Monitoring (MRM) or Selective Reaction Monitoring (SRM) The first mass analyzer transmits only the target precursor ion with a fixed m/z. The second mass analyzer is set to transmit only the product ion fragments with specific m/z. A precursor ion and a particular product ion are referred to as transition. A precursor ion can give rise to several transitions with product ions of different m/z. More than one transition can be monitored in the same experiment. The SRM mode is the most common mode of operation for quantitative application. The reduction of chemical noise by isolating the precursor ion significantly increases the signal-to-noise ratio in the selective reaction monitoring (SRM) mode to enable detection of analytes, which may have been buried by the noise. Typically, the transition with the highest product ion intensity will be used for quantitation purpose and a second transition can be used for confirmation purpose.

In a QqQ system, the first quadrupole (Q1) and the third quadrupole (Q3) operate as mass analyzers to provide mass selection functions by varying the DC and the RF. The second quadrupole (q) that operates in RF only allows all ions to pass through functions as an ion containment region and as a collision cell for collision-induced fragmentation. All four types of scan can be performed by a QqQ system.

The Q-TOF instruments are another popular hybrid instruments combining two types of mass analyzer. The quadrupole mass analyzer provides the initial mass selection to isolate the precursor ions. Usually, there is another quadrupole operating in RF only to function as a collision cell. The TOF mass analyzer with reasonably good resolution and mass accuracy and high sensitivity is used to scan the product ions. The duty cycle of a Q-TOF instrument is much faster than a QqQ instrument. The standard scan mode for a Q-TOF instrument is the product ion scan. Other scan modes such as precursor ion scan and neutral loss scan are not directly applicable.

11.3 EXAMPLES OF QUALITATIVE AND QUANTITATIVE APPLICATIONS

11.3.1 Structural Identification

LC-MS has played an important role in the identification of related substance in active pharmaceutical ingredient because of its superb combination of separation capability, selectivity, sensitivity, much improved mass resolution, and accuracy [7]. The characterization and identification of related substances of a potent corticosteroid mometasone furoate using an Orbitrap is a good example to illustrate the usefulness of the powerful LC-MS combination [5]. In this application, two of the impurities

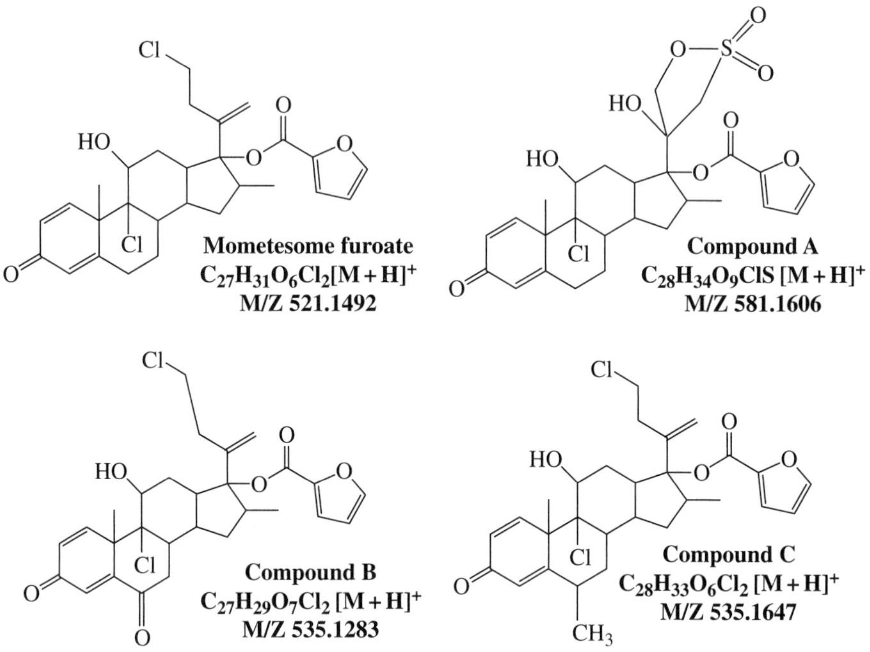

FIGURE 11.8 Mometasone furoate and its related impurities.

related to mometasone furoate with m/z of 535 and 581 were found to coelute in the total ion chromatogram before the mometasone furoate peak. The high-resolution and high-mass-accuracy capabilities of the Orbitrap enable the determination of elemental composition of the coeluting impurities based on isotopic patterns and the elemental combinations. The best possible elemental composition for the [M+H]$^+$ ions with m/z of 581.1606 with a mass accuracy of 0.18 ppm was determined to be $C_{28}H_{34}O_9ClS$ (Compound A, Fig. 11.8). Comparing the elemental composition of mometasone furoate, the additional moiety corresponds to CH_3O_3S with the removal of on chlorine atom. Further high-resolution LC-MS/MS experiments establish that the sulfur moiety is inserted at the 20 keto position. The formation of the sulfur-containing impurity is properly related to the action with the reagent CH_3O_2Cl used in the synthetic processes.

The best possible elemental composition for the coeluting [M+H]$^+$ ions with m/z of 535.1283 with a mass accuracy of 0.41 ppm was determined to be $C_{27}H_{29}O_7Cl_2$, which corresponds to the 6 keto structure (Fig. 11.8, Compound B). Interestingly, there is a late-eluting impurity after the mometasone furoate peak with the same nominal m/z of 535. The best possible elemental composition for the late-eluting impurity ([M+H]$^+$ ions with m/z of 535.1647 with a mass accuracy of 0.41 ppm) was determined to be $C_{28}H_{33}O_6Cl_2$, which corresponds to a methyl substitution at the 6 keto position (Fig. 11.8, Compound C). The difference in structure between the two impurities with the same nominal mass was differentiated using accurate mass data.

11.3.2 Bioanalytical Applications

Tandem LC-QqQ systems with API interfaces have been the workhorse in quantitative analysis in pharmacokinetics and metabolism studies. The SRM mode is the method of choice for its specificity, sensitivity, and high throughput. The signal-to-noise ratio is greatly enhanced using the SRM mode to improve the limit of detection and limit of quantitation (LOQ). The analytes of interest are ionized by the soft atmospheric ionization techniques such as ESI or APCI depending on the polarities. The unfragmented targeted analyte ions are selected as precursor ions in Q1. The selected ions are fragmented in Q2 by collision-induced dissolution into product ions and transmitted to Q3. More than one transition can be monitored in Q3. The transition that produces the highest ion intensity is selected in Q3 for detection and quantitation. Other transitions can be used to confirm the identity of the targeted analyte for specificity.

The development of a bioanalytical method involves the development of sample extraction and cleaning procedure, chromatographic conditions to separate the analyte of interest from potential interferences, and optimization of the MS detection of the analytes of interests [10, 21]. A good sample preparation procedure enables good and consistent recovery of the analytes from the complex biological matrixes such as urine, blood, plasma, and tissues. Depending on the applications, the analytes of interest may not necessary be completely resolved in the chromatographic separation as the mass spectrometer can provide orthogonal separation according to the m/z ratio of the analytes. However, it is certainly useful to achieve an adequate level of chromatographic separation up front using a solvent mixture and pH that facilitate ionization by the API interface for good sensitivity. Experimental parameters such as temperature and voltage of the API interface, the flow of evaporation gas to assist the vaporization of the droplets from the spray tip, various potentials of the electrical lens along the ion optics, the collision gas pressure, and the detector voltage have to be optimized for the formation and detection of precursor ions and product ions.

The composition of the biological samples can vary significantly between individuals and species. One of the major factors affecting the accuracy and precision of the analysis of biological samples is the matrix effect [22]. The ionization of the analytes are either suppressed or enhanced by the presence of the matrix components such as endogenous compounds and metabolites, which result in increase or decrease in signal intensity. Matrix effect is more pronounced in ESI than APCI, and it affects the early-eluting compounds more than late-eluting compounds. Several mechanisms for ion suppression caused by the matrix components such as competition of excess charges on the ESI droplets, changes in surface tension and viscosity of the droplets, and neutralization of the analyte ions have been proposed [23]. The matrix effect can be evaluated by comparing the response obtained from a neat standard solution and a postextraction sample spiked with the analyte of interest. Another way to evaluate matrix effect is the postcolumn infusion of the analyte solution into the ion source while the blank matrix extract is eluting from a column. The postcolumn infusion method can provide information about the retention time or the chromatography region affected by the matrix components in the chromatogram. Due to the unpredictability of the matrix effect, it is a major concern in quantitative analysis using LC-MS with ESI and APCI. Evaluation of matrix effect is required in the validation of bioanalytical method by the U.S. FDA Guidance for Industry on Bioanalytical Method Validation [24].

Various approaches can be taken to avoid or reduce the detrimental matrix effect. The first approach is to explore different APIs such as ESI, APCI, and atmospheric pressure photoionization (APPI) techniques to see what ionization technique for the analyte is less prone to matrix effect. The second approach is to chromatographically separate the analyte from the matrix interference. The postcolumn infusion technique can provide useful information on the retention of the interfering matrix components. The third approach is to remove the interfering matrix components in the sample preparation stage by protein precipitation or solid phase extraction (SPE). In case all these approaches cannot effectively clean up the interfering matrix, the use of matrix match standards can be considered provided the matrix effect is similar among same type of samples. However, the use of stable isotopic-labeled internal standard is the best way to compensate the variability caused by the matrix effect on the accuracy and precision of the quantitative LC-MS/MS method for bioanalysis. The stable isotopic-labeled internal standard with a sufficient number of ^{13}C or ^{15}N incorporated will have the same protein bonding characteristics, recovery, ionization efficiency, response, and retention as the unlabeled analyte [25]. For LC-MS analysis, the stable isotopic-labeled internal standard does not have to be separated chromatographically from the analyte as long as they have separate mass. The stable isotopic-labeled internal standard should be added to the sample prior to sample pretreatment so that both matrix effect and extraction loss can be compensated.

The assay of melamine in milk is used as example to illustrate the use of stable isotopic-labeled internal standard for quantitative analysis. The nonlabeled melamine has a molecular weight of 126. The stable isotopic-labeled melamine has three ^{13}C and three ^{15}N substitutes with a molecular weight of 132 as shown in Figure 11.9. For the nonlabeled melamine, the precursor ion [MH$^+$] has a m/z of 127. The product ions to be monitored for the quantitation have a m/z of 85 and 68. For the stable isotope-labeled melamine, the precursor ion [MH$^+$] has a m/z of 133. The products to be monitor for the quantitation have a m/z of 91 and 74 (Fig. 11.9b). A fix quantity of an internal standard is introduced to each standard and sample preparation. The extracted ion chromatograms of the labeled and the nonlabeled melamine and the fragment ions are shown in Figure 11.9c. A calibration curve is then constructed based on the ratio of the responses of analyte and internal standard versus the analyte concentration for the quantification of the unknown level of melamine in the samples.

11.3.3 Protein and Peptide Analysis

In life science research, the ability to identify and quantify any protein or set of proteins of interest in various physiological states is essential to advance the understanding of biological systems. The qualitative and/or qualitative analysis of all proteins in a tissue and cell in both temporal and spatial terms is referred to as

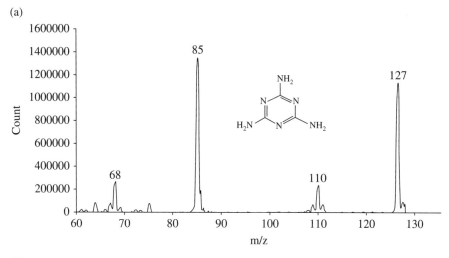

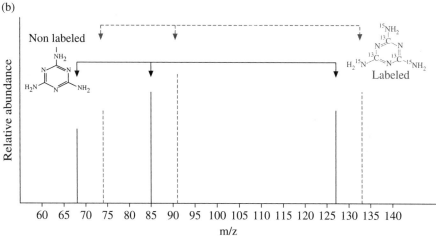

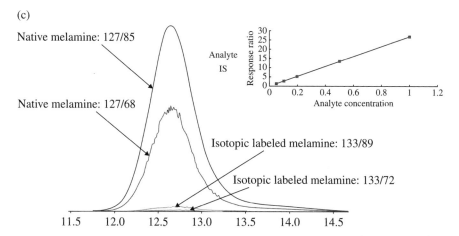

FIGURE 11.9 (a) MS spectrum of nonlabeled melamine. (b) MS spectrum of nonlabeled and labeled melamine. (c) Extracted ion chromatograms of the transitions of nonlabeled and labeled melamine and a calibration curve constructed based on the area ratio of ion nonlabeled and labeled melamine ion fragments.

proteomics. Since many small-molecule drugs and biologics act on protein targets, proteomics has been used in drug discovery and preclinical studies to try to understand the effects of drug on the protein targets [10, 26–29]. For a long time, the use of MS for protein analysis had been restricted by the lack of suitable ionization techniques to produce intact gas phase ions for large biomolecules using traditional ionization methods. Large biomolecules are broken up into random fragments during the vaporization and ionization processes. The development of ESI and matrix-assisted laser desorption/ionization (MALDI) has revolutionized the protein and peptide analysis to make them amendable for MS. ESI can be coupled with various mass filters and detectors in tandem for qualitative and quantitative proteomics. MALDI in general is not amendable for LC-MS application because of the sample preparation steps required prior to the laser desorption.

11.3.3.1 Qualitative Analysis The identification of proteins in a complex biological matrix is sometimes referred to as shotgun proteomics. The workflow for shotgun proteomics involved in the LC-MS/MS analysis of peptides obtained from tryptic digestion of the cell or tissues is shown in Figure 11.10 [26]. The peptide mixture first undergoes a capillary LC or nanoflow LC separation to reduce the complexity of the mixture before the peptides in the eluent are ionized by ESI. The number of peptides in the sample is expected to be far exceeding peak capacity for complete separation; multiple peptides are likely to be coeluding within a fraction at a given time. Peptide coelution can lead to ion suppressions during the ionization process and under sampling of peptides in the MS analysis. The use of multidimensional LC that harnesses the separation power of different column chemistries to increase the separation power and peak capacity has gain popularity in shotgun proteomics [11, 12].

In a data-dependent approach, the mass spectrometer can be programmed to perform a scan of the peptides in the first mass detector, usually a quadrupole, to select few peptides with the highest signal intensity for collision-induced dissociation in Q2. The resulting fragments are analyzed in the second mass detector with good mass accuracy and resolution to enable better identification of the mass fragments from very similar peptides. The MS analysis can also be conducted in a direct precursor selection approach to select peptides with specific masses and LC retention time if prior information is available [30]. The low-intensity peptide ions can be analyzed in the direct precursor selection mode.

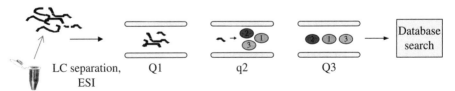

FIGURE 11.10   Work for shotgun proteomics.

The spectra of the peptides and fragment ions are then searched against known peptide spectral libraries or theoretical peptide fragment lists in sequence databases [31]. The process of peptide/protein identification is called spectral library searching. A peptide spectral library is a curated, annotated, and nonredundant collection/ database of LC-MS/MS peptide spectra built by consensus. For a sequence database search, the experimental spectral information of the peptide ion fragments is searched against the calculated spectra of all putative peptide candidates in the given setting (proteolytic enzymes, miscleavages, posttranslational modifications (PTM)) to find a sufficiently close match with the experimental mass spectra, which serves as the basis for peptide/protein identification. Since the sequence of amino acids in a protein is very unique, its identification can be inferred by the presence of one or more unique characteristic peptides of the protein from the tryptic digest.

11.3.3.2 *Quantitative Analysis*

Different levels of a particular protein or set of proteins can be present in different physiological states that correspond to different stages of a disease or in drug interactions [28, 29, 32, 33]. LC-MS/MS analysis using stable isotope labels plays a key role in the relative and absolute quantification of proteins in proteomics research. Among the many different protein quantification techniques involving stable isotope labeling, the isobaric tags for relative and absolute quantification (iTRAQ™) reagents are useful tools to monitor relative changes in protein and PTM abundance across perturbed biological systems allowing comparison of normal, disease, or drug-treated states [34].

The iTRAQ reagents consist of a set of four or eight isobaric reagents to label the primary amines of peptides and proteins. For simplicity, an iTRAQ reagent set with four isobaric label reagents each consisting of an *N*-methyl piperazine reporter group with mass equal to 114, 115, 116, or 117; a balance group with mass equal to 31, 30, 29, or 28; and an *N*-hydroxysuccinimide ester group that is reactive with the primary amines of peptides is used to illustrate the workflow of the iTRAQ technique for protein and peptide quantification (Fig. 11.11). The balance groups present in each of the iTRAQ reagents render the labeled peptides from each sample isobaric (same mass). Each isobaric tag has a *m/z* of 145.

A typical iTRAQ workflow is illustrated in Figure 11.12. Samples to be quantified are prepared and digested using an enzyme, such as trypsin, to generate proteolytic peptides. Each peptide digest is labeled with a different iTRAQ reagent. The labeled digests from different samples are then combined into one sample mixture. The combined peptide mixture is analyzed by LC-MS/MS for both identification and quantification. Since the chemical composition of each iTRAQ reagent is the same, the labeled peptide fragments from a particular protein of interest from different samples with different iTRAQ tags are chemically identical and with the same characteristics in LC separation, ionization, and MS analysis. The sequence of a peptide is determined from the product ions of the peptide fragments that are generated from cleavage of the peptide bonds as discussed previously. The relative abundance of the peptides from the same protein in different samples can be determined by comparing the intensities of reporter ion signals in the MS/MS scan.

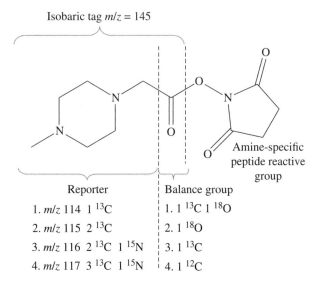

FIGURE 11.11 Chemical structure of the iTRAQ reagents and the stable isotope combinations.

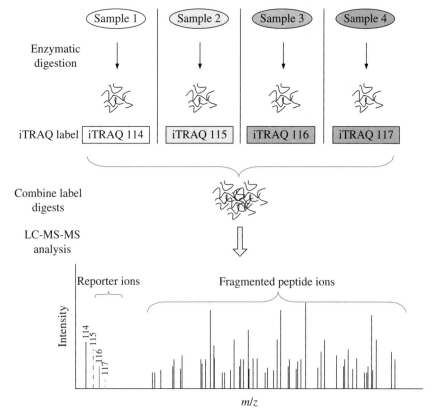

FIGURE 11.12 Workflow of an iTRAQ experiment for protein quantitation.

11.4 SUMMARY

The rapid advancements of LC-MS/MS instrument and methodology with high-throughput capability, super selectivity, exquisite sensitivity, and unprecedented mass accuracy make LC-MS/MS an indispensable tool for pharmaceutical and biological research and development.

REFERENCES

1. Lim CK, Lord G. Current developments in LC-MS for pharmaceutical analysis. Biol Pharm Bull 2002;25(5):547–557.
2. Lee MS. *LC-MS Applications in Drug Development*. New York: John Wiley & Sons; 2002.
3. Rossi DT, Sinz MW. *Mass Spectrometry in Drug Discovery*. New York: Marcel Dekker; 2002.
4. Korfmacher W. Principles and applications of LC-MS in new drug discovery. Drug Discov Today 2005;10(20):1357–1367.
5. Chen G, Pranmanik B, Liu Y, Mitza U. Applications of LC/MS in structure identifications of small molecules and proteins in drug discovery. J Mass Spectrom 2007;42:279–287.
6. Gillespie TA, Winger BE. Mass spectrometry for small molecule pharmaceutical product development: a review. Mass Spectrom Rev 2011;30:479–490.
7. Pramanik BN, Chen MS, Chen GD. Characterization of Impurities and Degradant Using Mass Spectrometry. Hoboken: Wiley; 2011.
8. Lee MS, Zhu M. *Mass Spectrometry in Drug Metabolism and Disposition*. Hoboken: Wiley; 2011.
9. Espada A, Molina-Martin M, Dage J, Kuo MS. Application of LC/MS and related techniques to high-throughput drug discovery. Drug Discov Today 2008;13:417–423.
10. Zhou S, Song Q, Tang Y, Weng N. Critical review of development, validation, and transfer for high throughput bioanalytical LC-MS/MS methods. Curr Pharm Anal 2005;1:3–14.
11. Qian WJ, Jacobs JM, Liu T, Camp D, Smith R. Advances and challenges in liquid chromatography-mass spectrometry-based proteomics profiling for clinical applications. Mol Cell Proteomics 2006;5:1727–1744.
12. Zhang X, Fang A, Riley C, Wang M, Regnier F, Buck C. Multi-dimensional liquid chromatography in proteomics – a review. Anal Chim Acta 2010;664:101–113.
13. Siu SO, Lam MP, Lau E, Kong RP, Lee SM, Chu IK. Fully automatable two-dimensional reversed-phase capillary liquid chromatography with online tandem mass spectrometry for shotgun proteomics. Proteomics 2011;11:2308–2319.
14. Lam P, Siu SO, Lau E, Mao X, Sun P, Chiu CN, Yeung W, Cox D, Chu IK. Online coupling of reverse-phase and hydrophilic interaction liquid chromatography for protein and glycoprotein characterization. Anal Bioanal Chem 2010;398:791–804.
15. Wilm M. Principles of electrospray ionization. Mol Cell Proteomics 2011;10(7): M111.009407.
16. Harris G, Nyadong L, Fernandez F. Recent developments in ambient ionization techniques for analytical mass spectrometry. Analyst 2008;133:1297–1301.

17. de Hoffman E, Stroobant V. *Mass Spectrometry Principles and Applications.* Chichester/ Hoboken: John Wiley; 2007.

18. *Basics of LC/MS Primer (5988-2045EN).* Santa Clara: Agilent Technologies; 2001.

19. *Mass Spectrometry Primer.* Milford: Waters. Available at http://www.waters.com/waters/ partDetail.htm?partNumber=715001940. Accessed March 1, 2014.

20. Perry RH, Cooks RG, Noll RJ. Orbitrap mass spectrometry: instrumentation, ion motion and applications. Mass Spectrom Rev 2008;27:661–99.

21. Yu R, Fan L, Rieser M, El-Shourbagy T. Recent advances in high through quantitative bioanalysis by LC-MS-MS. J Pharm Biomed Anal 2007;44:342–355.

22. Van Eeckhaut A, Lanckmans K, Sarre S, Smolders I, Michotte Y. Validation of bioanalytical LC–MS/MS assays: evaluation of matrix effects. J Chromatogr B 2009;877(23): 2198–2207.

23. Cote C, Berferon A, Mess J, Furtado M, Garofolo F. Matrix effect elimination during LC-MS/MS bioanalysis method development. Bioanalysis 2009;1:1243–1257.

24. FDA. 2001. Guidance for industry bioanalytical method validation. Available at http:// www.fda.gov/downloads/Drugs/GuidanceComplianceRegulatoryInformation/Guidances/ ucm070107.pdf.

25. Nilsson L, Eklund G. Direct quantification in bioanalytical LC–MS/MS using internal calibration via analyte/stable isotope ratio. J Pharm Biomed Anal 2007;43:1094–1099.

26. Domon B, Aebersold R. Mass spectrometry and protein analysis. Science 2006;312(5771): 212–217.

27. Khan M, Bennett M, Jumper C, Percy A. Proteomics by mass spectrometry – go big or go home. J Pharm Biomed Anal 2011;55:832–841.

28. Lee J, Han J, Altwerger G, Kohn E. Proteomics and biomarkers in clinical trials for drug development. J Proteomics 2011;74:2632–2641.

29. Schirle M, Bantscheff M, Kuster B. Mass spectrometry-based proteomics in preclinical drug discovery. Chem Biol 2012;19:72–84.

30. Schmidt A, Claassen M, Aebesold R. Direct mass spectrometry: towards hypothesis-driven proteomics. Curr Opin Chem Biol 2009;13:510–517.

31. Lam H, Deutsch EW, Eddes J, Eng J, King K, Stein S, Aebersold R. Development and validation of a spectral library searching method for peptide identification from MS/MS. Proteomics 2007;7:655–667.

32. Bantscheff M, Schirle M, Sweetman G, Rick J, Kuster B. Quantitative mass spectrometry proteomics: a critical review. Anal Bioanal Chem 2007;389:1017–1031.

33. Ohtsuki S, Uchida Y, Kubo Y, Terasake T. Quantitative targeted absolute proteomics-based ADME research as a new path to drug discovery and development: methodology, advantages, strategy, and prospects. J Pharm Sci 2011;100:3547–3559.

34. Zieske L. A perspective in the use of iTRAQ reagent technology for protein complex and profiling studies. J Exp Bot 2006;57:1501–1508.

12

BIORELEVANT DISSOLUTION TESTING

May Almukainzi, Nádia Araci Bou-Chacra, Roderick B. Walker, and Raimar Löbenberg

12.1 BACKGROUND

12.1.1 Dissolution Testing Definition

Dissolution is a state of transforming a solid into solution, which is an essential step for a drug product in order to be orally absorbed [1]. Pharmaceutical dissolution testing is the *in vitro* measurement of the release of an active pharmaceutical ingredient (API) from a dosage form. It includes the rate as well as the extent of drug dissolution. Dissolution testing has important applications throughout a product's life cycle [2] from the early development stages to routine batch releases of marketed products.

12.1.2 History of Dissolution

The foundation of dissolution experiments is based on Noyes and Whitney equation published in 1897 [3]. Arthur A. Noyes and Willis R. Whitney studied the solubility of benzoic acid and lead chloride; they found that the rate of dissolution is proportional to the difference between the concentrations of the bulk solution (not saturated) and the solution in saturated state. This can be mathematically expressed as $(dC/dt) = k(C_s - C)$, where C is the concentration, t is the time, k is a constant, and C_s is the saturated solubility of the substance under observation. *In vitro* dissolution in pharmaceutical sciences started to grow in the early 1950s [4]. In 1970, dissolution

Therapeutic Delivery Solutions, First Edition. Edited by Chung Chow Chan,
Kwok Chow, Bill McKay, and Michelle Fung.
© 2014 John Wiley & Sons, Inc. Published 2014 by John Wiley & Sons, Inc.

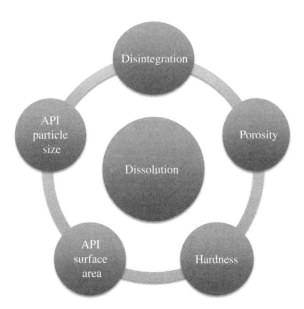

FIGURE 12.1 Factors that might impact drug dissolution. Reproduced from Ref. 7.

testing was introduced officially in the United States Pharmacopeia (USP) 18 and National Formulary (NF) as a quality control (QC) test in six monographs [4]. Over the years, dissolution testing has gained other valuable applications; one of them is the ability to predict the *in vivo* performance of drug products, and this will be discussed further.

Dissolution testing is used in formulation development to scan the impact of excipients such as disintegrates, lubricants, and binders or physical properties such as hardness or powder compression on a drug's dissolution [5]. Excipients might not always have a significant impact on immediate-release (IR) dosage forms; however, they are crucial in the case of extended-release dosage forms [6]. Furthermore, dissolution testing is a valuable test to assess a drug's stability at different storage conditions and to monitor any aging of a formulation at different temperatures over time. Additionally, intrinsic dissolution testing can be used to optimize a drug's particles size and size distribution or surface area to achieve a desired rate and extend of dissolution [5]. Figure 12.1 summarizes formulation factors that impact dissolution testing.

12.2 DISSOLUTION APPARATUS

12.2.1 Apparatus Types

Different apparatuses were introduced over the past decades to test *in vitro* the performance of pharmaceutical dosage forms. However, only few of them are listed in a major pharmacopoeia. USP is one of the most important pharmacopoeias.

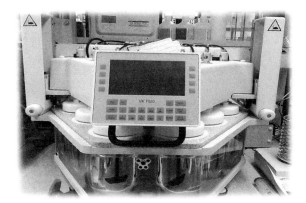

FIGURE 12.2 Apparatus 1 and 2 dissolution test.

Therefore, this chapter will focus on USP apparatuses and will only review some other important methods. The first apparatus described in USP is USP apparatus 1, which was introduced in 1970 [5, 8, 9]. Eight years later, USP apparatus 2 was added to the USP. USP adopted apparatus 3 in 1991 followed by apparatus 4, which was first described in 1995 [2]. Today, USP 34 describes seven apparatuses used for dissolution testing in its general chapters <711> and <724>. Four dissolution apparatuses are mainly used for oral dosage forms: apparatus 1(basket), apparatus 2 (paddle), apparatus 3 (reciprocating cylinder), and apparatus 4 (flow-through cell). The other three apparatuses are usually used for transdermal delivery systems—apparatus 5(paddle over disk), apparatus 6 (cylinder), and apparatus 7 (reciprocating holder)—but they can also be used for other dosage forms such as suspensions, patches, or drug-eluting stents for which apparatus 4 has gained importance.

Apparatuses 1 and 2 are the most widely used test apparatuses in industry because they are easy to operate and well established as performance tests for regulatory submissions [10] (Fig. 12.2). Apparatuses 1, 2, 3, and 4 settings and specifications are described in detail in USP chapter <711>, whereas apparatuses 5, 6, and 7 are described in chapter <724> of USP.

Franz cells are currently not listed in the USP [11], but they are commonly used since 1975 for semisolid dosage forms such as creams and ointments [12] (Fig. 12.3). A stimulus article was published by USP, PF Vol. 35(3) [May–June 2009], to define Franz cells and their usage for topical dosage forms [13]. Franz cells were also used in the past to develop buccal tablets [14].

It can be expected that this kind of diffusion cell will be listed one day in USP as apparatus to assess the performance of pharmaceutical dosage forms.

The individual drug product monographs in the USP specify parameters like media, pH, and sample time points, apparatus settings like rotation speed, and acceptance criteria.

Dissolution tests are performed usually at 37°C, and buffers like simulated gastric fluid (SGF) or simulated intestinal fluid (SIF) can be used; the volume in the apparatus should be set to maintain sink conditions [5]; however, just because these

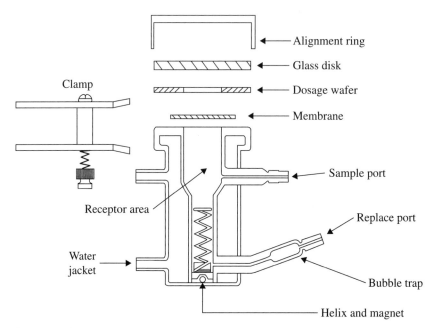

FIGURE 12.3 Franz cells as described in *Pharmacopeial Forum*. Reproduced from *Pharmacopeial Forum*, May-June 2009, with permission of the US Pharmocopeial Convention.

sets of conditions mimic some physiological conditions, it is wrong to assume that dissolution testing is a good predictor of *in vivo* dissolution. This chapter will discuss how dissolution tests can be made more predictive in regard to the *in vivo* dissolution of oral dosage forms.

Over the years, there were different concerns about dissolution equipment and its calibration and its reliability. Physical calibration procedures issued by USP [15], Food and Drug Administration (FDA) [16], and ASTM [17] try to minimize any source of variation in dissolution test apparatuses and aim to eliminate them as much as possible. For example, switching dissolution vessels or paddle wobble is a source of variability [18]. Therefore, dissolution test procedures and apparatus require validation and continuous calibration of physical parameters. USP has made reference standards available, which can be used for a performance verification test [15]. Similar procedures were outlined for Franz cells [19].

12.2.1.1 Dosage Forms and Related Dissolution Test Each dosage form requires a specific dissolution apparatus. Table 12.1 summarizes typical dissolution apparatuses used for common dosage form.

As mentioned before, dissolution testing is used as a QC test or as *in vitro* performance test for oral dosage forms. However, it has to be pointed out that *in vitro* dissolution does not necessary reflect the *in vivo* dissolution. While *in vivo* dissolution

TABLE 12.1 Apparatus typically used for special dosage forms

Dosage from	Dissolution apparatus
Solid oral dosage forms (conventional)	Basket, paddle, reciprocating cylinder, or flow-through cell
Oral suspensions	Paddle
Orally disintegrating tablets	Paddle
Chewable tablets	Basket, paddle, or reciprocating cylinder with glass beads
Transdermalpatches	Paddle over disk
Topicalsemisolids	Franz cell diffusion system
Suppositories	Paddle, modified basket, or dual-chamber flow-through cell
Chewing gum	Special apparatus
Powders and granules	Flow-through cell (powder/granule sample cell)
Microparticulate formulations	Modified flow-through cell
Implants	Modified flow-through cell

Reproduced from Ref. 20, with permission of Springer Science and Business Media.

is a prerequisite for drug absorption in the human body, the *in vitro* dissolution testing is not necessarily a predictor of *in vivo* dissolution or *in vivo* performance of a drug product. However, there are some examples where *in vitro* dissolution was predictive for *in vivo* performance. Blume et al. compared the quality of 128 glibenclamide tablets in 28 countries. They reported that some glibenclamide IR tablet products that had less than 80% drug release in more than 10 min were not bioequivalent to a reference product and thus they were not interchangeable with each other [21]. This example shows that in some cases QC methods can pick up therapeutic relevant performances. The common conclusion that this has to be right for any dissolution method is wrong. *In vitro* QC methods are in most cases overdiscriminating, and industry struggled for many years with the application and link between *in vitro* dissolution and *in vivo* performance of drug products [7]. This is especially true for IR products, which have low solubility such as Biopharmaceutics Classification System (BCS) class II or IV drugs.

A different picture can be seen for controlled-release drug products. Here, dissolution over an extended time period controls drug absorption. The FDA issued an industry guidance in 1997, which describes how *in vitro* dissolution and *in vivo* performance can be linked to each other [22]. Details will be discussed later in this chapter under *in vitro/in vivo* correlations (IVIVCs). The FDA recommends that at least three data points are defined in a dissolution profile of extended-release products: The first one shows that no dose dumping has occurred; therefore, not more than 30% of the drug should be dissolved within, for example, 3 h. The second time point should capture about 50% of the drug released, and the last time point should show that over 80% of the dose is dissolved, for example, within at least 12 or 24 h [23].

12.3 MEDIA USED IN DISSOLUTION TESTING

12.3.1 Simple Media

Dissolution testing is used in QC as a performance test to ensure that defined drug product specifications and lot-to-lot consistency are met [7]. Therefore, simple media, including compendial dissolution media, are used in most cases [5]. Water is the simplest medium for dissolution testing; using it alone can make analytical assay work easier. However, one has to have in mind that the pH of water may vary according to its source [10]. Other sources of variability are pH changes triggered by dosage form excipients or environmental factors like the absorption of carbon dioxide, which can impact the pH of the medium [24]. In such cases, the use of buffers as dissolution media is recommended to stabilize the pH of the media or to avoid any negative impact of pH changes on the drug's stability or the performance of the analytical assay. The pH of the media can also be used to simulate gastric or intestinal pH conditions by using USP SGF or SIF [25]. Both media usually are used without enzymes. The composition of commonly used simple media is listed in Table 12.2.

Phosphate buffers are commonly used as simple media; however, phosphate buffers are not physiologically relevant because bicarbonate buffers play a major role in the gut. Lui et al. showed that enteric-coated tablets had a delayed drug dissolution in phosphate buffer compared to pH 6.8 bicarbonate buffer [27]. The *in vitro* bicarbonate dissolution results showed a better fit with the *in vivo* observed data [27]. However, Boni et al. emphasized that such a result can only be achieved as long as the bicarbonate buffer is freshly prepared; otherwise, the dissolution profile will not be reproducible [28]. Another important consideration is the nature of the buffers used in performance testing [29]. Almukainzi reported that the use of sodium and potassium buffers resulted in different disintegration times of capsules [30]. Differences in disintegration will presumably trigger differences in dissolution times. This is an important observation because different pharmacopoeias use different salts to make buffers. SIF and USP buffer pH 6.8 is made with K+, while the International Pharmacopoeia (Ph. Int.) uses Na+ to make pH 6.8 buffer [31].

TABLE 12.2 Simple media used in dissolution testing

Media	Composition	References
SGF USP	2 g/l NaCl 7 ml/l HCl 1 l H_2O 3.2 purified pepsin	[25]
SIF USP	KH_2PO_4 6.8 g NaOH 77 ml Pancreatin 10 g Deionized water to 1 l	[25]
SLS	Concentration used (0.1–3%)	[26]
Tween 80 media	Concentration used (0.1–5%)	[26]

In vivo drug dissolutions are sensitive to many factors such as pH, food effects, transit time, motility, volume of coadministered water, bacteria, age, gender, and disease states [32, 33]. Additionally, the gastrointestinal (GI) system contains solubilizing agents such as bile salts, which can enhance a drug's solubility, and this might impact its dissolution behavior. Therefore, simple media for poorly soluble drugs often contain sodium lauryl sulfate (SLS) or Tween 80 to increase the solubility of the drug [34]. Surfactant purity is another important factor. Crison et al. investigated the intrinsic dissolution rate and solubility of carbamazepine using SLS prepared with two different purity grades, 95% and 99% [35]. They showed that the commonly used SLS grade, which has at least a purity of 95%, had a lower dissolution rate compared to highly purified SLS. Relatively small differences in purity had a significant impact on the dissolution results [35]. They concluded that in order to get a reliable and reproducible result, attention must be given to the purity of SLS.

12.3.2 Biorelevant Media

One of the major challenges in dissolution testing is to make *in vitro* results relevant to clinically observed data. For example, there are reports that the *in vivo* dissolution of a product can vary between individuals and within the same individual [36]; additional factors that can complicate the dissolution rate are the location, for example, stomach or intestine, and the fasted/fed state [37]. Food can have an impact on drug absorption by changing the physiological environment of the GI environment [38]. Gastric emptying is prolonged in the fed state and the pH in the stomach is increased, while the pH in the small intestine is decreased. This may impact a drug's dissolution behavior. For example, the dissolution of weak acidic drugs may be enhanced after a meal, which might increase absorption, whereas the opposite might happen for weak basic drugs [38]. Food also increases the secretion of bile into the small intestine, and the hepatic blood flow is increased, which can impact metabolism.

Media that can account for such factors are desirable and are needed in order to closely mimic the *in vivo* dissolution situation [36].

As outlined earlier, the demands for a medium mimicking the *in vivo* environment lead to developed biorelevant dissolution media (BDM) [32]. The goal of such media is to be an *in vitro* dissolution surrogate for *in vivo* release [32]. The development of these media started with simple buffers in combination with naturally occurring surfactants and has evolved throughout the years to more complex media that are more predictive of the *in vivo* environment. Today, the most common biorelevant media are fasted-state simulated intestinal fluid (FaSSIF) and fed-state simulated intestinal fluid (FeSSIF), which were first introduced in 1998 by Dressman [39]. FaSSIF contains taurocholate and lecithin, which are natural surfactants. These physiologically occurring compounds assist solubilizing lipophilic drugs into micelles [10]. Research has shown that the purity of these surfactants has a high impact on the dissolution and solubility of certain drugs.

In 2005, Vertzoni et al. proposed a fasted-state simulated gastric fluid (FaSSGF) [40] followed by fed-state simulated gastric fluid (FeSSGF).

TABLE 12.3 SGF composition

Composition	FaSSGF	FeSSGF Early	FeSSGF Middle	FeSSGF Late
Sodium taurocholate (μM)	80	—	—	—
Lecithin (μM)	20	—	—	—
Pepsin (mg/ml)	0.1	—	—	—
Sodium chloride (mM)	34.2	148	237.02	122.6
Acetic acid (mM)	—	—	17.12	—
Sodium acetate (mM)	—	—	29.75	—
Orthophosphoric acid (mM)	—	—	—	5.5
Sodium dihydrogen phosphate	—	—	—	32
Milk/buffer	—	1:0	1:1	1:3
Hydrochloric acid/sodium hydroxide q.s.	pH 1.6	pH 6.4	pH 5	pH 3
Properties				
pH	1.6	6.4	5	3
Osmolality (mOsm/kg)	120.7±2.5	559	400	300
Buffer capacity (mmol/l/pH)	—	21.33	25	25
Surface tension (mN/m)	42.6	—	—	—

Reproduced from Ref. 41, with permission of Dissolution Technologies.

FeSSGF can be differentiated into an early-, middle-, and late-stage composition after ingestion, and its composition is shown in Table 12.3. Additionally, milk was used in combination with FeSSGF to simulate an ingested breakfast meal in the stomach [10]. The compositions of most of these media have slightly changed since they were introduced to better mimic the physiological environment [42, 43]. Table 12.3, Table 12.4, Table 12.5, and Table 12.6 summarize the most commonly used composition of BDM.

The following examples will explain the different uses of BDM. Wei and Löbenberg reported that the dissolution of glyburide, a BCS class II drug, was increased when low purity grades of taurocholate and lecithin were used [44]. Okumu et al. reported the opposite behavior for montelukast sodium, which is much more lipophilic compared to glyburide. This drug was more soluble in higher-purity bile salt/lecithin media compared to the lower purity grades [45]. Figure 12.4 shows the reported solubility of both drugs in high- and low-purity BDM.

In another study, Hammed et al. studied the impact of the chemical purity on the solubility of different steroidal drugs such as estradiol, prednisolone, and progesterone and compared it with the solubility of benzodiazepines such as clonazepam, tetrazepam, diazepam, and lorazepam [46]. They showed that the solubility in the surfactants depended on the drug nature; hence, steroidal drugs showed a higher affinity for bile salt than benzodiazepines as these steroids have better affinity to micelles [46].

These studies show that the quality of the individual components is an important consideration when BDMs are used. Other factors like the preparation methods used to make BDM might also impact the performance of the media. Kloefer et al. studied

TABLE 12.4 FaSSIF composition

Composition	FaSSIF	FaSSIF
Sodium taurocholate (mM)	3	—
Lecithin (mM)	0.2	—
Maleic acid (mM)	19.12	—
Sodium hydroxide (mM)	34.8	q.s. pH 6.5
Sodium chloride (mM)	68.62	2 g
Glyceryl monocholate (mM)	—	—
Sodium oleate (mM)	—	—
Deionized water q.s.	—	1 L
24-Phosphonobile acid (mM)	—	3
Sodium dihydrogen phosphate (M)	—	0.025 M
Properties		
pH	6.5	6.5
Osmolality (mOsm/kg)	180 ± 10	140
Buffer capacity (mmol/l/pH)	10	13
Surface tension (N/m²)	—	35

Reproduced from Ref. 41, with permission of Dissolution Technologies.

TABLE 12.5 FeSSIF composition

Composition	FeSSIF	FeSSIF Early	FeSSIF Middle	FeSSIF Late	FeSSIF
Sodium taurocholate (mM)	—	10	7.5	4.5	10
Lecithin (mM)	—	3	2	0.5	2
Maleic acid (mM)	—	28.6	44	58.09	55.02
Sodium hydroxide (mM)	—	52.5	65.3	72	81.65
Sodium chloride (mM)	5 g	145.2	122.8	51	125.5
Glyceryl monocholate (mM)	—	6.5	5	1	5
Sodium oleate (mM)	—	40	30	0.8	0.8
24-Phosphonobile acid (mM)	15	—	—	—	—
Sodium acetate (M)	0.05	—	—	—	—
Acetic acid q.s.	pH 5	—	—	—	—
Deionized water q.s.	1 l	—	—	—	—
Properties					
pH	5	6.5	5.8	5.4	5.8
Osmolality (mOsm/kg)	272	400 ± 10	390 ± 10	240 ± 10	390 ± 10
Buffer capacity(mmol/l/pH)	29	25	25	15	25
Surface tension (N/m²)	30	—	—	—	—

Reproduced from Ref. 41, with permission of Dissolution Technologies.

different manufacturing procedures to make BDMs [47]. They examined how FeSSIF and FaSSIF media can be reproducibly prepared from commercial powders. The study looked at the physical, chemical, and microbiological stability of media along with the dissolution behavior of three poorly water-soluble drugs: ketoconazole, dipyridamole,

TABLE 12.6   Simulated colonic fluid 1 (SCoF1)

Composition	Amount (g/l)
Potassium chloride	0.20
Sodium chloride	8.00
Potassium phosphate monobasic	0.24
Sodium phosphate dibasic	1.44
Properties	**Value**
pH	7.00

Reproduced from Ref. 41, with permission of Dissolution Technologies.

(a)

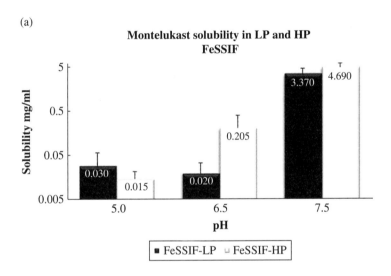

(b)

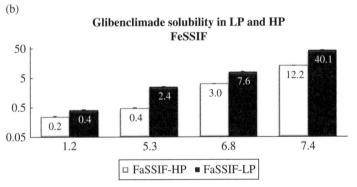

FIGURE 12.4   Comparison of the solubility of montelukast (a) and glyburide (b) between high and low purity grades of taurocholate and lecithin.

and phenytoin [47]. The study concluded that using a standard preparation method to prepare FeSSIF and FaSSIF resulted in reproducible dissolution profiles; however, differences observed in the micelle sizes of the differently prepared media might impact the dissolution behavior of other drugs [47]. Today, although biorelevant media were intensively studied and their composition was further developed over the years, many of their properties are still not well known. For example, FaSSIF and FeSSIF contain a trihydroxy acid as the only bile salt. In fact, it represents only 20% of the *in vivo* bile salt content. *In vivo* bile salt contains additionally dihydroxy acid [33]. Moreover, the lecithin used to make these media is not the exact phospholipid, which is present in small intestine. Lysolecithin is not included in the composition of FaSSIF and FeSSIF, which is the naturally occurring phospholipid in small intestine [33].

Some drugs are intended to be delivered to the colon to have a local effect as in the case of ulcerative colitis [48]. Therefore, drugs intended for colonic absorption need to be protected from gastric and intestinal absorption by designing a drug delivery system, which can achieve a localized drug release. One dosage form used for this purpose contains partially digested polysaccharides. This delivery system consisted of drug-layered pellets, which were coated first with two pH-independent polymers, EUDRAGIT RL and RS, and then with an outer layer of a pH-dependent polymer, EUDRAGIT FS 30D, which dissolves rapidly at pH > 7 [49]. To simulate the colonic environment, simulated colonic fluid (SCoF) as biorelevant medium was introduced. Marques et al. have reviewed the most common compositions of this medium, which is shown in Table 12.6 and Table 12.7 [41]. The medium has a neutral pH and can contain enzymes such as galactomannanase, which digest polysaccharide films or matrices, which is used as mechanism for drug releases [48].

Vertzoni et al. and colleagues developed and evaluated a fasted- and fed-state colonic fluid [50]. In a study, they compared the solubility of model drugs ketoconazole, danazol, and felodipine using human colonic fluids and gastric and intestinal fluids with the solubility in fasted-state simulated colonic fluid (FaSSCoF) and fed-state simulated colonic fluid (FeSSCoF) and plain buffers. Figure 12.5 shows the FaSSCoF and FeSSCoF and human colonic fluids. The study concluded that these biorelevant media have a solubility capacity closer to human colonic fluid than plain buffers [50]. Table 12.8 shows the properties of FaSSCoF and FeSSCoF.

TABLE 12.7 Simulated colonic fluid 2 (SCoF2)

Composition	Concentration (mM)
Acetic acid	170.0
Sodium hydroxide	157.0
Properties	**Value**
pH	5.80
Osmolality (mOsmol/kg)	295.00
Buffer capacity (mEq/l/pH)	29.10
Ionic strength	0.16

Reproduced from Ref. 41, with permission of Dissolution Technologies.

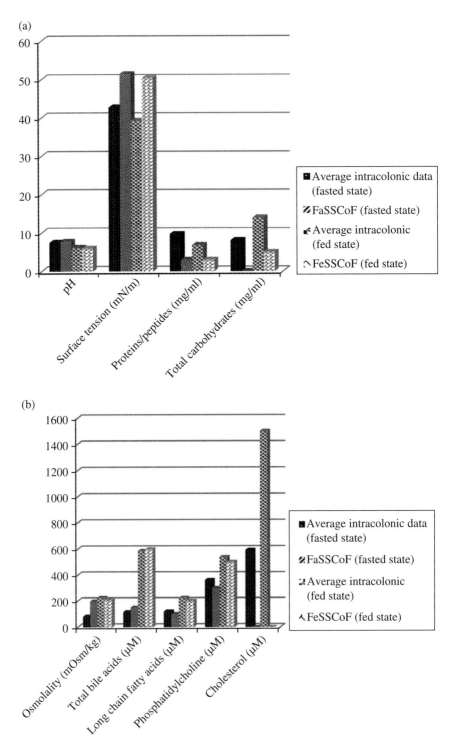

FIGURE 12.5 (a) pH, surface tension, protein content, and total carbohydrate concentration of FaSSCoF and FeSSCoF compared to human colonic fluids. (b) Osmolarity, total bile, long-chain fatty acids, phosphatidylcholine, and cholesterol of FaSSCoF and FeSSCoF compared to human colonic fluids. Reproduced from Refs. 41 and 50.

TABLE 12.8 Properties of FaSSCoF and FeSSCoF

Properties	Fasted state FaSSCoF	Fed state FeSSCoF
pH	7.8	6.0
Buffer species	Tris/maleates	Tris/maleates
Buffer capacity (mmol/l/ΔpH)	16/26	15/14
Osmolality (mOsm/kg)	196	207
Surface tension (mN/m)	51.4	50.4
Proteins/peptides (mg/ml)	3	3 total carbohydrates (mg/ml)
Total carbohydrates (mg/ml)	0	14
Total bile acids (μM)	150	600
Long-chain fatty acids (μM)	100	200
Phosphatidylcholine (μM)	300	500
Cholesterol (μM)	0	0

Reproduced from Ref. 50, with permission of Springer Science and Business Media.

To mimic the juices of the stomach, simple media such as SLS were tested at low pH. However, these media have some problem at this pH level. For example, SLS interacts with gelatin at pH values less than pH 5, or they interfere with salts of weak bases [40]. The introduction of FaSSGF media avoided all of these problems.

Different drug products have sometimes different *in vitro* dissolution behaviors due to the nature of excipients used in the formulation. Buch et al. investigated the dissolution of fenofibrate using different formulations [51]. Some of these formulations contained SLS. They found that formulations that contain SLS can cause a decrease in a drug's solubility when FaSSIF as dissolution medium was used. This effect seems to be due to an interaction between SLS and FaSSIF surfactants, which affects the solubilization of the mixed micelle systems [51].

Dissolution media containing synthetic surfactants are cheaper, easier to prepare, and presumably more stable. Therefore, some studies compared the *in vitro* drug dissolution behavior in media containing SLS versus BDM and assessed their ability to correlate the results with *in vivo* observed data. Lehto et al. compared SLS alone with FaSSIF for three BCS class II drugs: danazol, spironolactone, and a phase I compound (N74) [52]. FaSSIF was composed out of the higher purity grades of lecithin and sodium taurocholate, while SLS and Tween 80 were of analytical grade and used in concentration of 0.02%, 0.1%, 0.2%, 0.014%, 0.035%, and 0.07%. The author concluded that all tested media were generally able to reflect some of the *in vivo* dissolution. However, an *in vivo* study for a phase I product showed that 0.1% (v/v) SLS media were the best predictive media for this drug [52].

Similarly, Taupitz and Klein investigated using SLS and Tween 80 as the dissolution media in comparison to FaSSIF and FeSSIF [34]. Tamoxifen was used as a model drug. 0.1%, 0.175%, 0.25%, 0.5%, 0.75%, 1.0%, and 1.5% (w/v) of surfactant were used in the two buffers. They found that SLS and Tween 80 can be used as substitutes for BDMs to develop predictive and discriminatory methods in early formulation development [34]. The study results are promising, but they might be due to the nature of the drug used, the composition of the tested formulation, or the purity and concentration of the used surfactants. It is still recommended to investigate each drug in regard to its dissolution behavior in different media until an optimal medium is found. The recommendation is also supported by Zoeller and Klein, who investigated simplified media containing simple surfactants and compared them to conventional FaSSIF and FeSSIF [53]. Ketoconazole, glyburide, and tamoxifen were used as the model drugs in USP apparatus 2. They found that in some cases the dissolution profiles were comparable, whereas in other cases they were not. Synthetic surfactants are more cost effective compared to BDMs. They can be used in QC and formulation screening and showed in some of the studied cases a good predictability of the *in vivo* dissolution; however, in general, biorelevant media were more reliable in predicting *in vivo* release compared to SLS [53]. Therefore, the right media selection is the key in designing biorelevant dissolution methods, and different media are available that have to be tested for the drug under investigation. A case-by-case determination is required to get the best result and to define the best dissolution conditions.

12.4   *IN VITRO/IN VIVO* CORRELATION (IVIVC)

12.4.1   What is IVIVC

IVIVC is the establishment of a mathematical relationship between *in vivo* observed pharmacokinetic parameters and *in vitro* data, which are usually dissolution profiles [8]. The FDA has defined an IVIVC as "a predictive mathematical model which describes the relationship between *in vitro* properties of a dosage form and its *in vivo* response" [22]. USP (33 or 34) defined the IVIVC as "the establishment of a rational relationship between a biological property, or a parameter derived from a biological property produced from a dosage form, and a physiochemical property of the same dosage form" [54].

There are two general approaches that can be used to establish IVIVC. Traditionally, deconvolution of pharmacokinetic data was used to establish an IVIVC. This can be achieved by correlating the fraction dose dissolved versus fraction dose absorbed. The fraction dose absorbed is calculated by deconvolution from the observed data. Pharmacokinetic models like the one-compartment model of Wagner–Nelson or the multicompartment model of Loo–Riegelman can be used to extract the fraction dose absorbed or fraction dose dissolved [55]. This step is followed by correlating the fraction dose absorbed versus the *in vitro* fraction dose dissolved. This process is a two-stage process and can be successfully applied to most extended-release products if dissolution is the rate-limiting step as outlined in the FDA guidance for extended-release products [56].

For IR products, the traditional approach using deconvolution technique mostly fails. Polli et al. investigated the cause of this failure and found that the relationship between the *in vivo* and *in vitro* fraction dose absorbed is in most cases nonlinear [57].

This is the main reason why an IVIVC cannot be established for this kind of products. However, when the dissolution is the rate-limiting step, an IVIVC will produce a straight line, and a correlation will be possible. In contrast, when absorption is the time-limiting factor, then the plotted data will produce a "reverse L"-shaped graph. When both dissolution and permeability are limited, a more "hockey stick"-shaped graph will be the result rather than an L shape [57]. Therefore, in order to build IVIVC for these drugs, Dunne et al. suggested to use a nonlinear mixed-effects modeling approach [58].

In some other cases, IR products required a scale factor between *in vitro* and *in vivo* data. Löbenberg et al. investigated how to establish an IVIVC for IR glibenclamide tablets [59]. They found if a scale factor was used for the *in vitro* release data, then they can correlate with the *in vivo* fraction dose absorbed data [59].

Dunne et al. and Gaynor et al. reviewed deconvolution approaches to establish IVIVC. Their article lists many limitations for this method. For example, using a linear model for IVIVC might lead to biased predictions when the real data do not follow a straight line. In addition, many other statistical concerns are mentioned and summarized in Dunne et al.'s review [60]. Therefore, the authors suggested using convolution models to avoid these problems.

One of the first examples of a convolution was given by Gillespie using a fast-, medium-, and slow-release dosage forms of diltiazem [61]. He reported that the relationship between *in vitro* release and plasma drug concentrations can directly be modeled in a single stage rather than via an indirect two-stage approach [61].

As reviewed by Buchwald, the equation that forms the basis of these approaches [55, 61–65] relies on a convolution-type integral transform:

$$c(t) = \int_{0}^{t} c_\delta(t-u) r_{\text{abs}}(u) du$$

where $(c(t))$ is the plasma concentration resulting from the absorption rate time course (r_{abs}). The function c_δ represents the concentration time course that would result from the instantaneous absorption of a unit amount of drug and can be estimated from either IV bolus data or oral solution, suspension, or rapidly releasing (*in vivo*) IR dosage forms [56].

Costello et al. evaluated using a time scaling approach to develop IVIVCs using a convolution-based technique [66]. This approach seems to be successful for drugs that have a disparity in their *in vitro* and *in vivo* time scales. They found that incorporation of a time scale factor into the convolution-based model gave more accurate and robust predictions. In addition, this method accounts for variability that can come from the subjects or the dosage forms [66].

Another way to construct IVIVC is to use physiologically based pharmacokinetic (PBPK) models. These models are composed out of a series of compartments that represent different organs. Yu and Amidon described this in their Compartmental Absorption and Transit (CAT) Model [67]. Figure 12.6 shows the improved version

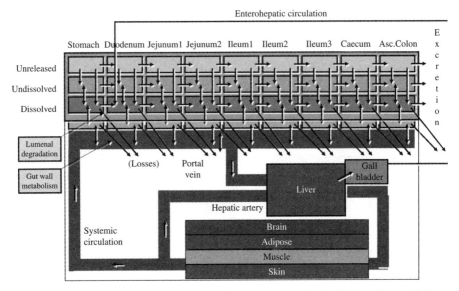

FIGURE 12.6 A schematic of a PBPK model. Adopted simulation plus workshop. Reproduced from Ref. 68, with permission of Simulations Plus, Inc.

of this model known as Advanced Compartmental Absorption and Transit (ACAT) developed by Simulations Plus.

These models require species-specific and compound-specific input parameters. The parameters have to be taken from the appropriate sources in the literature [69]. The combination of all input parameters allows for the prediction of plasma and tissue concentration time profiles in animals or humans; without that, an experiment has to be performed [69]. As reviewed by Jones et al., there are various simulation packages available for the prediction of the rate and extent of oral absorption in human and preclinical species, for example, GastroPlus™, SimCYP, PK-SIM®, and ChloePK® [69].

These models are complex and depend on the accuracy of the input parameters. However, independent of the input parameters, any model used for pharmaceutical product development needs to be validated before it can be part of a submission package.

The current FDA guidelines are based on deconvolution methods to establish an IVIVC, but they allow alternative approaches to develop IVIVC and suggest convolution procedures that model the relationship between *in vitro* dissolution and plasma concentration in a single step as outlined earlier.

The FDA guideline states that convolution-based IVIVC needs to be validated using a prediction error. Similarly, IVIVCs established by deconvolution are normally assessed using the regression coefficient between the plotted *in vivo* and *in vitro* data. Besides, the FDA guideline allows an internal and an external validation of these models. The internal validation is performed with formulations, which release the drug at different rates, while the external validation uses convolution models that were successfully used for a variety (seven sets) of other drugs [56]. The guideline states:

- Plasma concentrations predicted from the model and those observed are compared directly.
- An IVIVC should be evaluated to demonstrate that predictability of *in vivo* performance of a drug product from its *in vitro* dissolution characteristics is maintained over a range of *in vitro* dissolution release rates and manufacturing changes.
- Since the objective of developing an IVIVC is to establish a predictive mathematical model describing the relationship between an *in vitro* property and a relevant *in vivo* response, the proposed evaluation approaches focus on the estimation of predictive performance or, conversely, prediction error.
- Depending on the intended application of an IVIVC and the therapeutic index of the drug, evaluation of prediction error internally and/or externally may be appropriate. Evaluation of internal predictability is based on the initial data used to define the IVIVC model.
- Evaluation of external predictability is based on additional test data of seven sets. Application of one or more of these procedures to the IVIVC modeling process constitutes evaluation of predictability.

The mechanistic knowledge of the rate-limiting factor (dissolution or absorption controlled) is essential for the establishment of an IVIVC. It can be used to decide which approach—convolution or deconvolution—should be used. This can then be the basis to develop a meaningful dissolution method to establish an IVIVC.

Figure 12.7 shows the two general mechanisms in which a drug can be absorbed. In the first case, dissolution happens faster than absorption. In this case, absorption is the rate-limiting factor. If dissolution occurs slower than absorption, then dissolution is the rate-controlling step. An important differentiation needs to be added to these two general mechanistic pathways of oral absorption, which is the difference between the *in vitro* and *in vivo* dissolution. Both might or might not be similar. Only if they can be linked to each other, then the establishment of an IVIVC is possible.

Dissolution testing might be in this case a surrogate for *in vivo* dissolution [36, 70]. However, as shown in Figure 12.8, the chance to establish an IVIVC based on dissolution data requires that dissolution controls the absorption process. If dissolution is faster than absorption, only a minimum dissolution requirement can be defined (see Fig. 12.8). This minimum dissolution requirement is given when absorption and dissolution happen at the same rate.

Any dosage form, which has a faster dissolution than the minimum required dissolution, will have the same bioavailability because absorption and not dissolution is the rate-controlling mechanism. Differences in dissolution will presumably not affect bioavailability for these formulations. In a quality-by-design (QbD) approach, such dissolution specifications might be used to define the lower limit of a dissolution design space[67].

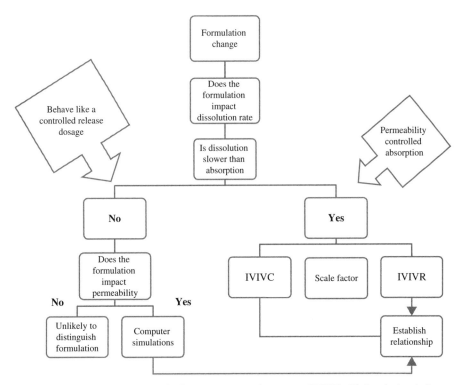

FIGURE 12.7 Biopharmaceutical parameters used to assess IVIVCs. If dissolution is faster than absorption, then the process is permeability controlled. If dissolution controls drug absorption, the dissolution testing might be a surrogate for *in vivo* dissolution. Reproduced from Ref. 7.

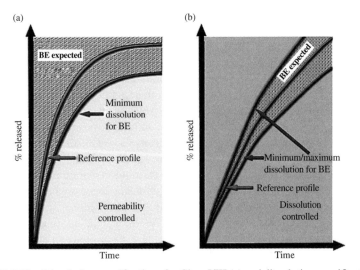

FIGURE 12.8 Dissolution specifications for Class I/III (a) and dissolution specifications for Class II/IV bioequivalence (BE) is expected in case (a) if a minimum required dissolution is achieved, in case (b) a minimum and maximum dissolution is required for BE.

Another situation is given when dissolution controls absorption. Here, the dosage form is the rate-limiting step in regard to the absorption process. It can be expected that *in vitro* dissolution methods will be able to differentiate between formulations and their *in vivo* performance. However, the prerequisite is that the *in vitro* dissolution is a surrogate for *in vivo* dissolution. It is obvious that biorelevant media might be needed in developing a dissolution test that mimics the *in vivo* environment. As mentioned before, only if *in vitro* dissolution can be linked to the *in vivo* dissolution, an IVIVC can be established. The challenge however is that the *in vivo* dissolution cannot be determined without either invasive procedures or fancy noninvasive techniques. Endoscopy is the most common invasive technique, but it is uncomfortable to subjects and sometimes requires a sedation of the subjects [71]. Noninvasive techniques such as roentgenography or fluoroscopy are used to follow the *in vivo* dissolution of a dosage form, but both techniques use high radiation doses. The gamma scintigraphy is the most common noninvasive technique, which is used to evaluate the *in vivo* behavior of pharmaceutical formulations [71, 72]. However, the costs of such studies are high, and the gained knowledge does not justify using these methods on a regular basis. Therefore, the search for methods, which can mimic the *in vivo* dissolution, is still an ongoing field of research.

12.4.2 Levels of IVIVC

The FDA has developed IVIVC guidelines for IR and extended-release dosage forms. The FDA guideline for extended-release products lists three possible levels of IVIVC: level A, level B, and level C [56]. For regulatory purposes, level A correlations are acceptable to waive bioequivalence studies. However, all other levels like levels B and C cannot be used to apply for a biowaivers, but they are useful for formulation development and to validate the *in vitro* dissolution test conditions [5, 73]. Level D correlations are not mentioned in the FDA guidance, but this rank-order correlation is commonly used in research to rank formulations according to their *in vitro* dissolution behavior. This is in most cases performed without any knowledge of the *in vivo* performance [74]. All other correlations require mathematical calculations.

Level A is a point-to-point correlation, for example, between the *in vitro* fraction dose dissolved and the *in vivo* fraction dose absorbed; ideally, it is a linear correlation [5]. However, nonlinear correlations are discussed in literature [58, 75]. Similarly, level A correlations can be achieved using convolution techniques. Examples are given later in the chapter. Level A is the most predictive level and can be used for regulatory filing to waive bioequivalence studies when formulation changes are needed. The main purpose of establishing an IVIVC is to waive bioequivalence studies.

An example of level A IVIVC is given by Rekhi and Jambhekar. They correlated the fraction of propranolol absorbed against the percentage of the drug released *in vitro* [76]. The *in vitro* data were generated in a USP 1, whereas the *in vivo* study was performed in healthy individuals. An IVIVC between *in vitro* and *in vivo* fraction dose absorbed/dissolved was established. This is the traditional way to establish IVIVC, and many examples exist in literature using this method to establish level A IVIVC [65, 77–79].

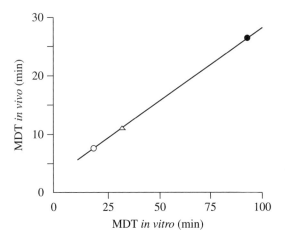

FIGURE 12.9   Level B correlation between the *in vitro* MDT and *in vivo* MRT. Reproduced from Ref. 80, with permission of Springer Science and Business Media.

Level B correlations are statistical moment analysis between, for example, the *in vitro* mean dissolution time (MDT) and the *in vivo* mean residence time (MRT) [6].

Graffner et al. correlated the *in vitro* dissolution of alaproclate extended-release solid oral dosage forms [80]. All dissolution studies were performed using USP apparatus 2 at 50 rpm in 500 ml of water at pH 1.2. The *in vivo* pharmacokinetic parameters were determined in healthy volunteers, who received 200 mg alaproclate hydrochloride. All profiles were linearized and a level B correlation between the *in vitro* MDT and *in vivo* MRT was obtained (Fig. 12.9) [80].

Level C correlations compare single-point relationships between *in vitro* and *in vivo* data. For example, the amount of drug dissolved at a particular time such as $t_{50\%}$ and *in vivo* parameters such as C_{max} or AUC are correlated [6, 56].

Balan et al. evaluate three metformin extended-release capsules [62]. The *in vitro* dissolution data were obtained using the USP apparatus 2 (paddle method) at 100 rpm in 1000 ml of phosphate buffer pH 6.8. The *in vivo* study was conducted in eight healthy volunteers. A linear regression was used to establish a level C IVIVC by correlating the mean C_{max} and AUC0-∞ values to the percent dissolved at 2 and 4 h [62] (Fig. 12.10). Table 12.9 summarizes all parameters required to establish IVIVCs of levels A to C.

12.4.3   Examples of IVIVC

The following section will review some examples of how IVIVC was established using conventional or slightly adapted dissolution apparatuses.

Shono et al. conducted a series of studies aimed to investigate the utility of using biorelevant media like FaSSGF, FeSSGF, FaSSIF-V2, and FeSSIF-V2 and conventional USP buffers in combination with a software called STELLA. The software allows the user to build his own PBPK models. In the first study, the author investigated

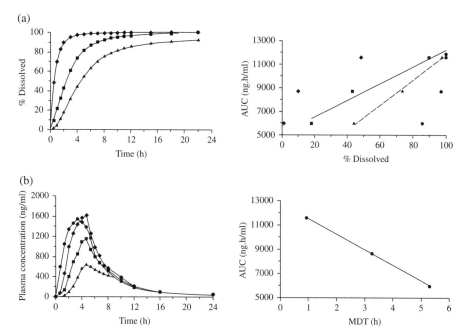

FIGURE 12.10 Left: mean *in vitro* dissolution data (a) and *in vivo* plasma concentration data (b) for metformin. Right: level C IVIVC for metformin. Mean observed AUC as a function of *in vitro* dissolution at 0.5, 2, and 4 h (a). AUC plotted versus the *in vitro* MDT (b). Reproduced from Ref. 62, with permission of Wiley-Liss, Inc. and the American Pharmaceutical Association.

TABLE 12.9 Parameters used in IVIVC corresponding to the level

Level	*In vitro*	*In vivo*
A	Dissolution curve	Input (absorption, plasma) curves
B	Statistical moments: MDT	Statistical moments: MRT, MAT, etc.
C	Disintegration time, time to have 10%, 50%, and 90% dissolved	C_{max}, T_{max}, K_a, time to have 10%, 50%, and 90% absorbed, AUC (total or cumulative)
	Dissolution rate	
	Dissolution efficiency	

Reproduced from Ref. 6, with permission from Dissolution Technologies.

the impact on fed and fasted conditions on celecoxib 200 mg capsules. They measured the solubility and dissolution profiles in biorelevant and compendial media. This study found that the PBPK approach corresponded much better to the food effect observed for celecoxib *in vivo*. Additionally, point estimates of AUC and C_{max} as well as a comparison of the predicted versus the observed data demonstrated a clear advantage of using results in biorelevant rather than compendial media.

In a second study, the previous model needed to be optimized for particle size effects on the fed- and fasted-state absorption of aprepitant, a poorly soluble antiemetic drug. The new model simulated the *in vivo* profiles for aprepitant well in both

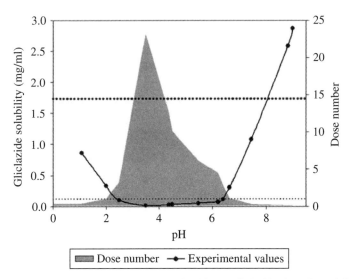

FIGURE 12.11 pH-dependent calculated Do for 80 mg gliclazide dose. Dotted lines represent the critical Do for GLK (conservative upper limit of Do=1 and the calculated value of Do = 14.5). Reproduced from Ref. 82, with permission of Springer Science and Business Media.

prandial states. It also predicted the dependency of the observed pharmacokinetic profiles on the dose and the particle size. The study concluded that the software model combined with dissolution results in biorelevant media successfully forecasts the *in vivo* performance of both nanosized and micronized formulations in the fed and fasted states.

In a third study, nelfinavir mesylate was used as a model compound. For this drug, it was found that the model needed to be adjusted for drug precipitation occurring in the GI tract in the fasted state. Without this adjustment, the model was not able to predict the observed data. However, in the fed state, no drug precipitation seemed to occur, and no adjustment for precipitation was necessary [81].

These simulations demonstrated that an *in silico* approach can be a powerful tool to test the impact of the fasted- and fed-state media on the absorption of certain low-solubility drugs. However, adjustments of the model are needed to accommodate compound-specific behaviors.

Grbic et al. used gliclazide as a model of a poorly soluble drug. Figure 12.11 shows that this drug has an unfavorable solubility in the small intestinal pH range or pH 5.0–6.5. The dose number (Do) is very high within this pH range. The fraction dose absorbed can be calculated as

$$Fa = \frac{2An}{Do} \tag{12.1}$$

where Fa is the fraction of drug absorbed, and An is the absorption number calculated (An = 7.27).The critical value of Do was calculated under the assumption that the drug is completely absorbed (Fa = 1). The authors calculated the critical Do as 14.5. Only Do of one or less fulfills the solubility requirements of BCS.

The study used USP apparatus 2 at $37 \pm 0.5°C$ at a rotational speed of 100 rpm and 900 ml of various dissolution media (media pH 1.2, 4.0, 4.5, 6.8, 7.2, and 7.4) [82]. The dissolution profiles were used as input into GastroPlus™. Other needed data like the volume of distribution, clearance, and solubility at pH 4.3 or pK_a were extracted from literature. Then, the software was used to estimate the drug plasma profiles of a clinical study. After the model was successfully established, the percent of drug absorbed was plotted against the percent of drug dissolved *in vitro*. In this study, a level A IVIVC was established using both a deconvolution and a convolution approach [82]. This demonstrates that PBPK models are important software solutions to establish IVIVCs. Other studies showed similar results [44, 83–85].

As shown earlier, if a convolution model has an acceptable degree of predictability, then it can be used to establish IVIVC. However, a model is only as good as its input data, which is especially true for PBPK models. Therefore, further modifications of the dissolution test condition to mimic the GI passage of a drug have proven to be beneficial when convolution approaches are used. The *in vivo* dissolution environment is dynamically changing in volume, composition, and pH. For example, in the fasted state, the small intestinal fluid volume may fluctuate from 45 to 319 ml, whereas in the fed state the volume may vary from 20 to 156 ml [43]. Then, there is the pH gradient from the stomach to the small intestine to the colon [59]. This is an especially important consideration for drugs that are sensitive to pH changes in GI and may precipitate or increase in their solubility when the pH changes. Using a two-stage pH technique, as used for enteric-coated dosage forms, is one way to mimic pH changes in a dissolution apparatus. A more advance technique used more than two pH values for the dissolution test. Wei and Löbenberg showed for glyburide that changing the pH during the dissolution test resulted in a different drug-release profile compared to a single pH test. They used the USP paddle apparatus and created a stepwise pH gradient from pH 6.0 to pH 7.5 and back to pH 5.0 as it occurs in the small and large intestine [44]. Using these dissolution data and GastroPlus™ as simulation software, they were able to predict *in vivo* observed data. In contrast, data acquired from a single pH condition was not able to predict the drug plasma time curves.

Okumu et al. showed in a study that a transfer model was needed to obtain dissolution profiles that were predictive of the *in vivo* dissolution [86]. They tested etoricoxib, which is a weak base. The drug dissolves fast and completely in the acetic environment of the stomach; however, its solubility decreases in the small intestinal pH range of 5.0–6.5. Using GastroPlus™, the measured solubility at pH 6.8 gave poor estimates of the plasma time course of a clinical study [86]. However, when they transferred the dissolved drug into SIF, they obtained a supersaturated solution at pH 6.8. The higher solubility value of this solution was able to achieve a good prediction of the clinically observed data.

A more convenient way to introduce dynamic pH changes while a dissolution test is performed is by using an open flow-through cell system. This apparatus allows to mimic dynamic pH changes of the *in vivo* environment by sequentially perfusing

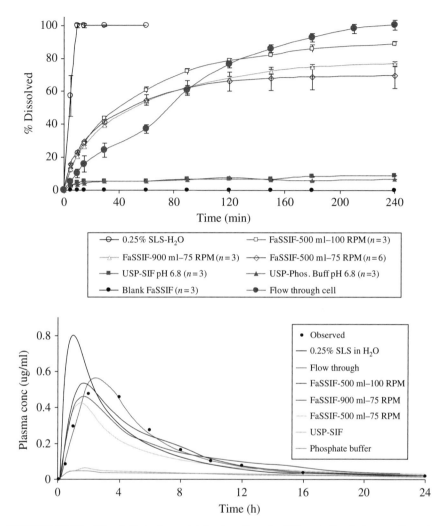

FIGURE 12.12   Dissolution of montelukast sodium in different dissolution media using USP apparatus 2 and a flow-through cell. The profiles were used in GastroPlus to simulate the *in vivo* performance. Reproduced from Ref. 45, with permission of Elsevier.

different biorelevant media. Okumu et al. showed that using a flow-through dissolution apparatus with dynamic pH changes can be used to predict the *in vivo* performance of montelukast sodium, a low-solubility drug [45]. The dynamic dissolution data were able to establish an IVIVC using GastroPlus™ as simulation software. Simulation data obtained with a single pH release profiles failed to generate a good fitting IVIVC (Fig. 12.12).

This is another example for how important the use of appropriate input data for PBPK simulation software is. Mimicking the GI environment using BDM combined with a dynamic dissolution protocol can provide dissolution data that reflect the *in vivo* dissolution. In such cases, simulation software can be used to establish strong IVIVCs.

USP apparatus 4 seems to be one of the most promising standard apparatuses to mimic the *in vivo* dissolution [36]. This is because this apparatus has the ability to supply continuously fresh media, remove dissolved materials, and maintain sink conditions over the duration of the study [36]. The performance of this apparatus was investigated by different studies [87–89]. Fang et al. investigated the ability of USP apparatus 4 with biorelevant media (SGF, SIF, FeSSIF, and FaSSIF) to predict the *in vivo* drug release in early-stage drug development [36]. The study investigated IR formulations containing BCS class II drugs. The impact of flow rates between 2 and 20 ml/min, the diameter of the flow-through cells, and the impact of using a sample holder were tested. The study found that the cumulative drug-release rate generated with USP apparatus 4 had a high prediction power of observed *in vivo* release rates when biorelevant media were used.

As mentioned earlier, level A correlations can be used to waive bioequivalence studies and will be more and more required by regulatory agencies in the future. It is therefore important to know how IVIVC can be established. However, the available standard dissolution apparatuses might not be suitable for every drug. There are some more complex dissolution protocols available that might be a solution for drugs where the normal apparatus will not provide predictive dissolution results.

McAllister reviewed more complex dissolution apparatuses [43]. Such approaches use, for example, multicompartment dissolution models like the artificial stomach–duodenum (ASD) model. Here, the drug is transferred from a stomach chamber to a duodenum chamber.

Carino reported that this model was able to simulate the fasted- and fed-state drug plasma curves of carbamazepine obtained in dogs [90].

Other models simulate the physical stress forces of the GI tract [91]. There are also combined dissolution and absorption models like the Flo Vitro system developed by Dow Chemicals. This system uses a chamber transfer model combined with an absorptive compartment. Finally, there are complex *in vitro* digestion models that aim to mimic the entire GI tract including enzymes and the bacterial flora. The TNO TIM is the most famous and complex model in this family [92].

There are many more apparatuses described in literature that focus on very specialized applications. Their successful use depends on the nature of the drug, the drug-release mechanisms, and whether the drug is actively or passively absorbed. However, it has to be pointed out that a dissolution method has to be developed for each drug individually.

12.5 SUMMARY

Dissolution as drug product performance test and drug development tool has been moved from being an *in vitro* test for assessing different dosage forms or formulations and batch-to-batch consistency to a test that can be used to predict the *in vivo* performance of a drug. Modern dissolution methods should be able to link the *in vitro* performance with the *in vivo* performance of a dosage form. The introduction of the BCS by Amidon in 1995 changed the way modern drug development is performed. The BCS is the mechanistic basis of oral drug absorption. The fundamental knowledge if a drug is dissolution or absorption limited in its bioavailability is important in any stage of the drug development process. This mechanistic understanding can be used to optimize a dosage form and to develop a strategy how to establish an IVIVC. In most cases, a dissolution test has to be developed for a specific purpose. In QC, a very sensitive and overdiscriminating dissolution test might be used for regulatory filing. However, such tests often do not reflect the *in vivo* dissolution, and they fail to predict the *in vivo* performance of a drug product. Therefore, the development of two separate dissolution tests might be in some cases a practical approach to capture the needs of routine batch releases and the need to have an IVIVC for life cycle management if production or formulation changes occur.

As shown, the currently available pharmacopoeial apparatuses are in many cases a good starting point to develop biorelevant dissolution methods. The important message to remember is that any biorelevant method must mimic the *in vivo* dissolution if the absorption process is dissolution limited. The use of computer models and here especially PBPK models will be in the future an integral part of drug development. This need will be mostly driven by QbD approaches that require predictive models to define design spaces and to set clinically relevant specifications for dosage form attributes like dissolution.

REFERENCES

1. Nagabandi V, Santosh Kumar M, Prasad G, Someshwar K, Varaprasad A. Comparative dissolution studies of marketed preparations and treatment of data by using ANOVA. *Int J Adv Pharm Sci* 2011;1(2):142–146.
2. Shiko G, Gladden LF, Sederman AJ, Connolly PC, Butler JM. MRI studies of the hydrodynamics in a USP 4 dissolution testing cell. *J Pharm Sci* 2011;100(3):976–991.
3. Noyes AA, Whitney WR. The rate of solution of solid substances in their own solutions. *J Am Chem Soc* 1897;19(12):930–934.
4. Dokoumetzidis A, Macheras P. A century of dissolution research: from Noyes and Whitney to the biopharmaceutics classification system. *Int J Pharm* 2006;321(1–2):1–11.
5. Gray V, Kelly G, Xia M, Butler C, Thomas S, Mayock S. The science of USP 1 and 2 dissolution: present challenges and future relevance. *Pharm Res* 2009;26(6):1289–1302.
6. Cardot JM, Beyssac E, Alric M. In vitro–in vivo correlation: importance of dissolution in IVIVC. *Dissolut Technol* 2007;14(1):15–19.
7. Tong C, Lozano R, Yun M, Mirza T, Löbenberg R, Nickerson B, Gray V, Wang Q. The value of in vitro dissolution in drug development. *Pharm Technol* 2009;33(4):52–64.
8. Bai G, Wang Y, Armenante PM. Velocity profiles and shear strain rate variability in the USP dissolution testing apparatus 2 at different impeller agitation speeds. *Int J Pharm* 2011; 403(1–2):1–14.
9. D'Arcy DM, Corrigan OI, Healy AM. Evaluation of hydrodynamics in the basket dissolution apparatus using computational fluid dynamics—dissolution rate implications. *Eur J Pharm Sci* 2006;27(2–3):259–267.

10. Klein S. The use of biorelevant dissolution media to forecast the in vivo performance of a drug. *AAPS J* 2010;12(3):397–406.

11. Hanson R. A primer on release-rate testing of semisolids. *Dissolut Technol* 2010;17:33–35.

12. Franz TJ. Percutaneous absorption on the relevance of in vitro data. *J Invest Dermatol* 1975;64(3):190–195.

13. Pharmacopeial Forum. Product test for transdermal drug products. In: *Pharmacopeial Forum.* Volume 35, Rockville: United States Pharmacopeia; May–June 2009.

14. Ahuja A, Dogra M, Agarwal SP. Development of buccal tablets of diltiazem hydrochloride. *Indian J Pharm Sci* 1995;57(1):26–30.

15. United States Pharmacopeia, Center for Drug Evaluation and Research. *Mechanical Calibration and Performance Verification Test Apparatus 1 and Apparatus 2.* 2nd ed. Rockville: United States Pharmacopeia; 2010. p 16.

16. United States Pharmacopeia, Center for Drug Evaluation and Research. *The Use of Mechanical Calibration of Dissolution Apparatus 1 and 2 – Current Good Manufacturing Practice (CGMP).* Rockville: United States Food and Drug Administration; 2010. p 1–8.

17. ASTM International. *ASTM E2503-07, Standard Practice for Qualification of Basket and Paddle Dissolution Apparatus.* West Conshohocken: ASTM International; 2007. p 1–4.

18. Deng G, Ashley AJ, Brown WE, Eaton JW, Hauck WW, Kikwai LC, Liddell MR, Manning RG, Munoz JM, Nithyanandan P, Glasgow MJ, Stippler E, Wahab SZ, Williams RL. The USP performance verification test, part I: USP lot P prednisone tablets: quality attributes and experimental variables contributing to dissolution variance. *Pharm Res* 2008; 25(5):1100–1109.

19. Hauck WW, Abernethy DR, Koch WF, Cecil TL, Brown W, Williams RL, USP (2008). USP responses to comments on stimuli article, "proposed change to acceptance criteria for dissolution performance verification testing". *Pharm Forum* 34(2): 474–476.

20. Siewert M, Dressman J, Brown C, Shah V, Aiache J-M, Aoyagi N, Bashaw D, Brown C, Brown W, Burgess D, Crison J, DeLuca P, Djerki R, Dressman J, Foster T, Gjellan K, Gray V, Hussain A, Ingallinera T, Klancke J, Kraemer J, Kristensen H, Kumi K, Leuner C, Limberg J, Loos P, Margulis L, Marroum P, Moeller H, Mueller B, Mueller-Zsigmondy M, Okafo N, Ouderkirk L, Parsi S, Qureshi S, Robinson J, Shah V, Siewert M, Uppoor R, Williams R. FIP/AAPS guidelines to dissolution/in vitro release testing of novel/special dosage forms. *AAPS PharmSciTech* 2003;4(1):43–52.

21. Blume H, Ali SL, Siewert M. Pharmaceutical quality of glibenclamide products a multi-national postmarket comparative study. *Drug Dev Ind Pharm* 1993;19(20):2713–2741.

22. United States Pharmacopeia, United States Food and Drug Administration, Center for Drug Evaluation and Research. *Dissolution Testing of Immediate Release Solid Oral Dosage Forms.* Rockville: United States Food and Drug Administration; 1997. p 1–17.

23. United States Pharmacopeia, United States Food and Drug Administration, Center for Drug Evaluation and Research. *Bioavailability and Bioequivalence Studies for Orally Administered Drug Products—General Considerations.* Rockville: Food and Drug Administration; 2003.

24. FIP guidelines for dissolution testing of solid oral products. *Pharm Ind* 1981;43:334–343.

25. United States Pharmacopeia, United States Food and Drug Administration, Center for Drug Evaluation and Research. *Test Solutions.* Volume 1, Rockville: United States Food and Drug Administration; 2010.

26. United States Pharmacopeia, United States Food and Drug Administration, Center for Drug Evaluation and Research. *Dissolution Methods: List of all Drugs in the Database.*

Rockville: Food and Drug Administration; 2011. Available at http://www.accessdata.fda.
gov/scripts/CDER/dissolution/dsp_SearchResults_Dissolutions.cfm?PrintAll=1.
Accessed March 1, 2014.

27. Liu F, Merchant HA, Kulkarni RP, Alkademi M, Basit AW. Evolution of a physiological
pH 6.8 bicarbonate buffer system: application to the dissolution testing of enteric coated
products. *Eur J Pharm Biopharm* 2011;78(1):151–157.

28. Boni JE, Brickl RS, Dressman J. Is bicarbonate buffer suitable as a dissolution medium?
J Pharm Pharmacol 2007;59(10):1375–1382.

29. Donauer N, Löbenberg R. A mini review of scientific and pharmacopeial requirements for
the disintegration test. *Int J Pharm* 2007;345(1–2):2–8.

30. Almukainzi M, Salehi M, Araci Bou-Chacra N, Löbenberg R. Investigation of the
performance of the disintegration test for dietary supplements. *AAPS J* 2010;
12(4):602–607.

31. Stippler E, Kopp S, Dressman JB. Comparison of US Pharmacopeia simulated intestinal
fluid TS (without pancreatin) and phosphate standard buffer PH 6.8, TS of the International
Pharmacopoeia with respect to their use in *in vitro* dissolution testing. Dissolut. *Technol*
2004;11(2):6–10.

32. Fotaki N, Vertzoni M. Biorelevant dissolution methods and their applications in *in vitro-in
vivo* correlations for oral formulations. *Open Drug Deliv J* 2010;4(2):2–13.

33. McConnell EL, Fadda HM, Basit AW. Gut instincts: explorations in intestinal physiology
and drug delivery. *Int J Pharm* 2008;364(2):213–226.

34. Taupitz T, Klein S. Can biorelevant media be simplified by using SLS and Tween 80 to
replace bile compounds? *Open Drug Deliv J* 2010;4:30–37.

35. Crison JR, Weiner ND, Amidon GL. Dissolution media for in vitro testing of water-insoluble
drugs: effect of surfactant purity and electrolyte on in vitro dissolution of carbamazepine in
aqueous solutions of sodium lauryl sulfate. *J Pharm Sci* 1997;86(3):384–388.

36. Fang JB, Robertson VK, Rawat A, Flick T, Tang ZJ, Cauchon NS, McElvain JS.
Development and application of a biorelevant dissolution method using USP apparatus 4
in early phase formulation development. *Mol Pharm* 2010;7(5):1466–1477.

37. Juenemann D, Bohets H, Ozdemir M, de Maesschalck R, Vanhoutte K, Peeters K, Nagels
L, Dressman JB. Online monitoring of dissolution tests using dedicated potentiometric
sensors in biorelevant media. *Eur J Pharm Biopharm* 2011;78(1):158–165.

38. Jones HM, Parrott N, Ohlenbusch G, Lavé T. Predicting pharmacokinetic food effects
using biorelevant solubility media and physiologically based modelling. *Clin Pharmacokinet*
2006;45(12):1213–1226.

39. Dressman JB, Amidon GL, Reppas C, Shah VP. Dissolution testing as a prognostic tool for
oral drug absorption: immediate release dosage forms. *Pharm Res* 1998;15(1):11–22.

40. Vertzoni M, Dressman J, Butler J, Hempenstall J, Reppas C. Simulation of fasting gastric
conditions and its importance for the in vivo dissolution of lipophilic compounds. *Eur J
Pharm Biopharm* 2005;60(3):413–417.

41. Marques M, Loebenberg R, Almukainzi M. Simulated biological fluids with possible
application in dissolution testing. *Dissolut Technol* 2011;18(3):15–28.

42. Jantratid E, Dressman JB. Biorelevant dissolution media simulating the proximal human
gastrointestinal tract: an update. *Dissolut Technol* 2009;16(3):21–25.

43. McAllister M. Dynamic dissolution: a step closer to predictive dissolution testing?
Mol Pharm 2010;7(5):1374–1387.

44. Wei H, Löbenberg R. Biorelevant dissolution media as a predictive tool for glyburide a class II drug. *Eur J Pharm Sci* 2006;29(1):45–52.

45. Okumu A, DiMaso M, Löbenberg R. Dynamic dissolution testing to establish in vitro/in vivo correlations for montelukast sodium, a poorly soluble drug. *Pharm Res* 2008; 25(12):2778–2785.

46. Hammad MA, Muller BW. Increasing drug solubility by means of bile salt-phosphatidylcholine-based mixed micelles. *Eur J Pharm Biopharm* 1998;46(3):361–367.

47. Kloefer B, van Hoogevest P, Moloney R, Kuentz M, Leigh MLS, Dressman J. Study of a standardized taurocholate-lecithin powder for preparing the biorelevant media FeSSIF and FaSSIF. *Dissolut Technol* 2010;17(3):6–13.

48. Libo Y. Biorelevant dissolution testing of colon-specific delivery systems activated by colonic microflora. *J Control Release* 2008;125(2):77–86.

49. Bott C, Rudolph MW, Schneider ARJ, Schirrmacher S, Skalsky B, Petereit HU, Langguth P, Dressman JB, Stein J. In vivo evaluation of a novel pH- and time-based multiunit colonic drug delivery system. *Aliment Pharmacol Ther* 2004;20(3):347–353.

50. Vertzoni M, Diakidou A, Chatzilias M, Söderlind E, Abrahamsson B, Dressman J, Reppas C. Biorelevant media to simulate fluids in the ascending colon of humans and their usefulness in predicting intracolonic drug solubility. *Pharm Res* 2010;27(10):2187–2196.

51. Buch P, Holm P, Thomassen JQ, Scherer D, Branscheid R, Kolb U, Langguth P. IVIVC for fenofibrate immediate release tablets using solubility and permeability as in vitro predictors for pharmacokinetics. *J Pharm Sci* 2010;99(10):4427–4436.

52. Lehto P, Kortejärvi H, Liimatainen A, Ojala K, Kangas H, Hirvonen J, Tanninen VP, Peltonen L. Use of conventional surfactant media as surrogates for FaSSIF in simulating in vivo dissolution of BCS class II drugs. *Eur J Pharm Biopharm* 2011; 78(3):531–538.

53. Zoeller T, Klein S. Simplified biorelevant media for screening dissolution performance of poorly soluble drugs. *Dissolut Technol* 2007;8–13.

54. United States Pharmacopeia. *United States Pharmacopeial Convention. Rockville: Food and Drug Administration* 2010;1(33):610.

55. Buchwald P. Direct, differential equation-based in-vitro–in-vivo correlation (IVIVC) method. *J Pharm Pharmacol* 2003;55(4):495–504.

56. United States Pharmacopeia, United States Food and Drug Administration, Center for Drug Evaluation and Research (CDER). *Guidance for industrial: Extended Release Oral Dosage Forms: Development, Evaluation and Application of In Vitro/In Vivo Correlations.* Rockville: Food and Drug Administration; 1997.

57. Polli JE, Crison JR, Amidon GL. Novel approach to the analysis of in vitro–in vivo relationships. *J Pharm Sci* 1996;85(7):753–760.

58. Dunne A, O'Hara T, Devane J. Level A in vivo–in vitro correlation: nonlinear models and statistical methodology. *J Pharm Sci* 1997;86(11):1245–1249.

59. Löbenberg R, Krämer J, Shah VP, Amidon GL, Dressman JB. Dissolution testing as a prognostic tool for oral drug absorption: dissolution behavior of glibenclamide. *Pharm Res* 2000;17(4):439–444.

60. Dunne A, Gaynor C, Davis J. Deconvolution based approach for level A in vivo/in vitro correlation modelling: statistical considerations. *Clin Res Regul Aff* 2005;22(1):1–14.

61. Gillespie WR. Convolution-based approaches for in vivo–in vitro correlation modeling. *Adv Exp Med Biol* 1997;423:53–65.

62. Balan G, Timmins P, Greene DS, Marathe PH. In vitro–in vivo correlation (IVIVC) models for metformin after administration of modified-release (MR) oral dosage forms to healthy human volunteers. *J Pharm Sci* 2001;90(8):1176–1185.

63. Modi NB, Lam A, Lindemulder E, Wang B, Gupta SK. Application of in vitro–in vivo correlations (IVIVC) in setting formulation release specifications. *Biopharm Drug Dispos* 2000;21(8):321–326.

64. O'Hara T, Hayes S, Davis J, Devane J, Smart T, Dunne A. In vivo–in vitro correlation (IVIVC) modeling incorporating a convolution step. *J Pharmacokinet Pharmacodyn* 2001;28(3):277–298.

65. Veng-Pedersen P, Gobburu JV, Meyer MC, Straughn AB. Carbamazepine level-A in vivo–in vitro correlation (IVIVC): a scaled convolution based predictive approach. *Biopharm Drug Dispos* 2000;21(1):1–6.

66. Costello C, Rossenu S, Vermeulen A, Cleton A, Dunne A. A time scaling approach to develop an in vitro–in vivo correlation (IVIVC) model using a convolution-based technique. *J Pharmacokinet Pharmacodyn* 2011;38(5):519–539.

67. Yu LX, Amidon GL. A compartmental absorption and transit model for estimating oral drug absorption. *Int J Pharm* 1999;186(2):119–125.

68. Lukacova V. A Generic PBPK Model for Predictive DMPK (GastroPlus), PBPK Modeling in Drug Development and Evaluation; April 8, 2009; Alexandria, VA. Available at http://www.simulations-plus.com. Accessed on March 1, 2014.

69. Jones HM, Gardner IB, Watson KJ. Modelling and PBPK simulation in drug discovery. *AAPS J* 2009;11(1):155–166.

70. Vitková Z, Vitko A, Zabka M, Cizmárik J. Alternative and generalized approach to in vitro–in vivo correlation. *Acta Pol Pharm* 2011;68(3):417–421.

71. Wilson CG, Washington N. Assessment of disintegration and dissolution of dosage forms in vivo using gamma scintigraphy. *Drug Dev Ind Pharm* 1988;14(2–3):211–281.

72. Billa N, Yuen KH, Khader MAA, Omar A. Gamma-scintigraphic study of the gastrointestinal transit and in vivo dissolution of a controlled release diclofenac sodium formulation in xanthan gum matrices. *Int J Pharm* 2000;201(1):109–120.

73. Sirisuth N, Eddington ND (2002). In-vitro–in-vivo correlation definitions and regulatory guidance. Int J Generic Drugs. Available at http://www.iagim.org/pdf/ivivc-01.pdf. Accessed March 1, 2014.

74. Emami J. In vitro–in vivo correlation: from theory to applications. *J Pharm Pharm Sci* 2006;9(2):169–189.

75. Akbor MM, Sultana R, Ullah A, Azad MAK, Latif AHMM, Hasnat A. In vitro–in vivo correlation (IVIVC) of immediate release (IR) levofloxacin tablet. *Dhaka Univ J Pharm Sci* 2007;6(2):113–119.

76. Rekhi GS, Jambhekar SS. Bioavailability and in-vitro/in-vivo correlation for propranolol hydrochloride extended-release bead products prepared using aqueous polymeric dispersions. *J Pharm Pharmacol* 1996;48(12):1276–1284.

77. Z-q Li, He X, Gao X, Y-y Xu, Wang Y-f, Gu H, Ji, RF, Sun, SJ. Study on dissolution and absorption of four dosage forms of isosorbide mononitrate: level A in vitro–in vivo correlation. *Eur J Pharm Biopharm* 2011;79(2):364–371.

78. Philip AK, Pathak K. Wet process-induced phase-transited drug delivery system: a means for achieving osmotic, controlled, and level A IVIVC for poorly water-soluble drug. *Drug Dev Ind Pharm* 2008;34(7):735–743.

79. Sankalia JM, Sankalia MG, Mashru RC. Drug release and swelling kinetics of directly compressed glipizide sustained-release matrices: establishment of level A IVIVC. *J Control Release* 2008;129(1):49–58.

80. Graffner C, Nicklasson M, Lindgren JE. Correlations between in vitro dissolution rate and bioavailability of alaproclate tablets. *J Pharmacokinet Biopharm* 1984;12(4):367–380.

81. Shono Y, Jantratid E, Dressman JB. Precipitation in the small intestine may play a more important role in the in vivo performance of poorly soluble weak bases in the fasted state: case example nelfinavir. *Eur J Pharm Biopharm* 2011;79(2):349–356.

82. Grbic S, Parojcic J, Ibric S, Djuric Z. In vitro–in vivo correlation for gliclazide immediate-release tablets based on mechanistic absorption simulation. *AAPS PharmSciTech* 2011; 12(1):165–171.

83. Crison JR, Timmins P, Keung A, Upreti VV, Boulton DW, Scheer BJ. Biowaiver approach for biopharmaceutics classification system class 3 compound metformin hydrochloride using in silico modeling. *J Pharm Sci* 2012;101(5):1773–1782.

84. Kovačević I, Parojĉić J, Homŝek I, Tubić-Grozdanis M, Langguth P. Justification of bio-waiver for carbamazepine, a low soluble high permeable compound, in solid dosage forms based on IVIVC and gastrointestinal simulation. *Mol Pharm* 2009;6(1):40–47.

85. Tsume Y, Amidon GL. The biowaiver extension for BCS class III drugs: the effect of dissolution rate on the bioequivalence of BCS class III immediate-release drugs predicted by computer simulation. *Mol Pharm* 2010;7(4):1235–1243.

86. Okumu A, DiMaso M, Löbenberg R. Computer simulations using GastroPlus™ to justify a biowaiver for etoricoxib solid oral drug products. *Eur J Pharm Biopharm* 2009;72(1):91–98.

87. Eaton JW, Tran D, Hauck WW, Stippler ES. Development of a performance verification test for USP apparatus 4. *Pharm Res* 2012;29(2):345–351.

88. Fotaki N. Flow-through cell apparatus (USP apparatus 4): operation and features. *Dissolut Technol* 2011;18(4):46–49.

89. Stippler ES. Review of research paper: development of a performance verification test for USP apparatus 4. *Dissolut Technol* 2011;18(4):44.

90. Carino SR, Sperry DC, Hawley M. Relative bioavailability estimation of carbamazepine crystal forms using an artificial stomach-duodenum model. *J Pharm Sci* 2006;95:116–125.

91. Garbacz G, Kelin S, Weitschies W. A biorelevant dissolution stress test device. *Expert Opin Drug Deliv* 2010;7(11):1251–1261.

92. Blanquet S, Zeijdner E, Beyssac E, Meunier JP, Denis S, Havenaar R, Alric M. A dynamic artificial gastrointestinal system for studying for behavior of orally administered drug dosage forms under various physiological conditions. *Pharm Res* 2004;21(4):585–589.

13

ICH QUALITY GUIDELINES: THEIR GLOBAL IMPACT

SULTAN GHANI

13.1 INTRODUCTION

By the end of the nineteenth century, America had seen tremendous progress in mechanization and was leading in innovations. The close of the century witnessed tremendous economic activity in the United States. As a result of political and socioeconomical progress, stability was achieved in the later part of nineteenth century and the early twentieth century; the pace of the drug development as well as regulatory requirement were enhanced. Industrial expansion and growth were noted in all sectors including the pharmaceutical sector. Some important drugs discovered before the twentieth century include nitrous oxide, ether, nitroglycerin, barbiturates, and aspirin.

A new consciousness and an invigorated awareness started to surface, contributing toward improved quality in the pharmaceutical field. Substandard and poor-quality drugs had created an awareness about protecting consumers, putting pressure on governments. As such, both developing and developed countries started regulating drugs and pharmaceuticals.

Regulations to control the quality of pharmaceuticals did exist in some form or the other at almost all times. However, the latest research in the scientific, medical, and pharmaceutical fields and the availability of more sophisticated automated manufacturing and analytical equipments put an onus on manufacturers, who were pressured by stricter and more stringent demands from the regulatory bodies to produce better-quality products. In essence, this would not have been possible if economic growth had not been commensurate to support the prohibitive expenses required by government regulations.

Therapeutic Delivery Solutions, First Edition. Edited by Chung Chow Chan,
Kwok Chow, Bill McKay, and Michelle Fung.
© 2014 John Wiley & Sons, Inc. Published 2014 by John Wiley & Sons, Inc.

Laws are developed to regulate or monitor the activity of people. Drug laws, like all the other laws, are not different in this respect. They exert a significant influence on the public starting from the point of manufacture to distribution and eventually to sale. They have social and economical implications, and they influence both national and international trade particularly where free enterprise exists and where there is competition between businesses.

The drug act and regulations are designed to provide effective protection to consumers from ineffective and unsafe medicines. They provide working documents of specifications for regulators in both developed and developing countries and must be fully understood by both the manufacturer and distributor of drugs. Regulatory requirements for pharmaceuticals or medicinal products are not straightforward and differ around the world. Manufacturers are usually successful in getting approval for their products if all international, national, regional, and local requirements and guidelines are met.

Major differences exist among countries with regard to data requirements, storage conditions (especially temperature and humidity conditions), and standard testing requirements. These differences and uncertainties make the development of globally recognized standards challenging and difficult. The lack of harmonized approaches increases cost and poses barriers to timely international access to drug products. This impact is significant especially for developing countries with limited resources and training, thus leading to differences in regulatory experience and decisions.

13.2 GLOBAL HARMONIZATION FORUMS

The global activities for harmonization of drug standards are wide-ranging and encouraging, thus providing many opportunities. Awareness of global harmonization and a thorough understanding of underlying issues are necessary for achieving success in the developing countries. These provide opportunities for the developing countries to become a part of the global community to further enhance their regulatory capabilities and to take enforcement action, ensuring that the medical products marketed in their jurisdiction comply with global international standards that are based on science and experience. In the global context, this provides a certain degree of acceptability for the free movement of such products when harmonized standards are met.

The purpose of the International Conference on Harmonization of Technical Requirements for Registration of Pharmaceuticals for Human Use (ICH) is to harmonize the interpretation and applications of technical guidelines. Harmonization and multilateral relation representatives act as a liaison for ICH steering committee experts as well as the ICH secretariat.

The Pan American Network for Drug Regulatory Harmonization (PANDRH) was established in 1999 to support processes on drug regulatory harmonization. Participants in PANDRH include regulatory authorities from America's various pharmaceutical interest groups, industry, and academia.

Other similar organizations were conceived to respond to the growing need for international harmonization in the regulation of pharmaceutical products. These organizations provide a forum in which official representatives of national regulatory bodies working with manufacturers and other organizations possessing the relevant expertise can harmonize global approaches to regulating the safety, efficacy, and quality of pharmaceutical products.

The globalization approach utilizes the expertise of international regulatory bodies and research-based industry of the member countries and seeks their help to document a single set of technical requirements for the registration of pharmaceutical products, thus streamlining the drug development process. There are still many challenges for the effective, timely implementation of harmonized procedures and guidance documents.

The globalization approach greatly influences the requirement related to the existing products especially regarding the quality of generic drug products and drug substances. This has a direct bearing on not only the generic industry but also the developing countries where drug use is based on well-established products and linked with many countries for producing the generic version of essential drugs.

The Global Cooperation Group (GCG) was established as a subcommittee of the ICH steering committee to garner support for the use of ICH guidelines by non-ICH countries (developing countries). It also supports various regional harmonization initiatives such as Asia Pacific Economic Cooperation (APEC), the Association of Southeast Asian Nations (ASEAN), the Gulf Cooperation Council (GCC), and the PAN American Network of Drugs (PANDRH).

13.3 ICH PROCESS

The ICH involves the three regulatory authorities of Europe, Japan, and the United States and their official industry organizations. The experts from these six groups discuss the scientific and technical requirements for product registration and monitoring.

The purpose of setting up these expert groups is to develop harmonized technical guidelines and requirements and thus reduce or eliminate the need to duplicate the testing carried out during the research and development of medicines.

There are obvious advantages associated with such types of harmonization from both the human resources and animal resources perspectives. At the same time, unnecessary delays in the global development and availability of medicines are eliminated while the key attributes needed of new medicines, such as safety, quality, and efficacy, are maintained and the regulatory obligations for public health protection are met.

The mission of such harmonization is a more economical use of human, animal, and material resources and the elimination of unnecessary delays in the global development and availability of new medicines while maintaining standards of quality, safety, and efficacy and the regulatory obligations to protect public health.

13.3.1 Structure

This is a joint initiative involving regulators and industry as partners in the scientific and technical discussion on various aspects of medicine that are essential to ensure and assess the safety, quality, and efficacy of medicines. Although the focus of the

ICH is on new drugs, its principles and standards have been accepted by many countries for generic products. The sole objective behind establishing the ICH was the fact that most of these new molecules are developed in Western Europe, Japan, and the United States.

The ICH consists of six parties, three observers, and the International Federation of Pharmaceutical Manufacturers Association (IFPMA). The six parties are founders of ICH and represent regulatory bodies and research-based industry of the European Union (EU), Japan, and the United States. The observers are the World Health Organization (WHO), the European Free Trade Association (EFTA), and Canada. Canadians are represented by Health Canada.

13.3.2   European Union

In Europe, the Commission represents the European member states and these member states have been working for the past so many years toward harmonization of technical requirements and procedures to achieve a single market in pharmaceuticals that would allow free movement of products throughout the EU.

The European Medicine Agency (EMEA) has been established by the Commission to support the scientific and technical activities of the ICH. The European Federation of Pharmaceutical Industries and Associations (EFPIA) is a representative of pharmaceutical industry associations involving pharmaceutical companies and is working toward research and development and manufacturing of medicinal products in Europe for human use.

In Japan, the Ministry of Health, Labour and Welfare has the responsibility for approval and administration of drugs, medical devices, and cosmetics. The Japan Pharmaceutical Manufacturer Association (JPMA) represents a large number of pharmaceutical manufacturers in Japan, and it also represents the industry's point of view through a specialized committee of experts and working group. The JPMA represents the industry's interest in international issues and promotes international standards among its members.

The U.S. Food and Drug Administration (FDA), which is responsible for the management of a wide range of drugs, biologicals, medical devices, and cosmetics, is also responsible for the premarket approval of drugs in the United States. Technical and scientific advisors of the FDA work with the ICH toward developing harmonized standards.

The Pharmaceutical Research and Manufacturers of America (PhRMA) represents the research-based industry of the United States. The member companies are involved in drug discovery and development and the manufacture of prescription drugs. The PhRMA coordinates the technical and scientific input to ICH through its scientific and regulatory committees.

The ICH observers include the WHO, the EFTA, which is currently represented by Swissmatic, Switzerland, and Canada, which is represented by Health Canada.

The IFPMA is a nonprofit, nongovernmental organization representing industry associations and companies from both developed and developing countries.

Many member companies of IFPMA are research-based pharmaceutical and biotech companies. The IFPMA has been closely associated with the ICH since its inception in 1990 to ensure contact with research-based industry outside the ICH region.

13.3.3 ICH Steering Committee

The administration of the ICH is conducted through a Steering Committee (SC), which is supported by its Secretariat. This Committee was established in 1990 and works as per the ICH Terms of Reference and determines policies and procedures for ICH, selects topics for harmonization, and monitors the progress of harmonization. Each of the six cosponsors occupies two seats on the ICH SC. The IFPMA provides the Secretariat services and participates as a nonvoting member in the SC. The observers also have nonvoting participation in the SC meetings.

For smooth running of the ICH, each of the six cosponsors acts as a coordinator to establish a contact point with the ICH Secretariat in order to ensure that the ICH documents are shared with the appropriate persons within their area of jurisdiction. Each party also establishes a contact network of experts within their own organization to ensure that in the discussion they reflect the views and policies of the cosponsors they represent.

The Secretariat operates from its office in Geneva and is basically responsible for documentation, meetings, the coordination of Expert Working Groups (EWGs), and other administrative support services. It is also responsible for organizing and arranging the ICH conferences and maintaining all contact with relevant bodies.

13.3.4 Conferences and Workshops

One of the key objectives of the ICH was to organize international conferences and workshops on harmonization activities and initiatives so as to keep the harmonization activities more transparent and to have an open forum to discuss the various ICH decisions and recommendations. Until now, six international conferences on harmonization have been organized by the ICH in different parts of three regions—the United States, Europe, and Japan. The first conference was organized in 1991 and the last in 2003. Other workshops and meetings on various ICH topics were organized by different organizations.

13.3.5 ICH Harmonization Activities

The ICH harmonization activities are given in the following Table 13.1. A stepwise progression of guideline procedures is followed, which also includes maintenance activities.

TABLE 13.1   ICH Harmonization Activities

S. No.	Procedure	Scientific discussion group	Description	Remarks
1	ICH procedure	EWG	Development of a new guideline	M5
2	Q&A procedure	Implementation Working Group (IWG)	Q&As to assist in the implementation of guidelines	Common Technical document (CTD)–IWG
3	Revision procedure	EWG	Revision and/or modification of guidelines	E2B (R3)
4	Maintenance procedure	EWG	Addition of standards to existing guidelines and/or recommendations	Q3C (R3) M2 recommendations

The suggestions for harmonization initiatives (the suggestion or the topic) come from various sources. However, the formal proposal for ICH action should be channelized through one of the six parties of ICH or one of the ICH observers on the ICH SC.

A concept paper is initiated or triggered indicating the type of harmonization action proposed, the perceived problems, issues to be resolved, references, and the recommendation of the type of EWG required. This concept paper is presented to the SC, and, after the endorsement from the SC, the ICH Working Group is established. Each of the six official ICH parties nominates official representatives (usually two officials per Working Group). The SC officially designates a reporter; sometimes a corapporteur may be appointed. The observers also nominate their representatives in the EWG.

13.3.6   ICH Procedure

Basically, there are four key steps in the ICH process for the development of guidelines. This is a streamlined procedure; however, in certain situations, the SC may decide to follow some accelerated procedures for new topics. The steps are as follows:

13.3.6.1   Step 1—Consensus Building   When a new topic is accepted, the process of consensus building begins. As specified in the concept paper, an EWG is established. The rapporteur prepares an initial draft guideline based on the objective set out in the concept paper and in consultation with experts designated to the EWG. The initial draft goes through successive revisions and is circulated within the EWG. Consultation on the draft within the EWG is carried out by correspondence, fax, and e-mails. The face-to-face meeting of the EWG normally takes place during biennial steering committee meetings. If any formal additional meetings are required, the EWG is required to receive permission from the SC. When consensus is reached among the working group members of all the six parties, the

EWG signs the Step 2 sign-off sheet. When consensus within the six parties has not been reached, a report is made to the SC indicating the extent of agreements and disagreements, and, in this situation, experts from all parties represented on the EWG have an opportunity to explain their position to the SC. Based on the discussion between the SC and the EWG, the SC may give an extension to the EWG to accomplish the consensus or may decide to suspend the harmonization project.

13.3.6.2 Step 2—Confirmation of Six-party Consensus

When sufficient scientific consensus on technical issues for the draft guideline is accomplished and accepted at the SC, the consensus text approved by the SC proceeds to the next stage of regulatory consultation. The consensus text approved by the SC is signed off by the SC as a Step 2 final document.

13.3.6.3 Step 3—Regulatory Consultation and Discussion

Step 3 involves wide-ranging regulatory consultation in three regions. Also, there is an opportunity for industry associations and regulatory authorities of non-ICH regions to comment on the draft consultation document, which is distributed using IFPMA and WHO contact lists.

After obtaining all regulatory consultation results, the EWG, which organizes the discussion for consensus building, reviews all the comments received during the consultation period. If both regulatory and industry parties of EWGs are satisfied, the consensus achieved at Step 2 is not substantially altered. The Step 3 expert document is signed by the EWG regulatory experts. This sign-off is called Step 3 Expert Sign-Off. In case complete consensus has not been achieved, a report is made to the SC highlighting the differences between the parties. The EWG representatives have an opportunity to explain their position to the SC.

13.3.6.4 Step 4—Adoption of an ICH Harmonized Guideline

This step is reached when the SC agrees on the basis of the report provided by the rapporteur of the EWG. This is the confirmation that sufficient scientific consensus on the technical issues has been accomplished. The document is signed by three regulatory parties to the ICH confirming that the guidelines or document is considered to be satisfactory for adoption by the regulatory parties of three regions. In case of any major objection to adoption due to some new issues, the regulatory parties may agree to revise and submit it for further consultation. In such a situation, the EWG discussion restarts.

The final document is signed off by the SC signatories for regulatory parties of ICH as an ICH Harmonized Tripartite Guideline at Step 4 of the ICH process.

The Step 4 Harmonized Guideline is moved to the final step of the process, which is regulatory implementation. This step of regulatory implementation is carried out according to the national and regional procedures in the EU, Japan, and the United States. The final guideline is published by the ICH Secretariat on the ICH website.

To support the implementation of guidance, additional information in the form of questions and answers (Q&A) is usually developed. The development and adoption of Q&A has an established process that has to be followed. The Q&A process is intended as a mechanism by which questions received from the stakeholders are collected, analyzed, reformulated, and ultimately used as model questions for which standard answers are developed and posted on the website.

Any questions sent to the mailbox of the ICH website or raised by any of the six official ICH parties are brought to the attention of the appropriate implementation working group.

The regional questions and issues are handled by the regulatory party of the concerned region and then shared and evaluated within the IWG; the final proposed answer is presented to the SC for approval and endorsement before its publication on the website. The IWG rapporteur sends questions to members of his/her IWG thereafter, and, based on this information, the IWG prepares model questions and their responses for presentation at the SC meeting. Based on the level of guidance given by the answer, the IWG assesses whether the Q&A document should be a Step 2 document and published for comments or a Step 4 document and published as final.

Each IWG presents its draft Q&A to a SC meeting and makes recommendations on the status of the documentation. The SC concurs with the Q&A document and its step status. For Step 2, the Expert and SC members sign the Q&A document, and, for Step 4, the Regulatory Experts and the Regulatory SC members sign the Q&A document.

This procedure is intended to provide results quickly and efficiently by using the minimum amount of resources consistent with the achievement of scientifically valid results.

The revision procedure of ICH guidelines is almost identical to the formal ICH procedures. The only difference compared to ICH procedure category 1 is the final outcome. In the revision procedure (category 3), the final outcome is a revised version of the current existing guideline, whereas, in category 1, the final outcome is a new guideline. The revision of a guideline is designated by the letter R1 after the usual denomination of the guidelines. However, when a guidance or guideline is revised more than once, the document is renamed R2, R4, and so on.

The maintenance procedure (category 4) is based on updating new information. This has been harmonized by all parties. The maintenance procedure follows almost on the same lines as the other procedures. However, instead of having many face-to-face meetings, the regulatory rapporteurs rely on correspondence or teleconferencing to avoid unnecessary travel.

13.3.7   ICH Global Cooperation Group

The ICH GCG was formed in 1999 as a subcommittee of the ICH SC. The mission of the GCG is to promote mutual understanding of regional harmonization initiatives in order to facilitate the harmonization process related to ICH guidelines globally and to facilitate the capacity of drug regulatory authorities and the industry to utilize them.

The GCG comprises one representative from each of the six parties on the ICH SC plus the ICH Secretariat. The ICH observers (WHO, Canada, EFTA) are also part of the GCG. The group includes other Regional Harmonization Initiatives (RHIs). These

RHIs include APEC, ASEAN, the GCC, the Pan American Network of Drug Regulatory Harmonization (PANDRH), and Southern African Development Community (SADC) as permanent representatives to GCG.

The purpose of the GCG is to set up principles intended to guide harmonization activities and to answer requests for information by non-ICH regulators and industry. It is interesting to note that some of the key principles of the GCG include the following:

1. The ICH will serve as a resource for information and data and not seek to impose its views on any country, region, or company.
2. Documents related to the GCG initiative will be provided to non-ICH member countries or companies without charge.
3. The GCG will work closely with the WHO and other international organizations in order to achieve these goals.

13.4 IMPACT OF ICH GUIDELINES

Drugs have been most extensively regulated among all consumer products. Extensive testing requirements and meeting high-quality standard guidelines with added regulatory close monitoring and supervision and every aspect of development, manufacturing, and testing are creating a major issue of affordability and accessibility to drugs for consumers of both developed and developing countries.

Besides these factors, many regulatory authorities are not applying the risk assessment and management model in their evaluation/review of product applications particularly generic or branded generic products. Furthermore, some review processes require information from the perspectives of "nice to have with no added value to the quality and safety of product." In some other cases a host of additional administrative information is a part of the drug review process, thus creating unnecessary delays in the approval of drug products. Although the ICH process for formulating globalization and harmonization guidelines has significant benefits but, at the same time, there are significant difficulties particularly in non-ICH countries regarding generic drugs.

In 2001, the WHO issued a report with a host of recommendations on the impact of implementation of ICH guidelines in non-ICH countries. These are as follows:

1. The WHO should maintain its position as an observer with the ICH SC adopting a more proactive role.
2. The WHO should maintain a position as an observer in the ICH Global Expert Group, making it clear that participation should not be considered as automatic endorsement of ICH guidelines or procedures.
3. With regard to the ICH process, the WHO should establish a consultation procedure for assessing the usefulness of new ICH guidelines for the pharmaceutical industries and drugs regulatory authorities in non-ICH countries as early as possible.

4. The WHO should establish a mechanism to review and build on ICH guidelines in order to establish WHO guidelines. It should also assess the benefits and risks of implemented selected ICH drugs quality guidelines and the possible impact of pharmacopoeia and harmonization activities in the ICH region on standards for the manufacturer of generic products in non-ICH countries.

5. In order to improve access to essential drugs of assured quality especially in developing countries, there is an urgent need for the WHO to intensify its efforts to further develop international standard/guidelines for the approval of generic products in consultation with the generic drug industry, related organizations, and national authorities.

6. The ICH should be encouraged to benefit from the work already carried out by the WHO in the area of pharmacovigilance, and all ICH countries should be encouraged to participate more actively in the WHO Program for International Drug Monitoring. Although a decade has passed since these recommendations, one should look at what progress has been made by the WHO with regard to the impact of ICH International quality guidelines on non-ICH countries, particularly on the accessibility and reasonable price of generic drugs.

Points 2, 3, and 4 are important from the point of view of usefulness of new ICH guidelines, additional benefits, and the risk of implementation of these guidelines as well as improved access to essential drugs of assured quality in developing countries. One can argue on this point that there are many ICH quality guidelines that are significantly beneficial especially for the generic drugs and their accessibility in the developing countries. The two guidelines (ICH Q3A and ICH Q3B) are most significant from the impact point of view on the quality and accessibility of drug at a reasonable price. With the exception of a few WHO guidelines for stability testing, the WHO did not build on ICH guidelines but rather endorsed all ICH guidelines in their prequalification program.

The development of international standards has definitely presented a challenge to the WHO as an intergovernmental organization because of the associated costs of setting international standards related to pharmaceuticals. "A resolution of 9th International Conference of Drugs Regulatory Authorities (ICDRA)" held in Berlin in April 1999 requested the WHO when participating in the ICH process to take into account the implication for non-ICH countries. It was also recognized that the harmonized standards developed by the ICH will impact non-ICH countries, where the largest number of people reside and where there is a need for generic drugs at a reasonable price. The most important group of guidelines is ICH Q3A to Q3D and Q7A. The most serious impact was created by ICH Q3A and Q3B as these guidelines along with the others has created two types of Active Pharmaceutical Ingredient (API) standards for generic drugs. One category of generic drugs is referred to as Drug Master File (DMF) grade and the other as non-DMF grade. There is no doubt that impurities that are inherently toxic and are known to be toxic or have side effects must be controlled and monitored. However, the regulatory authorities apply these guidelines without taking into consideration risk–benefit assessment and all API pharmaceutical synthetic materials. For ICH Q7A many regulatory authorities from developing countries have significantly accelerated inspection programs for including the WHO and European directorate of Quality Medicine (EDQM). These additional activities and requirements are the result of the two types of API grades, as stated earlier. The difference between the prices of the two grades (type) of API is significantly huge. For example, the cost of sitagliptin with an assay limit of 98.6% is $1000 per kg, while the same material of DMF grade with an assay limit of 99.4% is about $10,000 per kg. The important question is whether the difference of 0.8% impurity justifies the use of sitaglipton at $10,000 per kg. There are many similar examples and differences in the price of APIs of DMF and non-DMF grades. The principle of risk assessment of ICH Q9 can be used to justify that such minor differences of impurities could justify the cost. The new emerging markets and many developing countries using these guidelines cannot have generic drugs at a reasonable cost.

The driving force behind the harmonization initiatives has been the need to hasten access to pharmaceutical products and respond to the demands of international trade, but this is not the WHO principle of harmonization. Besides having consistent and uniform standards and avoiding unnecessary duplication of regulatory requirements, the harmonized standards should not be based on the state of the art but also on cost considerations.

13.5 CONCLUSION

There are positive impacts resulting from many ICH guidelines and in the interests of global harmonization. These guidelines should be adopted by all regulatory authorities including the WHO. A good example is the stability guidelines.

ICH Quality Guideline list

Stability Q1A–Q1F	
Code	Document title
Q1A (R2)	Stability testing of new drug substances and products
Q1B	Stability testing : photo-stability testing of new drug substances and products
Q1C	Stability testing for new dosage forms
Q1D	Bracketing and matrixing designs for stability testing of new drug substances and products
Q1E	Evaluation of stability data
Q1F	Stability data package for registration applications in climatic zones III and IV
Analytical validation Q2	
Code	Document title
Q2 (R1)	Validation of analytical procedures: text and methodology

(*Continued*)

ICH Quality Guideline list **(Cont'd)**

Impurities Q3A–Q3D

Code	Document title
Q3A (R2)	Impurities in new drug substances
Q3B (R2)	Impurities in new drug products
Q3C (R5)	Impurities: guideline for residual solvents
Q3D	Impurities: guideline for metal impurities

Pharmacopoeias Q4–Q4B

Code	Document title
Q4	Pharmacopoeias
Q4A	Pharmacopoeia harmonization
Q4B	Evaluation and recommendation of pharmacopoeial texts for use in the ICH Regions
Q4B	Annex 1R1 residue on ignition/sulphated ash general chapter
Q4B Annex 2R1	Test for extractable volume of parenteral preparations general chapter
Q4B Annex 3R1	Test for particulate contamination: sub-visible particles general chapter
Q4B Annex 4AR1	Microbiological examination of non-sterile products: microbial enumeration tests general chapter
Q4B Annex 4BR1	Microbiological examination of non-sterile products: tests for specified micro-organisms general chapter
Q4B Annex 4CR1	Microbiological examination of non-sterile products: acceptance criteria for pharmaceutical preparations and substances for pharmaceutical use general chapter
Q4B Annex 5R1	Disintegration test general chapter
Q4B Annex 6R1	Uniformity of dosage units general chapter
Q4B Annex 7R2	Dissolution test general chapter
Q4B Annex 8R1	Sterility test general chapter
Q4B Annex 9R1	Tablet friability general chapter
Q4B Annex 10R1	Polyacrylamide gel electrophoresis general chapter
Q4B Annex 11	Capillary electrophoresis general chapter
Q4B Annex 12	Analytical sieving general chapter
Q4B Annex 13	Bulk density and tapped density of powders general chapter
Q4B Annex 14	Bacterial endotoxins test general chapter

Quality of biotechnological products Q5A–Q5E

Code	Document title
Q5A (R1)	Viral safety evaluation of biotechnology products derived from cell lines of human or animal origin Q5A
Q5B	Analysis of the expression construct in cells used for production of r-DNA derived protein products
Q5C	Stability testing of biotechnological/biological products

(Continued)

ICH Quality Guideline list   (Cont'd)

Q5D	Derivation and characterisation of cell substrates used for production of biotechnological/biological products
Q5E	Comparability of biotechnological/biological products subject to changes in their manufacturing process

Specifications Q6A–Q6B

Code	Document title
Q6A	Specifications: test procedures and acceptance criteria for new drug substances and new drug products: chemical substances
Q6B	Specifications: test procedures and acceptance criteria for biotechnological/biological products

Good manufacturing practice Q7

Code	Document title
Q7	Good manufacturing practice guide for active pharmaceutical ingredients

Pharmaceutical development Q8

Code	Document title
Q8 (R2)	Pharmaceutical development
Q8/9/10 Q&As R4Q8/Q9/Q10	Implementation

Quality risk management Q9

Code	Document title
Q9	Quality risk management
Q8/9/10 Q&AsR4Q8/Q9/Q10	Implementation

Pharmaceutical quality system Q10

Code	Document title
Q10	Pharmaceutical quality system
Q8/9/10 Q&AsR4Q8/Q9/Q10	Implementation

Development and manufacture of drug substances Q11

Code	Document title
Q11	Development and manufacture of drug substances (chemical entities and biotechnological/biological entities)

FURTHER READING

1. Carman HJ. Social and Economic History of the United States: From Handicraft to Factory, 1500–1820. New York: Health and Co.; 1968.

2. Dato Che Mohd Zin Che Awang. The Sixth International Conference of Harmonization, New Horizon and Future Challenges; November 12–15, 2003;Osaka, Japan.

3. Guarino RA. New Drug Approval Process: Accelerating Global Registration. 4th ed. New York: M. Dekker; 2004.

4. Gulf Cooperation Council (GCC). Available at world/gulf/gcc.htm. Accessed March 11, 2014.

5. ICH. The process of harmonization. Available at http://www.ich.org/about/process-of-harmonisation/formalproc.html. Accessed March 15, 2014.

6. Javrongrit Y. ACSCO/PPWG 2006. Symposium of APEC Network on Pharmaceutical Regulatory Science; November 15–16, 2006; Ha Noi, Vietnam.

7. Jayasurya DC. Regulation of Pharmaceutical in Developing Countries – Legal Issues and Approach. Geneva: World Health Organization; 1985.

8. Mike W. The ICH Experience International Conference Affairs, Kathmandu, Nepal; 2001.

9. Silverman M, Lee PR. Pill, Profits and Politics. Los Angeles: University of California Press; 1974. p 3–4.

10. Thirty-fourth Session of Sub-Committee on Planning and Programming of the Executive Committee, Washington, DC; March 2000.

11. World Bank. Addressing the Challenges of Globalization: An Independent Evaluation of the World Bank Program. Washington, DC: World Bank; 2004.

12. World Health Organization (WHO). The impact of implementation of ICH guideline in Non-ICH countries. Report of a WHO Meeting; September 13–15, 2001; Geneva: WHO.

14

OUT OF SPECIFICATION/ATYPICAL RESULT INVESTIGATION

Yu-Hong Tse and Chung Chow Chan

14.1 BACKGROUND

In the development of therapeutic solutions, there are situations when out of specification (OOS) or aberrant data are obtained. OOS data can be any analytical laboratory-recorded result that is not within the stated specification of the tested material or official compendia (detailed definition of OOS data is listed in Appendix I). When this happens, an investigation must be carried out.

Workshops, seminars, consultants, and guidance are available to support laboratory personnel in overcoming difficulties and in having a consistent approach during an *OOS/atypical result* investigation. Most firms have developed a specific *OOS/atypical result* investigation document. Unfortunately, an inadequate OOS investigation is still a leading cause for warning letters being issued by the Food and Drug Administration (FDA) since the emphasis in FDA-483 laboratory citations has shifted from the lack of an adequate OOS standard operating procedure (SOP) to observations on details within the document or for not following the procedure without any scientific justification during the investigation [1]. The following excerpt from the FDA guidance document clearly summarizes the overall requirements for the OOS investigation:

> To be meaningful, the investigation should be thorough, timely, unbiased, well-documented, and scientifically sound. The first phase of such an investigation should include an initial assessment of the accuracy of the laboratory's data. Whenever possible, this should be done

Therapeutic Delivery Solutions, First Edition. Edited by Chung Chow Chan,
Kwok Chow, Bill McKay, and Michelle Fung.
© 2014 John Wiley & Sons, Inc. Published 2014 by John Wiley & Sons, Inc.

before test preparations (including the composite or the homogenous source of the aliquot tested) are discarded. This way, hypotheses regarding laboratory error or instrument malfunctions can be tested using the same test preparations. If this initial assessment indicates that no meaningful errors were made in the analytical method used to arrive at the data, a full-scale OOS investigation should be conducted. For contract laboratories, the laboratory should convey its data, findings, and supporting documentation to the manufacturing firm's quality control unit (QCU), who should then initiate the full-scale OOS investigation [2].

In this chapter, we will look at how sound scientific judgment and good documentation can lead to a successful *OOS/atypical result* investigation in a case study based on current guidance.

Successful OOS investigation in the laboratory = sound scientific judgment + good documentation

In the case study described here, *Analyst A* and *Supervisor B* followed the typical process flow for an *OOS/atypical result* investigation process discussed in Appendix II for their investigation.

During routine release testing, *Analyst A* followed the approved testing method to perform a high-performance liquid chromatography (HPLC) analysis in Cream A (material code: 101, batch #123), which contained methyl paraben (limit: 95.0–105.0% of label claim at release) and propyl paraben (limit: 95.0–105.0% label claim at release) as preservatives. The bulk samples from the top, middle, and bottom of the manufacturing tank were tested. The sample was prepared as follows:

- 1 g of Cream A was accurately weighed in a 25-ml centrifuge tube.
- 5 ml of methanol and 5 ml of the mobile phase were added in the tube.
- The tube was closed and the sample mixed well until the cream matrix dissolved.
- The sample preparation was transferred to the HPLC vial for analysis.

After completion of the system suitability test (SST), *Analyst A* analyzed the sample by injecting the sample preparation with bracketing of standard preparation to the HPLC system.

CASE STUDY   PART I—ORIGINAL TESTING RESULT

Test result (acceptance criteria: release limit: 95.0–105.0%)

Sample ID	Propyl paraben assay result (%)	Methyl paraben assay result (%)
Batch #123—tank top	96.1	99.8
Batch #123—tank middle	95.5	99.6
Batch #123—tank bottom	94.9	99.4

14.2   IDENTIFYING AND ASSESSING OOS TEST RESULTS— PHASE I: LABORATORY INVESTIGATION

14.2.1   Responsibility of *Analyst A*

14.2.1.1   Before Testing   The first and primary responsibility for *Analyst A* is to ensure that he/she has been trained on the appropriate testing procedure in order to achieve accurate laboratory testing results. *Analyst A* should document the evidence in his/her training record and get it approved by the supervisor to indicate that he/she has the confidence to perform the test. *Analyst A* should be aware of any potential problems that could occur during the testing process and should watch for problems that could create *OOS/atypical* results. *Analyst A* should not knowingly continue an analysis he/she expects to invalidate at a later time for an assignable cause (i.e., analyses should not be completed for the sole purpose of seeing what results can be obtained when obvious errors are known):

> In accordance with the CGMP regulations in §211.160 (b)(4), the analyst should ensure that only those instruments meeting established performance specifications are used and that all instruments are properly calibrated [2].

14.2.1.2   During Testing   *Analyst A* should ensure system suitability requirements (SST) in the analytical method, and systems not meeting such requirements should not be used:

> For example, in chromatographic systems, reference standard solutions may be injected at intervals throughout chromatographic runs to measure drift, noise, and repeatability. If reference standard responses indicate that the system is not functioning properly, all of the data collected during the suspect time period should be properly identified and should not be used. The cause of the malfunction should be identified and, if possible, corrected before a decision is made whether to use any data prior to the suspect period [2].

14.2.1.3   After Testing   Before discarding test preparations or standard preparations, *Analyst A* should check the data for compliance with specifications. *Analyst A* should retain the test preparations and glassware. *Analyst A* should inform the supervisor immediately and start an assessment of the accuracy of the results if any OOS/atypical results are obtained. *Analyst A* should also prevent the analytical parameters from being altered until the instrument has been investigated. If errors are obvious, such as the spilling of a sample solution or the incomplete transfer of a sample composite, *Analyst A* should immediately document what happened and repeat the testing at the original level of duplication with the approval of the supervisor.

14.2.2   Responsibilities of Supervisor B

Once *OOS/atypical results* have been obtained, *Supervisor B* should perform the initial assessment with *Analyst A* in an objective and timely manner. There should be no preconceived assumptions as to the cause of the *OOS/atypical result*.

FDA's OOS guidance document iterated expectations that analysts are appropriately trained, followed validated procedures, use appropriate glassware, the correct reagents and standards, and perform analysis on well-maintained & calibrated instrumentation. [2]

Supervisor B should assess the data promptly to ascertain if the results may be attributed to laboratory error or whether the results could indicate problems in the manufacturing process. An immediate assessment could include visual examination of the actual solutions, test units, and glassware used in the original measurements and preparations, which would allow more credibility to be given to laboratory error theories.

According to current guidance, *Supervisor B* should take at least the following steps as part of the assessment (Appendix III) [3]:

- Discuss the test method with *Analyst A*; confirm *Analyst A*'s knowledge of and performance of the correct procedure.
- Review and examine the raw data obtained with *Analyst A* in the analysis, including the chromatogram, and identify anomalous or suspect information.
- Confirm the equipment setting with *Analyst A* in front of the instrument.
- Verify that appropriate reference standards, solvents, reagents, and other solutions were used and that they meet quality control specifications.
- Evaluate the performance of the testing method and the performance of the instruments to ensure that it is performing according to the standard expected based on method validation data.
- Document and preserve evidence of this assessment.
- Inform the quality assurance unit immediately since appropriate action may be required from some regulatory authorities.

The assignment of a cause for the OOS/atypical result will be greatly facilitated if the retained sample preparations are examined promptly. Hypotheses regarding what might have happened (e.g., dilution error, instrument malfunction) can be tested. Examination of the retained solutions can be performed as part of the laboratory investigation [2].

14.2.3   Documentation Requirement

Information to be included in the documentation during investigation includes but is not limited to the following:

- The nonconformity number should be unique and should be used as a reference number during the whole investigation.
- The incident date should be clearly defined in the investigation process. For example, does the date refer to when OOS data was collected or the date when the OOS data is observed?

- The date OOS was initiated should be confirmed. Since the OOS investigation should be thorough, timely, unbiased, well-documented, and scientifically defensible, the time difference between the dates of OOS initiated and those of the incident should not be significantly different. A significant time difference indicates that the laboratory is out of control and also increases the difficulty in identifying the root cause of the OOS.
- Sample information to be included is as follows:
  - Description of the sample being tested as received by the laboratory.
  - Name of the material.
  - Type of sample: raw material, in process, finished product, stability, consumer complaint, and so on.
  - Batch number or other unique code.
  - Amount received.
  - Date on which it was received for testing.
  - Identification of the source or location from where the sample was obtained.
- Testing method information to be included is as follows:
  - Method reference number (use only the most recent, approved version).
  - Description of modifications, including reason(s) for the modification and data to verify the accuracy and reliability of the modified method.
- Testing limit(s) to be included is as follows:
  - Upper and lower limits.
  - Specification reference number.
- The conclusion of each investigation should be documented and the next step of the investigation step should be discussed based on whether laboratory error *is* or *is not* clearly identified.

14.2.4 Reanalysis

After initial assessment, *Supervisor B* can authorize *Analyst A* to perform reanalysis if the obvious determinate laboratory error is not identified. The reanalysis plan should be prepared and preapproved. The justification should be discussed and documented (Appendix IV).

According to the guidance, original sample preparation can be reinjected as part of an investigation where a transient equipment malfunction is suspected. This could occur if bubbles were introduced during an injection on a chromatographic system, while other tests indicated were proper. Such theories are difficult to prove. However, a reinjection can provide strong evidence that the problem should be attributed to the instrument rather than the sample problem or its preparation [2].

Supervisor B can also authorize further extraction of the active ingredient from the cream matrix to determine whether the active ingredient was fully extracted during the original analysis.

For release rate testing of certain specialized dosage form drugs that are not destroyed during testing, where possible, examination of the original dosage unit tested might determine whether it was damaged during laboratory handling in a way that affected its performance. Such damage would provide evidence to invalidate the OOS test result, and a retest would be indicated.

Further extraction of a dosage unit, where possible, can be performed to determine whether it was fully extracted during the original analysis. Incomplete extraction could invalidate the test results and should lead to questions regarding validation of the test method [2].

Supervisor B should ascertain not only the reliability of the individual value obtained, but also the significance of the OOS/atypical results represent in the overall quality assurance program.

When clear evidence of laboratory error exists, the OOS/atypical results should be invalidated. When evidence of laboratory error remains unclear, a full scale OOS investigation should be conducted by the manufacturing firm to determine what caused the unexpected results. It should not be assumed that OOS test results are attributable to analytical error without performing and documenting an investigation. Both the initial laboratory assessment and the following failure investigation should be documented fully [2].

CASE STUDY: PART II—REANALYSIS RESULT

Reanalysis result obtained: (acceptance criteria: release limit: 95.0–105.0%)

	Propyl paraben assay result			Methyl paraben assay result		
Sample ID	Original result (%)	Original HPLC vial (%)	Revial after remixing	Original result (%)	Original HPLC vial (%)	Revial after remixing
Batch #123—tank top	96.1	96.5	96.2%	99.8	99.9	99.6
		96.4	96.4%		100.5	100.0
Batch #123—tank bottom	94.9	95.3	95.8	99.4	100.4	100.7
		95.2	95.9		99.3	101.1

14.2.5 Reanalysis Result Discussion

14.2.5.1 Averaging In this case study, the average result from the original analysis (94.9%) and the reanalysis results from the original HPLC vial (95.3 and 95.2%) provide a result that passed the specification:

Appropriate uses: It should be noted that a test might consist of a specific number of replicates to arrive at a result. For instance, an HPLC assay result may be determined by averaging the peak responses from a number of consecutive, replicate injections from the same preparation (usually 2 or 3). The assay result would be calculated using the peak response average. This determination is considered one test and one result. This is a distinct difference from the analysis of different portions from a lot, intended to determine variability within the lot, and from multiple full analyses of the same homogenous sample. The use of replicates to arrive at a single reportable result, and the specific number of replicates used, should be specified in the written, approved test method. Acceptance limits for variability among the replicates should also be specified in the method. Unexpected variation in replicate determinations should trigger remedial action as required by §211.160(b)(4). If acceptance limits for replicate variability are not met, the test results should not be used. In some cases, a series of complete tests (full run-throughs of the test procedure), such as assays, are part of the test method. It may be appropriate to specify in the test method that the average of these multiple assays is considered one test and represents one reportable result. In this case, limits on acceptable variability among the individual assay results should be based on the known variability of the method and should also be specified in the test methodology. A set of assay results not meeting these limits should not be used. These appropriate uses of averaging test data should be used during an OOS investigation only if they were used during the original testing that produced the OOS result.

Inappropriate use: Reliance on averaging has the disadvantage of hiding variability among individual test results. For this reason, all individual test results should normally be reported as separate values. Where averaging of separate tests is appropriately specified by the test method, a single averaged result can be reported as the final test result. In some cases, a statistical treatment of the variability of results is reported. For example, in a test for dosage form content uniformity, the standard deviation (or relative standard deviation) is reported with the individual unit dose test results. Averaging can also conceal variations in different portions of a batch, or within a sample. For example, the use of averages is inappropriate when performing powder blend/mixture uniformity or dosage form content uniformity determinations. In these cases, testing is intended to measure variability within the product, and individual results provide the information for such an evaluation [2].

Since the initial assessment does not determine that laboratory error caused the *OOS/atypical* test result and testing results appear to be accurate, *Supervisor B* should inform other business units such as production and quality assurance to perform a full-scale failure investigation using a predefined procedure.

14.3   EXPANDED INVESTIGATING OOS/ATYPICAL TEST RESULTS

14.3.1   Review of Production

Supervisor B should also request for a review of manufacturing and packaging processes to ascertain if any deviation from these processes caused an *OOS/atypical* result for the preservative assay result. *Supervisor B* should also review the historical testing result for product *Cream A*. Reconfirmation of the assay result from the raw material can also help determine if sufficient amount of preservative *was* or *was not* added in the batch. *Supervisor B* should also review the validation of the testing method from the raw material and final product.

> The objective of such an investigation should be to identify the root cause of the OOS result and take appropriate corrective and preventative action. A full-scale investigation should include a review of production and sampling procedures, and will often include additional laboratory testing. Such investigations should be given the highest priority. Among the elements of this phase is evaluation of the impact of OOS result(s) on already distributed batches.
>
> The investigation should be conducted by the QCU and should involve all other departments that could be implicated, including manufacturing, process development, maintenance, and engineering. In cases where manufacturing occurs off-site (i.e., performed by a contract manufacturer or at multiple manufacturing sites), all sites potentially involved should be included in the investigation. Other potential problems should be identified and investigated [2].

14.3.2 Additional Laboratory Testing

Since the reanalysis cannot identify any determinate laboratory errors, *Supervisor B* should initialize expanded laboratory-followed approved procedures. A number of practices can be followed during the laboratory phase of an investigation. These include (1) retesting a portion of the original sample and/or (2) testing a specimen from the collection of a new sample from the batch.

14.3.2.1 Retest According to the guidance, *Supervisor B* should use the sample for the retesting from the same homogeneous material that was originally collected from the lot, tested, and yielded the OOS results. For a liquid, it may be from the original unit liquid product or composite of the liquid product; for a solid it may be an additional weighing from the same sample composite that had been prepared by *Analyst A. Supervisor B* should use the retesting situations to investigate the possibility of the testing instrument malfunctions or to identify a possible sample handling integrity problem, for example, a suspected dilution error. *Supervisor B* can authorize *Analyst A* and/or another trained analyst, *Analyst C*, when there is an increase in the level of duplication. *Supervisor B* can also authorize the retest performed with the "control sample" in order to provide valuable information to identify the product-related OOS (Appendix V).

> Generally, retesting is neither specified nor prohibited by approved applications or by the compendia. Decisions to retest should be based on the objectives of the testing and sound scientific judgment. The cGMP regulations require the establishment of specifications, standards, sampling plans, test procedures, and other laboratory control mechanisms (§211.160). The establishment of such control mechanisms for examination of additional specimens for commercial or regulatory compliance testing must be in accordance with "predetermined guidelines or sampling strategies" (United States Pharmacopeia 23, General Notices and Requirements, p. 9).

Some firms have used a strategy of repeated testing until a passing result is obtained (testing into compliance), then disregarding the OOS results without scientific justification. Testing into compliance is objectionable under the cGMPs. The number of retests to be performed on a sample should be specified in advance by the firm in the SOP.

The number may vary depending upon the variability of the particular test method employed, but should be based on scientifically sound, supportable principles. The number should not be adjusted depending on the results obtained. The firm's predetermined testing procedures should contain a point at which the testing ends and the product is evaluated.

If, at this point, the results are unsatisfactory, the batch is suspect and must be rejected or held pending further investigation (§211.165(f)). In the case of a clearly identified laboratory error, the retest results would substitute for the original test results. The original results should be retained, however, and an explanation recorded. This record should be initialed and dated by the involved persons and include a discussion of the error and supervisory comments.

If no laboratory or statistical errors are identified in the first test, there is no scientific basis for invalidating initial OOS results in favor of passing retest results. All test results, both passing and suspect, should be reported and considered in batch release decisions [2].

14.3.2.2 *Resample (If Required)*   If sufficient quantity of sample is not available, the original sample has been contaminated, and/or the sample is suspected to not be representative of the batch, then *Supervisor B* can request a resampling with the sample plan to ensure that the resamples are representative of the batch, and the establishment of control mechanisms for examination of additional specimens for commercial or regulatory compliance testing should be in accordance with predetermined procedures and sampling strategies (§211.165(c)) after Quality Assurance (QA)approval. In our case study, since the bulk cream has been packed into the finished pack, *Supervisor B* has decided to test the finish pack to confirm if there is any problem with the bulk sample during the first retest. However, the number of samples to be tested should be addressed if the homogeneity of the product during the packaging has not been established for this product (Appendix VI).

In some cases, when all data have been examined, it may be concluded that the original sample was prepared improperly and was therefore not representative of the batch (§211.160(b)(3)). A resampling of the batch should be conducted if the investigation shows that the original sample was not representative of the batch. This would be indicated, for example, by widely varied results obtained from several aliquots of the original composite (after determining there was no error in the performance of the analysis). Resampling should be performed by the same qualified, validated methods that were used for the initial sample. However, if the investigation determines that the initial sampling method was in error, a new accurate sampling method must be developed, qualified, and documented (§§211.160 and 165(c)) [2].

CASE STUDY: PART III—RETEST RESULT

Retest result obtained: (acceptance criteria: release limit: 95.0–105.0%)

Analyst	Sample ID	Propyl paraben (%)	Methyl paraben (%)
A	Batch #123—tank bottom 1	95.4	99.8
	Batch #123—tank bottom 2	95.3	99.7
	Batch #123—tank bottom 3	95.1	99.6
	Batch #123—tank bottom 4	95.2	99.6
	Batch #123—finished pack	95.5	101.0
	Control sample (batch #456)	99.1	99.7
C	Batch #123—tank bottom 1	95.8	100.1
	Batch #123—tank bottom 2	95.9	100.2
	Batch #123—tank bottom 3	95.4	99.9
	Batch #123—tank bottom 4	95.3	99.9
	Batch #123—finished pack	95.8	100.9
	Control sample (batch #456)	99.5	100.0

Since the retest results confirm the *atypical/OOS results* for propyl paraben assay and typical result is obtained for methyl paraben, *Supervisor B* has decided to investigate the raw material used in manufacturing as stated in the QA preapproved documented investigation plan (Appendix VII).

CASE STUDY: PART IV—RESAMPLE TESTING RESULT FOR PROPYL PARABEN

Acceptance criteria: Release limit: 99.0–101.0% using titration method

Batch number	Titration method (%)	HPLC method	
ABC	99.7	Sample #1:	95.10%
		Sample #2:	95.22%
		Average:	95.16%
CDE	99.5	Sample #1:	99.82%
		Sample #2:	99.10%
		Average:	99.46%

14.4   CONCLUDING THE INVESTIGATION

14.4.1   Interpretation of the Investigation Results

The expectation from the agency is clear on interpretation of the investigation results. However, it is critical for firms to understand the impact of the *OOS result/atypical result* on the batch quality and the impact on the patient.

An initial OOS result does not necessarily mean the subject batch fails and must be rejected. The OOS result should be investigated, and the findings of the investigation, including retest results, should be interpreted to evaluate the batch and reach a decision regarding release or rejection (§211.165).

In those instances where an investigation has revealed a cause, and the suspect result is invalidated, the result should not be used to evaluate the quality of the batch or lot. Invalidation of a discrete test result may be done only upon the observation and documentation of a test event that can reasonably be determined to have caused the OOS result.

In those cases where the investigation indicates an OOS result is caused by a factor affecting the batch quality (i.e., an OOS result is confirmed), the result should be used in evaluating the quality of the batch or lot. A confirmed OOS result indicates that the batch does not meet established standards or specifications and should result in the batch's rejection, in accordance with §211.165(f), and proper disposition. For inconclusive investigations—in cases where an investigation (1) does not reveal a cause for the OOS test result and (2) does not confirm the OOS result—the OOS result should be given full consideration in the batch or lot disposition decision. In the first case (OOS confirmed), the investigation changes from an OOS investigation into a batch failure investigation, which must be extended to other batches or products that may have been associated with the specific failure (§211.192) [2].

In our case study, the initial OOS result should be considered as a valid result based on the investigation finding. Most of the QA personnel will reject the batch since this action will avoid any potential finding during audit with assumption that the appropriate correction action and preventive action (CAPA) are in place to prevent the reoccurrence. On the other hand, *Supervisor B* has recommended batch release based on the following scientific discussion:

- Additional testing shows atypical result only during retest with increase of sample preparations.
- Stability trending for this product has been reviewed and indicates that there is only a slight drop in the propyl paraben assay result with proposed shelf life. The end of shelf life specification for propyl paraben has a wider limit of 90.0–110.0%.
- Both propyl paraben and methyl paraben can work as preservatives in products.
- USP antimicrobial effective test for preservatives on Cream A, lot #123, has been performed and passed.

14.4.2 Additional Reporting Requirement

For those products that are the subject of applications, regulations require submitting within three working days a field alert report (FAR in the United States) of information concerning any failure of a distributed batch to meet any of the specifications established in an application (21 Code of Federal Regulations (CFR) 314.81(b)(1)(ii). The *OOS* test results not invalidated on distributed batches or lots for this class of products are considered to be one kind of "information concerning any failure" described in this regulation. This includes *OOS* results that are considered to be discordant and of low value in batch quality evaluation. In these cases, an FAR should be submitted.

14.4.3   Tracking and Trending OOS/Atypical Results

Supervisor B should be especially alert in developing trends during investigation finding. To conclude the investigation, *Supervisor B* should evaluate the available results and determine the batch quality with quality assurance's approval regarding when a release decision should be made. The SOPs should be followed in arriving at this point. *Supervisor B* should also periodically review the *OOS/atypical result* with senior management to determine if any trend has been established. Each review should be documented to include proposed corrective actions for any trends observed (Appendix VIII).

Laboratory error should be relatively rare. Frequent errors suggest a problem that might be due to inadequate training of analysts, poorly maintained or improperly calibrated equipment, or careless work. Whenever laboratory error is identified, the firm should determine the source of that error and take corrective action to ensure that it does not occur again. To ensure full compliance with the cGMP regulations, the manufacturer also should maintain adequate documentation of the corrective action [2].

The tracking system used during the atypical/OOS investigation should also be established and used for the follow-up for corrective action and preventative measurements.

14.5   CORRECTIVE AND PREVENTIVE ACTION

CAPA has been a high-priority topic for many years. It is expected that a CAPA be implemented after each *OOS/atypical* result investigation, although this has not been discussed in detail in the guidance. Therefore it is critical that *Supervisor B* clearly identify the root cause of the *OOS/atypical* result.

14.5.1   Corrective Actions

In our case study, the impact of low potency of propyl paraben raw material should be investigated in the firm. This should be undertaken not only for *Cream A* but for other products in which the same raw material has been used. Extended investigation should also look at other in-house batches of propyl paraben raw material.

The stability limit for Cream A was reviewed. There was no significant change in the methyl paraben assay and about 2% drop in the propyl paraben assay after 24 months shelf life storage. The current USP antimicrobial effective test for preservatives for Cream A, lot #123, will be performed before it is recommended for product release.

14.5.2   Preventive Actions

The quality assurance unit in the firm should work with the vendor to determine the root cause of the problem to prevent reoccurrence. Additional procedures may be implemented.

APPENDIX I

Out of specification (OOS)   The term *OOS results* includes *all* suspect results that fall outside the specifications or acceptance criteria established in new drug applications, official compendia, or by the manufacturer.

Acceptance criteria   Predefined limits that are to be applied to the test results to determine if they meet the specification required with the approved methods.

Analyst   A trained individual performed the tests.

Assignable cause   An assignable cause is a clear and scientific reason for the OOS result.

Atypical result   Any result that is significantly different from previously obtained results for other batches of the same product or raw material but is within specification limits. Examples include (1) variability between sample preparations that exceeds the typical past performance of the method; (2) out of trend result; (3) atypical sample chromatogram.

Control sample   A sample that is generally from a batch of the same material that has previously given that result within specification.

Determinate error   An error that has unequivocally occurred within the laboratory (e.g., wrong size of glassware was used).

Expanded OOS investigation   Expanded OOS investigation included further assessment of the laboratory testing and an external investigation conducted by QA, which may include a department outside the laboratory.

Immediately   As soon as it is known; the same working day if it is possible.

Indeterminate laboratory error   An error that has been suspected, but not proven, to have occurred in the laboratory. Typically, this will occur where a result cannot be repeated (e.g., loss of drying).

Investigation   An investigation conducted to identify the reason for an OOS result and indicated both initial assessment and expanded OOS investigation.

Laboratory sample   A sample from a batch or lot of product or material—intended for inspection or testing—obtained according to a defined sampling plan.

Prepared sample   A laboratory sample that has been prepared for a testing according to the procedures documented in the approved method.

Reanalysis   Reanalysis involves remeasurement or rerunning of an original sample preparation provided these solutions were correctly prepared, properly stored, and within their expiration date.

Remeasure   Where the same final test solution, in the same final container, is measured again (e.g., sample in the HPLC vial).

Reprepared   Where a further portion of drug product is taken from the same laboratory sample and then treated to sample preparation and measurement.

Rerun   Where the same final test solution, in a different final container, is measured again (e.g., fresh HPLC vial is filled from the final solution).

Resample   Where a further, fresh laboratory sample is taken from the batch as stored in the warehouse and then treated to sample preparation and measurement.

Result   The value obtained when a defined test method is performed on a sample. The result may be an individual value, the mean of the values obtained from replicate measurements on the same prepared sample, or the mean of the values obtained for replicate sample preparations.

Retest   To restart or reinitiate a test by a test by repreparing a sample according to the test method using the original sample or a new laboratory sample obtained after resampling.

Sample   One or more portion of material derived from a larger amount in one item or one or more items taken from a larger number of items. Sample should be taken from a batch or lot of material according to a defined sampling plan.

SST   System suitability test.

System   Instrument, reagents, conditions, and procedural details constitute a system.

Test   A defined procedure performed to measure or classify a characteristic of a laboratory sample.

APPENDIX II   SUGGESTED OOS INVESTIGATION FLOWCHART

The charts summarizing the activities that will follow for the investigation of an OOS/atypical result are illustrated in the below figure.

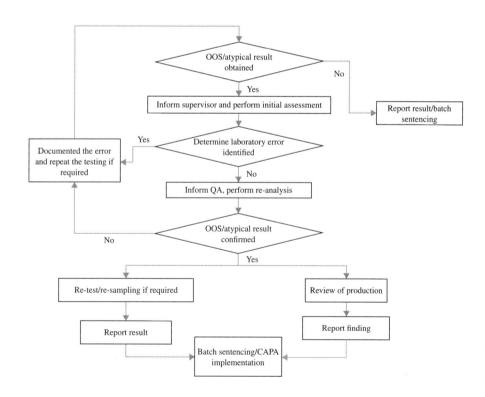

APPENDIX III   EXAMPLE OF INITIAL ASSESSMENT FORM

Document tracking information

Investigation tracking number:	2012–0001
Analyst(s)	Analyst A
Incident date:	OOS result was obtained on Aug 12, 2012, and observed on Aug 13, 2012
Date OOS initiated:	Aug 13, 2012

Sample information

Product name:	*Cream A*
Item code:	*101*
Batch number	*123*
Analytical record reference:	*1200001*
Sample ID number:	*20120812*
Vendor batch number:	*NA, in-house product*
Sample received date (amount):	*Aug 09, 2012 (3x100g, bulk—top, middle, and bottom)*
Type of OOS (OOS/atypical):	*OOS and atypical*
Sample storage condition:	*Room temperature at QC sample cabinet*
Test performed:	*HPLC preservative assay*
Limit and specification reference:	*95.0–105.0% (Cream A, item code: 101, version 2)*
Result obtained:	*94.9% on bottom of holding tank sample*
Method number:	*Method 013 (version 2)*
Testing purpose:	*Release testing*

Questions during assessment:	Answers:
Verify the analyst has enough experience for performing this technique?	*Analyst A has been trained on this method (reference: Analyst A training record on file) and has been testing this product for a few years.*
Was the correct method used?	*Method #013 (version 2) is the current version and has been verified for testing this batch. No change has been made on this version since 2004.*
Were the physical attributes of the laboratory sample as expected?	*The appearances for all samples received are typical for this product.*
Were appropriate reference standards, solvents, reagents, media, or other solutions used and prepared correctly?	*Reference standard, solvents, reagents, media, or other solutions used were checked and used as specified in Method #013 (version 2).*
Was the correct glassware used?	*All glassware was checked and corrected glassware was used.*

(*Continued*)

APPENDIX III   (Cont'd)

Was the instrument calibration current?	*HPLC001 was calibrated on July 03, 2012, and next calibration date was Jan 03, 2012. During the calibration, all testing results complied with the acceptance criteria in the SOP.*
Did the instrument perform correctly?	*% RSD of five injections for the standard was less than 1% (limit: 2%).*
Did the testing method perform as historically expected?	*Yes, except OOS result was obtained.*
Are the features of absorbance spectra consistent with expectations?	*Sample and standard chromatogram were reviewed, and they appeared to be typical. The response factor is typical based for the standard injection comparing with the historical data.*
Were all the arithmetic/ calculations correct?	*Yes, manual calculation confirmed that the result from the sample was low.*
Is chromatographic peak shapes/ intensity consistent with expectations?	*Sample peaks were smaller based on the expected sample size.*
Were the SST criteria defined in the method met?	*% RSD for five injections was less than 1% (limit: 2%)*
Were control sample results satisfactory?	*No control sample was performed.*
Is QA informed?	*Yes, QA Manager D has been contacted.*

Initial assessment of deviation:

Background information
- *On Aug 12, 2012, Analyst A followed method #013 (version 2) to perform the HPLC assay analysis on one batch of Cream A (Item code: 101, batch #123). The bulk samples from the top, middle, and bottom of the manufacturing tank were tested. Samples were prepared at 14:00 hours on August 12, 2006. The analysis was completed at 23:00 hours. All the chromatographic data was stored on file: 12AUG2012-1400.*
- *Analyst A has been testing this product for a few years.*
- *HPLC system used: HPLC001.*
- *Sample preparation: Weigh 1 g of Cream A in a 25-mL centrifuge tube. Add 5 mL of chloroform and 5 mL of mobile phase in the tube. Close the tube and mix the sample well until the ointment matrix dissolves completely. Wait 5 minutes until the separation of the layers. The top layer is the sample layer for analysis.*
- *After completion of the SST, the sample preparation was injected only once with bracketing of standard preparation.*
- *Testing resting for other samples:*
  - *Batch #123—tank top 96.1% (propyl paraben); 99.8 (methyl paraben)*
  - *Batch #12—tank middle 95.5% (propyl paraben); 99.6 (methyl paraben)*
  - *Batch #123—tank bottom 94.9% (propyl paraben); 99.4 (methyl paraben)*

Problem analysis:

What other tests indicate have been wrong but are not affected?	*No, methyl paraben assay results are typical for all samples analyzed.*
Size or extent of deviation:	*All QC samples.*
Is this deviation on a specific portion of material?	*No, all samples showed low propyl paraben assay results.*
Could this deviation have been on other portion but was not observed in this deviation?	*No, all samples showed low propyl paraben assay results.*
Is there any trend in this kind of deviation?	*No, OOS trending for Cream A was reviewed.*
Is this a recurring deviation?	*Yes, one OOS was observed in 2009, reference: 2009–013, and the root cause was due to incomplete extraction.*
How many samples/units have deviation?	*Only sample from bottom of the tank has OOS result but the top and middle samples are atypical compared with the product trending.*
Any recent changes in the laboratory /manufacturing process?	*Production manager indicated there was no change of manufacturing process.*
Could this change cause this deviation?	*NA*
List all probable causes of deviation:	*Incomplete extraction during sample preparation.* *Injector problem—equipment error* *Insufficient raw material is added in the batch.* *Inhomogeneous of the sample.*

Conclusion:

Laboratory error cannot be clearly identified during initial assessment. Reanalysis will be performed according to the reanalysis plan.

Prepared by:	*Analyst A, Aug 13, 2012*
Approved by:	*Supervisor B, Aug 13, 2012*

APPENDIX IV   EXAMPLE OF REANALYSIS PLAN FORM

Investigation tracking number: 2012–0001

Objective of analysis:

The original sample preparation for Cream A , tank bottom (item code: 101, batch #123) will be reanalyzed using method #013 on the same HPLC system (HPLC001) to determine if the OOS/atypical result is due to equipment error or sample preparation error.

Plan of analysis:

Original analyst will undertake the following tasks:

- *Duplicate injections will be performed on the original HPLC vial for batch #123 tank bottom samples to confirm there is no instrument error during the original analysis.*
- *Original sample preparation from batch #123 tank bottom will be remixed before reanalysis and revialed to confirm there is no sample preparation error.*
- *Duplicate injections will also be performed on the original HPLC vial for batch #123 bulk top sample preparation as control sample.*
- *Original sample preparation from batch #123 tank top sample will also be remixed before reanalysis and revialed as control.*

Acceptance criteria:

All results must meet specification limits.

Prepared by: *Analyst A*   Approved by: *Supervisor B*
Date: *Aug 13, 2012*   Date: *Aug 13, 2012*

Result from analysis:
Release limit: 95.0–105.0%

| | *Propyl paraben assay result* | | | *Methyl paraben assay result* | | |
Sample ID	*Original result*	*Original HPLC vial*	*Revial after remixing*	*Original result*	*Original HPLC vial*	*Revial after remixing*
Batch #123—	*96.1%*	*96.5%*	*96.2%*	*99.8%*	*99.9%*	*99.6*
tank–top		*96.4%*	*96.4%*		*100.5%*	*100.0*
Batch #123—	*94.9%*	*95.3%*	*95.8*	*99.4%*	*100.4%*	*100.7*
tank bottom		*95.2%*	*95.9*		*99.3%*	*101.1*

Conclusion from the reanalysis plan:

The reanalysis results are consistent with the initial result and indicate that the OOS was not due to the equipment error and sample mixing error. However, the elimination of preparation error cannot be established without further retesting. Original sample will be retested according to the new retest plan.

Prepared by: *Analyst A*   Approved by: *Supervisor B*
Date: *Aug 14, 2012*   Date: *Aug 14, 2012*

APPENDIX V   EXAMPLE OF RETEST PLAN I FORM

Investigation tracking number: 2012–0001

Objective of retest:
To determine if the cause of the OOS/atypical result for Cream A is due to sample preparation error or sample error.

Design of retest plan:
The samples will be prepared from the original sample. All the samples listed below will be tested according to Method #013 (Version 2).
- *Analyst A: Batch #123 tank-top*
- *Analyst A: Batch #123 tank-middle*
- *Analyst A: Batch #123 tank- bottom*
- *Analyst A: Batch #123 finished pack*
- *Analyst A: Control sample: Batch #789*
- *Analyst C: Batch #123 tank-top*
- *Analyst C: Batch #123 tank middle*
- *Analyst C: Batch #123 tank-bottom; four sample preparations*
- *Analyst A: Batch #123 finished pack*
- *Analyst C: Control sample : Batch #789*

Each analyst will prepare standard, chemicals separately. Two different HPLC will be used in the analysis.

Name of Analyst: *Analyst A and C*
Number of results obtained: *10*

Acceptance criteria:
All results must meet specification limits and show typical results.

Prepared by: *Analyst A*   Approved by: *Supervisor B*
Date: *Aug 14, 2012*   Date: *Aug 14, 2012*

Retest result obtained:
Release limit: 95.0–105.0%

Analyst	Sample ID	Propyl paraben	Methyl paraben
A	Batch #123—tank bottom 1	95.4%	99.8%
	Batch #123—tank –bottom 2	95.3%	99.7%
	Batch #123—tank bottom 3	95.1%	99.6%
	Batch #123—tank bottom 3	95.2%	99.6%
	Batch #123—finished pack	95.5%	101.0%
	Control sample (batch #789)	95.5%	99.7%
C	Batch #123—tank bottom 1	95.8%	100.1%
	Batch #123—tank bottom 2	95.9%	100.2%
	Batch #123—tank bottom 3	95.4%	99.9%
	Batch #123—tank bottom 4	95.3%	99.9%
	Batch #123—finished pack	95.8%	100.9%
	Control sample (batch #789)	99.5%	100.0%

Reference for all analytical data: Analytical record: *1200001, 1200002, and 1200003*

Conclusion from retest plan:
Although all reanalysis passed the acceptance criteria, the results for propyl paraben were atypically low and indicated that the cause of the original OOS was not laboratory-related error. Expanded investigation will be performed.

Prepared by: *Analyst A*   Approved by: *Supervisor B*
Date: *Aug 14, 2012*   Date: *Aug14, 2012*

APPENDIX VI   EXAMPLE OF RETEST PLAN II FORM

Investigation tracking number: 2012–0001

Objective of retest II:
Investigate the potency for propyl paraben (material code: RM101) used in manufacturing Cream A (material code: 101).
Design of retest:
Propyl paraben raw material (material code: RM101) from batch ABC, which was used in Cream A, batch #123, and batch CDE, which was used in other batch of Cream A with typical assay result will be tested using titration method and HPLC method 013 (version 2)
Name of analyst: *Analyst A*
Number of results obtained: *4*
Acceptance criteria:
All results must meet specification limits.

Prepared by: *Analyst A*   Approved by: *Supervisor B*
Date: *Aug 16, 2012*   Date: *Aug 16, 2012*

Result from retest:
Acceptance criteria: Release limit: 99.0–101.0% using titration method

Batch #	Titration method (%)	HPLC method	
ABC	99.7	Sample #1:	95.10%
		Sample #2:	95.22%
		Average:	95.16%
CDE	99.5	Sample #1:	99.82%
		Sample #2:	99.10%
		Average:	99.46%

Conclusion from retest result:
The assay results obtained from batch ABC using titration and HPLC methods are significantly different. Sample should be sent out for impurity analysis.

Prepared by: *Analyst A*   Approved by: *Supervisor B*
Date: *Aug 22, 2012*   Date: *Aug 22, 2012*

APPENDIX VII   EXAMPLE OF RETEST PLAN III FORM

Investigation tracking number: 2012–0001

Objective of retest III:
Investigate the cause of low potency for propyl paraben, material code: RM101, batch ABC using GC-MS and NMR spectroscopy

Design of retest:
*Sample from propyl paraben, material code: RM101, batches ABC and DEF
(as control) will be sent out to contract laboratory for impurity identification
using GC-MS and H NMR spectroscopy*
Name of analyst: *Analyst A*
Number of results obtained: *NA*
Acceptance criteria:
Report result and finding.

Prepared by: *Analyst A*   Approved by: *Supervisor B*
Date: *Aug 20, 2012*   Date: *Aug 20, 2012*

Result from retest:
*The impurities in propyl paraben (Batch ABC) were analyzed by gas chromatog-
raphy. Two impurity peaks at the retention time of 5.9 and 6.5 minutes were
observed. Using mass spectrometry coupled with gas chromatography, these
impurities were identified to be methyl paraben and ethyl paraben.*

*The ^1H NMR spectra for propyl paraben (batch ABC (purity: 93.89%) and
DEF (purity: 99.46)) were compared. Three additional NMR signals were
observed on the ^1H NMR spectrum for propyl paraben, Lot ABC. Using the
information obtained from GC-MS, all these additional peaks were identified and
assigned in Table III.*

Table III: Additional peaks observed on H NMR spectrum for propyl paraben (Lot ABC)

Chemical shift	Intensity	Number of protons	Assignment
1.34–1.37 (triplet)	7.87	3	CH_3-CH_2-$O(C=O)Ar$*
4.3 (quartet)	5.42	2	CH_3-CH_2-$O(C=O)Ar$*
3.84 (singlet)	2.39	3	CH_3-$O(C=O)Ar$*

**Ar = 4-hydroxyl-aryl; the NMR signals for the aromatic hydrogen on the methyl
paraben, ethyl paraben, and propyl paraben are not resolved due to the similar
chemical shift and low concentration for methyl and ethyl paraben.
Using the NMR data, the concentration of methyl, ethyl, and propyl paraben
calculated in mole ratio are 1.7%, 5.5%, and 92.7% respectively.*

Conclusion from retest result:
Two impurities, ethyl paraben and methyl paraben, were identified in batch ABC
and the amounts of these two impurities were about 5.5% and 1.7%, respectively.

Prepared by: *Analyst A*   Approved by: *Supervisor B*
Date: *Aug 22, 2012*   Date: *Aug 22, 2012*

APPENDIX VIII   EXAMPLE OF FINAL REPORTING FOR ATYPICAL/ OOS RESULT INVESTIGATION FORM

Investigation tracking number: 2012–0001

Final conclusion:
The potency for propyl paraben, RM101, in raw material form is determined through titration according to the current testing procedure (RM101-M1). A known amount of propyl paraben raw material reacts with propyl paraben with excess 1N sodium hydroxide. The unreacted sodium hydroxide is then backed-titrated with 1N H_2SO_4. However, the propyl paraben content in Cream A is determined by HPLC. The purity of propyl paraben in raw material RM101 (batch ABC) obtained by the titration method is 99.7%. However, purity of 95.05% was obtained when HPLC was used. These results suggest that the titration method lacks selectivity in determining the purity of propyl paraben since any alkyl paraben will react with sodium hydroxide just like propyl paraben does. Also, these results have explained the cause of atypical propyl paraben for Cream A, batch #123.

Result to be reported:
The average result obtained from all the sample preparations from the bottom-tank sample will be reported as the result from the bottom sample.
Investigation completed within predefined period (20 working days) _✓_ Yes or ___No
Evaluation of atypical/OOS result for trending:
Indeterminate laboratory error ___Yes or _✓_ No
Determinate laboratory error___Yes or _✓_ No
Inconclusive___Yes or _✓_ No
Valid Results (Material related) _✓_ Yes or ___No

Corrective action:
The remaining inventory for propyl paraben raw material, material code RM101, batch ABC has been restricted for use and it is the only batch currently in-house.
Investigation findings indicated that batch ABC has been used in Cream A (batch #123) only. Therefore, no other product has been impacted due to this deviation.
The stability limit for Cream A is reviewed. There is no significant change in methyl paraben assay and there is about 2% drop of the 24 months shelf life storage. The current USP antimicrobial effective test for preservatives on Cream A, lot #123, will be performed before recommendations for product release.

Preventative action:
The vendor was informed about this investigation finding. Vendor investigation to determine the root cause of the problem if required. *The potency determination using validated HPLC for propyl paraben RM101 will be added in the specification.*

Prepared by: *Analyst A*   Approved by: *Supervisor B*
Date: *Aug 25, 2012*   Date: *Aug 25, 2012*

REFERENCES

1. Hoinowski AM, Motola S, Davis RJ, McArdle JV. Investigation of out-of-specification results. *Pharm Technol* January 2002;26:40–51.

2. FDA. 1998. Guidance for industry: investigating out of specification (OOS) test results for pharmaceutical production, p 3–13. Available at http://www.fda.gov/downloads/drugs/guidancecomplianceregulatoryinformation/guidances/ucm070287.

3. FDA. 1993. Guide to inspections of pharmaceutical quality control laboratories. Available at http://www.fda.gov/ICECI/Inspections/InspectionGuides/ucm074918.htm.

INDEX

Therapeutic Delivery Solutions, First Edition. Edited by Chung Chow Chan,
Kwok Chow, Bill McKay, and Michelle Fung.
© 2014 John Wiley & Sons, Inc. Published 2014 by John Wiley & Sons, Inc.